Student Solutions Manual

Alison Hyslop
St. John's University

to accompany

CHEMISTRY

Matter and Its Changes

Fifth Edition

James E. Brady
St. John's University

Fred Senese
Frostburg State University

John Wiley & Sons, Inc.

COVER PHOTO: © Chris Ewels

To order books or for customer service, please call 1-800-CALL-WILEY (225-5945).

ISBN 978-0-470-18465-3

Printed in the United States of America

10 9 8 7 6 5 4 3 2 1

Printed and bound by Bind-Rite Graphics, Inc.

Table of Contents

Practice Exercises

1.1 $V = \frac{4}{3}\pi r^3$, the SI unit for radius, r, is meters, the numbers $\frac{4}{3}$ and π do not have units. Therefore, the SI unit for volume is meter^3 or m^3.

1.2 Force equals mass × acceleration ($F = ma$), and acceleration equals change in velocity divided by change in time ($a = \frac{\text{change in v}}{\text{change in t}}$), and velocity equals distance divided by time ($v = \frac{d}{t}$). Put the equations together:

$$F = m\left(\frac{\text{change in v}}{\text{change in t}}\right)$$

$$F = m\left(\frac{\text{change in }\frac{d}{t}}{\text{change in t}}\right) = m\left(\frac{\text{change in d}}{\text{change in t}^2}\right)$$

The unit for mass is kilogram (kg); the unit for distance is meter (m) and the unit for time is second (s). Substitute the units into the equation above:

Unit for force in SI base units = $\text{kg}\left(\frac{\text{m}}{\text{s}^2}\right)$ or kg m s^{-2}

1.3 $$t_F = \left(\frac{9\ ^\circ F}{5\ ^\circ C}\right)t_C + 32\ ^\circ F = \left(\frac{9\ ^\circ F}{5\ ^\circ C}\right)(86\ ^\circ C) + 32\ ^\circ F = 187\ ^\circ F$$

The tool that relates the two temperature scales is the Celsius - Fahrenheit conversion tool.

1.4 $$t_C = (t_F - 32\ ^\circ F)\left(\frac{5\ ^\circ C}{9\ ^\circ F}\right) = (50\ ^\circ F - 32\ ^\circ F)\left(\frac{5\ ^\circ C}{9\ ^\circ F}\right) = 10\ ^\circ C$$

To convert from °F to K we first convert to °C. In the

$$t_C = (t_F - 32\ ^\circ F)\left(\frac{5\ ^\circ C}{9\ ^\circ F}\right) = (68\ ^\circ F - 32\ ^\circ F)\left(\frac{5\ ^\circ C}{9\ ^\circ F}\right) = 20\ ^\circ C$$

$$T_K = (273\ ^\circ C + t_C)\left(\frac{1\text{ K}}{1\ ^\circ C}\right) = (273\ ^\circ C + 20\ ^\circ C)\left(\frac{1\text{ K}}{1\ ^\circ C}\right) = 293\text{ K}$$

1.5
(a) 21.0233 g + 21.0 g = 42.0233 g: rounded correctly to 42.0 g
(b) 10.0324 g / 11.7 mL = 0.8574 g / mL: rounded correctly to 0.857 g / mL
(c) $\frac{14.24\text{ cm} \times 12.334\text{ cm}}{(2.223\text{ cm} - 1.04\text{ cm})} = 148.57$ cm: rounded correctly to 149 cm

1.6
(a) 32.02 mL – 2.0 mL = 30. mL
(b) 54.183 g – 0.0278 g = 54.155 g
(c) 10.0 g + 1.03 g + 0.243 g = 11.3 g
(d) $43.4\text{ in} \times \left(\frac{1\text{ ft}}{12\text{ in.}}\right) = 3.62$ ft (1 and 12 are exact numbers)
(e) $\frac{1.03\text{ m} \times 2.074\text{ m} \times 3.9\text{ m} \times}{12.46\text{ m} + 4.778\text{ m}} = 0.48\text{ m}^2$

1.7 $\text{m}^2 = (124\ \text{ft}^2)\left(\frac{30.48\ \text{cm}}{1\ \text{ft}}\right)^2\left(\frac{1\ \text{m}}{100\ \text{cm}}\right)^2 = 11.5\ \text{m}^2$

1.8 (a) $\text{in.} = (3.00\ \text{yd})\left(\frac{3\ \text{ft}}{1\ \text{yd}}\right)\left(\frac{12\ \text{in.}}{1\ \text{ft}}\right) = 108\ \text{in.}$

(b) $\text{cm} = (1.25\ \text{km})\left(\frac{1000\ \text{m}}{1\ \text{km}}\right)\left(\frac{100\ \text{cm}}{1\ \text{m}}\right) = 1.25 \times 10^5\ \text{cm}$

(c) $\text{ft} = (3.27\ \text{mm})\left(\frac{1\ \text{m}}{1000\ \text{mm}}\right)\left(\frac{100\ \text{cm}}{1\ \text{m}}\right)\left(\frac{1\ \text{in.}}{2.54\ \text{cm}}\right)\left(\frac{1\ \text{ft}}{12\ \text{in.}}\right) = 0.0107\ \text{ft}$

(d) $\frac{\text{km}}{\text{L}} = \left(\frac{20.2\ \text{mile}}{1\ \text{gal}}\right)\left(\frac{1.609\ \text{km}}{1\ \text{mile}}\right)\left(\frac{1\ \text{gal}}{3.785\ \text{L}}\right) = 8.59\ \text{km/L}$

1.9 $\text{Density} = \frac{\text{mass}}{\text{volume}}$

Density of the object $= \frac{365\ \text{g}}{22.12\ \text{cm}^3} = 16.5\ \text{g/cm}^3$

The object is not composed of pure gold since the density of gold is 19.3 g/cm^3.

1.10 The density of the alloy is 12.6 g/cm^3. To determine the mass of the 0.822 ft^3 sample of the alloy, first convert the density from g/cm^3 to lb/ft^3, then find the weight.

Density in lb/ft^3 $= \frac{12.6\ \text{g}}{\text{cm}^3}\left(\frac{1\ \text{lb}}{453.6\ \text{g}}\right)\left(\frac{30.48\ \text{cm}}{1\ \text{ft}}\right)^3 = 787\ \text{lb/ft}^3$

Mass of sample alloy = (0.822 ft^3) (787 lb/ft^3) = 647 lb

1.11 density = mass/volume = $(1.24 \times 10^6\ \text{g})/(1.38 \times 10^6\ \text{cm}^3) = 0.899\ \text{g/cm}^3$

1.12 volume of one-carat diamond $= 200\ \text{mg}\left(\frac{1\ \text{g}}{1000\ \text{mg}}\right)\left(\frac{1\ \text{cm}^3}{3.52\ \text{g}}\right) = 0.0568\ \text{cm}^3$

Review Problems

1.26 (a) 0.01 m (b) 1000 m (c) 10^{12} pm
(d) 0.1 m (e) 0.001 kg (f) 0.01 g

1.28 (a) $t_F = \left(\frac{9\ °\text{F}}{5\ °\text{C}}\right)(t_C) + 32\ °\text{F} = \left(\frac{9\ °\text{F}}{5\ °\text{C}}\right)(50\ °\text{C}) + 32\ °\text{F} = 120\ °\text{F}$ when rounded to the proper number of significant figures.

(b) $t_F = \left(\frac{9\ °\text{F}}{5\ °\text{C}}\right)(t_C) + 32\ °\text{F} = \left(\frac{9\ °\text{F}}{5\ °\text{C}}\right)(10\ °\text{C}) + 32\ °\text{F} = 50\ °\text{F}$

(c) $t_C = \left(\frac{5\ °\text{C}}{9\ °\text{F}}\right)(t_F - 32\ °\text{F}) = \left(\frac{5\ °\text{C}}{9\ °\text{F}}\right)(25.5\ °\text{F} - 32\ °\text{F}) = -3.61\ °\text{C}$

(d) $t_C = \left(\frac{5\ °\text{C}}{9\ °\text{F}}\right)(t_F - 32\ °\text{F}) = \left(\frac{5\ °\text{C}}{9\ °\text{F}}\right)(49\ °\text{F} - 32\ °\text{F}) = 9.4\ °\text{C}$

(e) $T_K = (t_C + 273\ °\text{C})$ and the= 60 +273 = 333 K

(f) $T_K = (t_C + 273\ °C)\left(\frac{1\ K}{1\ °C}\right) = -30 + 273 = 243\ K$

1.30 $t_C = (t_F - 32\ °F)\left(\frac{5\ °C}{9\ °F}\right) = (103.5\ °F - 32\ °F)\left(\frac{5\ °C}{9\ °F}\right) = 39.7\ °C$

This dog has a fever; the temperature is out of normal canine range.

1.32 Range in Kelvins:

$$K = (10\ MK)\left(\frac{1 \times 10^6\ K}{1\ MK}\right) = 1.0 \times 10^7\ K$$

$$K = (25\ MK)\left(\frac{1 \times 10^6\ K}{1\ MK}\right) = 2.5 \times 10^7\ K$$

Range in degrees Celsius:

$$t_C = (T_K - 273\ K)\left(\frac{1\ °C}{1\ K}\right) = (1.0 \times 10^7\ K - 273\ K)\left(\frac{1\ °C}{1\ K}\right) \approx 1.0 \times 10^7\ °C$$

$$t_C = (T_K - 273\ K)\left(\frac{1\ °C}{1\ K}\right) = (2.5 \times 10^7\ K - 273\ K)\left(\frac{1\ °C}{1\ K}\right) \approx 2.5 \times 10^7\ °C$$

Range in degrees Fahrenheit:

$$t_F = \left(\frac{9\ °F}{5\ °C}\right)(°C) + 32\ °F = \left(\frac{9\ °F}{5\ °C}\right)(1.0 \times 10^7\ °C) + 32\ °F \approx 1.8 \times 10^7\ °F$$

$$t_F = \left(\frac{9\ °F}{5\ °C}\right)(°C) + 32\ °F = \left(\frac{9\ °F}{5\ °C}\right)(2.5 \times 10^7\ °C) + 32\ °F \approx 4.5 \times 10^7\ °F$$

1.34 $t_C = (T_K - 273\ K)\left(\frac{1\ °C}{1\ K}\right) = (4\ K - 273\ K)\left(\frac{1\ °C}{1\ K}\right) = -269\ °C$

1.36 (a) 4 significant figures
(b) 5 significant figures
(c) 4 significant figures
(d) 2 significant figures
(e) 4 significant figures
(f) 2 significant figures

1.38 (a) 0.72 m^2
(b) 84.24 kg
(c) 4.19 g/cm^3 (dividing a number with 4 sig. figs by one with 3 sig. figs)
(d) 19.42 g/mL
(e) 858.0 cm^2

1.40 (a) $km/hr = (32.0\ dm/s)\left(\frac{1\ m}{10\ dm}\right)\left(\frac{1\ km}{1000\ m}\right)\left(\frac{3600\ s}{1\ h}\right) = 11.5\ km/h$

(b) $\mu g/L = (8.2\ mg/mL)\left(\frac{1\ g}{1000\ mg}\right)\left(\frac{1 \times 10^6\ \mu g}{1\ g}\right)\left(\frac{1000\ mL}{1\ L}\right) = 8.2 \times 10^6\ \mu g/L$

(c) $kg = (75.3\ mg)\left(\frac{1\ g}{1000\ mg}\right)\left(\frac{1\ kg}{1000\ g}\right) = 7.53 \times 10^{-5}\ kg$

(d) $L = (137.5\ mL)\left(\frac{1\ L}{1000\ mL}\right) = 0.1375\ L$

(e) $\text{mL} = (0.025\ \text{L})\left(\dfrac{1000\ \text{mL}}{1\ \text{L}}\right) = 25\ \text{mL}$

(f) $\text{dm} = (342\ \text{pm})^2\left(\dfrac{1\times10^{-12}\ \text{m}}{1\ \text{pm}}\right)^2\left(\dfrac{10\ \text{dm}}{1\ \text{m}}\right)^2 = 3.42 \times 10^{-20}\ \text{dm}$

1.42 (a) $\text{cm} = (36\ \text{in.})\left(\dfrac{2.54\ \text{cm}}{1\ \text{in.}}\right) = 91\ \text{cm}$

(b) $\text{kg} = (5.0\ \text{lb})\left(\dfrac{1\ \text{kg}}{2.205\ \text{lb}}\right) = 2.3\ \text{kg}$

(c) $\text{mL} = (3.0\ \text{qt})\left(\dfrac{946.4\ \text{mL}}{1\ \text{qt}}\right) = 2800\ \text{mL}$

(d) $\text{mL} = (8\ \text{oz})\left(\dfrac{29.6\ \text{mL}}{1\ \text{oz}}\right) = 200\ \text{mL}$

(e) $\text{km/hr} = (55\ \text{mi/hr})\left(\dfrac{1.609\ \text{km}}{1\ \text{mi}}\right) = 88\ \text{km/hr}$

(f) $\text{km} = (50.0\ \text{mi})\left(\dfrac{1.609\ \text{km}}{1\ \text{mi}}\right) = 80.4\ \text{km}$

1.44 (a) $\text{cm}^2 = (8.4\ \text{ft}^2)\left(\dfrac{30.48\ \text{cm}}{1\ \text{ft}}\right)^2 = 7{,}800\ \text{cm}^2$

(b) $\text{km}^2 = (223\ \text{mi}^2)\left(\dfrac{1.609\ \text{km}}{1\ \text{mi}}\right)^2 = 577\ \text{km}^2$

(c) $\text{cm}^3 = (231\ \text{ft}^3)\left(\dfrac{30.48\ \text{cm}}{1\ \text{ft}}\right)^3 = 6.54\times10^6\ \text{cm}^3$

1.46 $\text{mL} = (4.2\ \text{qt})\left(\dfrac{946.35\ \text{mL}}{1\ \text{qt}}\right) = 4.0 \times 10^3\text{mL}$ (stomach volume)

$4.0 \times 10^3\ \text{mL} \div 0.9\ \text{mL} = 4{,}000$ pistachios (don't try this at home)

1.48 $\dfrac{\text{m}}{\text{s}} = \left(\dfrac{200\ \text{mi}}{1\ \text{hr}}\right)\left(\dfrac{5280\ \text{ft}}{1\ \text{mi}}\right)\left(\dfrac{30.48\ \text{cm}}{1\ \text{ft}}\right)\left(\dfrac{1\times10^{-2}\text{m}}{1\ \text{cm}}\right)\left(\dfrac{1\ \text{hr}}{60\ \text{min}}\right)\left(\dfrac{1\ \text{min}}{60\ \text{s}}\right) = 90\ \dfrac{\text{m}}{\text{s}}$

1.50 $\dfrac{\text{mi}}{\text{h}} = \left(\dfrac{2230\ \text{ft}}{1\ \text{s}}\right)\left(\dfrac{1\ \text{mi}}{5280\ \text{ft}}\right)\left(\dfrac{60\ \text{s}}{1\ \text{min}}\right)\left(\dfrac{60\ \text{min}}{1\ \text{hr}}\right) = 1520\ \dfrac{\text{mi}}{\text{hr}}$

1.52 $1\ \text{light year} = 1\ \text{y}\left(\dfrac{365.25\ \text{d}}{1\ \text{y}}\right)\left(\dfrac{24\ \text{h}}{1\ \text{d}}\right)\left(\dfrac{3600\ \text{s}}{1\ \text{h}}\right)\left(\dfrac{3.00\times10^8\ \text{m}}{1\ \text{s}}\right) = 9.47 \times 10^{15}\ \text{m}$

$\text{miles} = 8.7\ \text{light years}\left(\dfrac{9.47 \times 10^{15}\text{m}}{1\ \text{light year}}\right)\left(\dfrac{1\ \text{km}}{1000\ \text{m}}\right)\left(\dfrac{1\ \text{mi}}{1.609\ \text{km}}\right) = 5.1 \times 10^{13}\ \text{mi}$

1.54 $\text{meters} = 6033.5 \text{ fathoms}\left(\frac{6 \text{ ft}}{1 \text{ fathom}}\right)\left(\frac{1 \text{ yd}}{3 \text{ ft}}\right)\left(\frac{0.9144 \text{ m}}{1 \text{ yd}}\right) = 11{,}034 \text{ m}$

1.56 density = mass/ volume = 36.4 g/45.6 mL = 0.798 g/mL

1.58 $\text{mL} = 25.0\text{g}\left(\frac{1 \text{ mL}}{0.791 \text{ g}}\right) = 31.6 \text{ mL}$

1.60 $\text{g} = 185 \text{ mL}\left(\frac{1.492 \text{ g}}{1 \text{ mL}}\right) = 276 \text{ g}$

1.62 mass of silver = 62.00 g – 27.35 g = 34.65 g
volume of silver = 18.3 mL –15 mL = 3.3 mL
density of silver = (mass of silver)/(volume of silver) = (34.65 g)/(3.3 mL) = 11 g/mL

1.64 $\text{density} = \left(\frac{227{,}641 \text{ lb}}{385{,}265 \text{ gal}}\right) = 0.5909 \ {}^{\text{lb}}\!/_{\text{gal}}$

$\text{sp. gr. of liquid hydrogen} = \frac{0.5909 \ {}^{\text{lb}}\!/_{\text{gal}}}{8.34 \ {}^{\text{lb}}\!/_{\text{gal}}} = 0.07085$

$\text{density} = 0.07085\left(1.00 \ {}^{\text{g}}\!/_{\text{mL}}\right) = 0.0709 \ {}^{\text{g}}\!/_{\text{mL}}$

Practice Exercises

2.1 The first sample has a ratio of

$$\frac{1.25\text{ g Cd}}{0.357\text{ g S}}$$

Therefore, the second sample must have the same ratio of Cd to S:

$$\frac{1.25\text{ g Cd}}{0.357\text{ g S}} = \frac{x}{3.50\text{ g S}}$$

Cross-multiplication gives,

(1.25 g Cd)(3.50 g S) = x(0.357 g S)

$$\frac{(1.25\text{g Cd})(3.50\text{ g S})}{0.357\text{g S}} = x$$

x = 12.3 g Cd

2.2 Compare the ratios of the mass of the compound before heating and the mass of the iron after heating, if they are the same, the compounds are the same.

Sample	Ratio
A	$\frac{25.36\text{ g}}{16.11\text{ g}} = 1.574$
B	$\frac{15.42\text{ g}}{8.28\text{ g}} = 1.86$
C	$\frac{7.85\text{ g}}{4.22\text{ g}} = 1.86$
D	$\frac{11.87\text{ g}}{7.54\text{ g}} = 1.57$

Compounds A and D are the same, as are compounds B and C.

2.3 $^{240}_{94}\text{Pu}$

The bottom number is the atomic number, found on the periodic table (number of protons). The top number is the mass number (sum of the number of protons and the number of neutrons). Since it is a neutral atom, it has 94 electrons.

2.4 $^{35}_{17}\text{Cl}$ contains 17 protons, 17 electrons, and 18 neutrons.

2.5 We can discard the 17 since the 17 tells the number of protons which is information that the symbol "Cl" also provides. In addition, the number of protons equals the number of electrons in a neutral atom, so the symbol "Cl" also indicates the number of electrons. The 35 is necessary to state which isotope of chlorine is in question and therefore the number of neutrons in the atom.

2.6 $2.24845 \times 12\text{ u} = 26.9814\text{ u}$

2.7 Copper is $63.546\text{ u} \div 12\text{ u} = 5.2955$ times as heavy as carbon

2.8 $(0.198 \times 10.0129\text{ u}) + (0.802 \times 11.0093\text{ u}) = 10.8\text{ u}$

2.9
(a) 1 Ni, 2 Cl
(b) 1 Fe, 1 S, 4 O
(c) 3 Ca, 2 P, 8 O
(d) 1 Co, 2 N, 12 O, 12 H

2.10
(a) 2 N nitrogen, 4 H hydrogen, 3 O oxygen
(b) 1 Fe iron, 1 N nitrogen, 4 H hydrogen, 2 S sulfur, 8 O oxygen
(c) 1 Mo molybdenum, 2 N nitrogen, 11 O oxygen, 10 H hydrogen
(d) 6 C carbon, 4 H hydrogen, 1 Cl chlorine, 1 N nitrogen, 2 O oxygen

2.11 This is a balanced chemical equation, and the number of each atom that appears on the left is the same as that on the right: 1 Mg, 2 O, 4 H, and 2 Cl.

2.12 $Mg(OH)_2(s) + 2HCl(aq) \rightarrow MgCl_2(aq) + 2H_2O$

2.13 6 N, 42 H, 2 P, 20 O, 3 Ba, and 12 C are on both the products and the reactants sides of the equation, therefore the reaction is balanced.

2.14 The term "octa' means eight, therefore there are 8 carbon atoms in octane. The formula for an alkane is C_nH_{2n+2}, so octane has 8 carbons and $((2 \times 8) + 2) = 18$ H. The condensed formula is $CH_3CH_2CH_2CH_2CH_2CH_2CH_2CH_3$, and the structural format is:

```
  H H H H H H H H
  | | | | | | | |
H-C-C-C-C-C-C-C-C-H
  | | | | | | | |
  H H H H H H H H
```

2.15 The condensed formula is $CH_3CH_2CH_2CH_2CH_2CH_2CH_2CH_2CH_2CH_3$
The structural formula is:

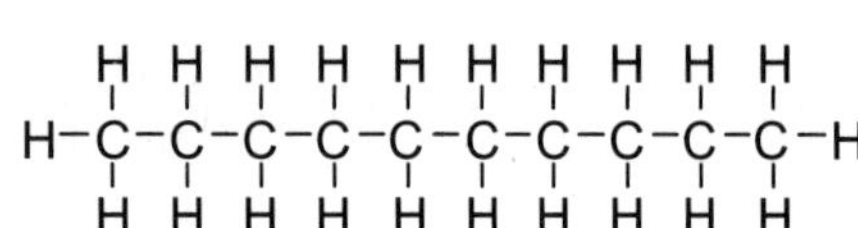
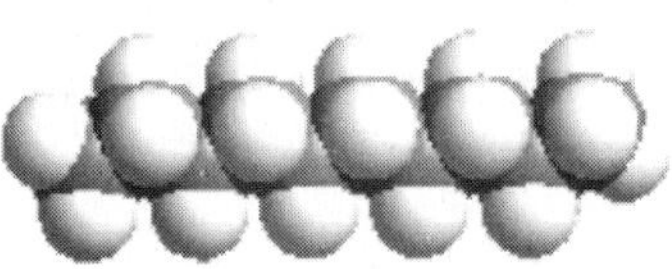

2.16 (a) Propanol: $CH_3CH_2CH_2OH$

```
  H H H
  | | |
H-C-C-C-O-H
  | | |
  H H H
```

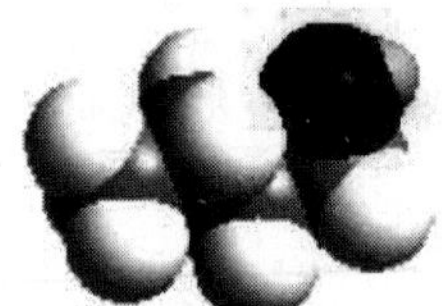

(b) Butanol: $CH_3CH_2CH_2CH_2OH$

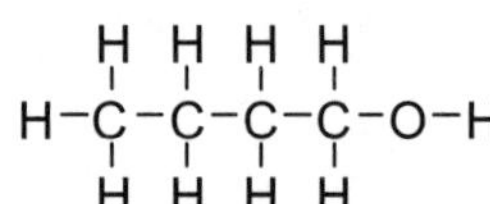
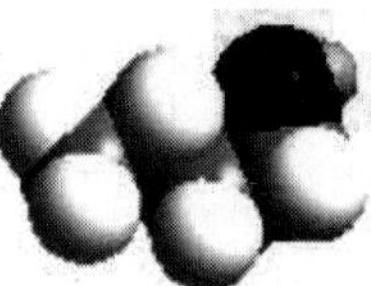

2.17
(a) Fe: 26 protons and 26 electrons
(b) Fe^{3+}: 26 protons and 23 electrons
(c) N^{3-}: 7 protons and 10 electrons
(d) N: 7 protons and 7 electrons

2.18 (a) O: 8 protons and 8 electrons
(b) O^{2-}: 8 protons and 10 electrons
(c) Al^{3+}: 13 protons and 10 electrons
(d) Al: 13 protons and 13 electrons

2.19 (a) NaF (b) Na_2O (c) MgF_2 (d) Al_4C_3

2.20 (a) Ca_3N_2 (b) $AlBr_3$ (c) Na_3P d) CsCl

2.21 (a) $CrCl_3$ and $CrCl_2$, Cr_2O_3 and CrO
(b) CuCl, $CuCl_2$, Cu_2O and CuO

2.22 (a) Au_2S and Au_2S_3, Au_3N and AuN
(b) SnS and SnS_2, Sn_3N_2 and Sn_3N_4

2.23 (a) $KC_2H_3O_2$ (b) $Sr(NO_3)_2$ (c) $Fe(C_2H_3O_2)_3$

2.24 (a) Na_2CO_3 (b) $(NH_4)_2SO_4$

2.25 (a) phosphorous trichloride
(b) sulfur dioxide
(c) dichlorine heptaoxide

2.26 (a) $AsCl_5$ (b) SCl_6 (c) S_2Cl_2

2.27 (a) K_2O (b) $BaBr_2$ (c) Na_3N (d) Al_2S_3

2.28 (a) aluminum trichloride (b) barium sulfide
(c) sodium bromide (d) calcium fluoride

2.29 (a) postassium sulfide
(b) magnesium phosphide
(c) nickel(II) chloride
(d) iron(III) oxide

2.30 (a) Al_2S_3 (b) SrF_2 (c) TiO_2 d) Au_2O_3

2.31 (a) lithium carbonate
(b) iron(III) hydroxide

2.32 (a) $KClO_3$ (b) $Ni_3(PO_4)_2$

2.33 diiodine pentaoxide

2.34 chromium(III) acetate

Review Problems

2.76 From the first ratio we see that there is a ratio of 9.33 g of nitrogen to 2.00 g of hydrogen. Multiplying the mass of nitrogen by 3.14 we see that for every 6.28 g hydrogen there will be 29.3 g nitrogen.

$$\frac{9.33 \text{ g nitrogen}}{2.00 \text{ g hydrogen}} = \frac{x \text{ g nitrogen}}{6.28 \text{ g hydrogen}}$$

$x = 29.3$ g nitrogen

2.78 5.54 g ammonia. From the ratio, Problem 2.76, above we see that for every 4.56 g nitrogen there needs to be 0.98 g hydrogen. According to the Law of Conservation of Mass 5.54 g ammonia will be produced.

$$\frac{9.33\text{ g nitrogen}}{2.00\text{ g hydrogen}} = \frac{4.56\text{ g nitrogen}}{x\text{ g hydrogen}}$$

$x = 0.98$ g hydrogen

mass of NH_3 = 4.56 g nitrogen + 0.98 g hydrogen = 5.54 g NH_3

2.80 2.286 g of O. The first nitrogen-oxygen compound has an atom ratio of 1 atom of N for 1 atom of O, and a mass ratio of 1.000 g of N for 1.143 g of O. The second compound has an atom ratio of 1 atom of N for 2 atoms of O; therefore, the second compound has twice as many grams of O for each gram of N, or 2.286 g of O.

2.82 Since we know that the formula is CH_4, we know that one fourth of the total mass due to the hydrogen atom constitutes the mass that may be compared to the carbon. Hence we have 0.33597 g H ÷ 4 = 0.083993 g H and 1.00 g assigned to the amount of C-12 in the compound. Then it is necessary to realize that the ratio 1.00 g C ÷ 12 for carbon is equal to the ratio 0.083993 g H ÷ X, where X equals the relative atomic mass of hydrogen.

$$\left(\frac{1.000\text{ g C}}{12\text{ u C}}\right) = \left(\frac{0.083993\text{ g H}}{\text{X}}\right) = 1.008\text{ u}$$

2.84 Regardless of the definition, the ratio of the mass of hydrogen to that of carbon would be the same. If C–12 were assigned a mass of 24 (twice its accepted value), then hydrogen would also have a mass twice its current value, or 2.01588 u.

2.86 $(0.6917 \times 62.9396\text{ u}) + (0.3083 \times 64.9278\text{ u}) = 63.55\text{ u}$

2.88

		neutrons	protons	electrons
(a)	Radium-226	138	88	88
(b)	^{206}Pb	124	82	82
(c)	Carbon-14	8	6	6
(d)	^{23}Na	12	11	11

2.90 1 Cr, 6 C, 9 H, 6 O

2.92 $MgSO_4$

2.94
(a) 2 K, 2 C, 4 O
(b) 2 H, 1 S, 3 O
(c) 12 C, 26 H
(d) 4 H, 2 C, 2 O
(e) 9 H, 2 N, 1 P, 4 O

2.96
(a) 1 Ni, 2 Cl, 8 O
(b) 1 Cu, 1 C, 3 O
(c) 2 K, 2 Cr, 7 O
(d) 2 C, 4 H, 2 O
(e) 2 N, 9 H, 1 P, 4 O

2.98
(a) 6 N, 3 O
(b) 4 Na, 4 H, 4 C, 12 O
(c) 2 Cu, 2 S, 18 O, 20 H

2.100 (a) 6 (b) 3 (c) 27

2.102 (a) K^+ (b) Br^- (c) Mg^{2+}
(d) S^{2-} (e) Al^{3+}

2.104 (a) NaBr (b) KI (c) BaO
(d) $MgBr_2$ (e) BaF_2

2.106 (a) KNO_3 (b) $Ca(C_2H_3O_2)_2$ (c) NH_4Cl
(d) $Fe_2(CO_3)_3$ (e) $Mg_3(PO_4)_2$

2.108 (a) PbO and PbO_2 (b) SnO and SnO_2 (c) MnO and Mn_2O_3
(d) FeO and Fe_2O_3 (e) Cu_2O and CuO

2.110 (a) silicon dioxide (b) xenon tetrafluoride
(c) tetraphosphorus decaoxide (d) dichlorine heptaoxide

2.112 (a) calcium sulfide (b) aluminum bromide
(c) sodium phosphide (d) barium arsenide
(e) rubidium sulfide

2.114 (a) iron(II) sulfide (b) copper(II) oxide
(c) tin(IV) oxide (d) cobalt(II) chloride hexahydrate

2.116 (a) sodium nitrite (b) potassium permanganate
(c) magnesium sulfate heptahydrate (d) potassium thiocyanate

2.118 (a) ionic chromium(II) chloride
(b) molecular disulfur dichloride
(c) ionic ammonium acetate
(d) molecular sulfur trioxide
(e) ionic potassium iodate
(f) molecular tetraphosphorous hexaoxide
(g) ionic calcium sulfite
(h) ionic silver cyanide
(i) ionic zinc(II) bromide
(j) molecular hydrogen selenide

2.120 (a) Na_2HPO_4 (b) Li_2Se (c) $Cr(C_2H_3O_2)_3$
(d) S_2F_{10} (e) $Ni(CN)_2$ (f) Fe_2O_3
(g) SbF_5

2.122 (a) $(NH_4)_2S$ (b) $Cr_2(SO_4)_3 \cdot 6H_2O$ (c) SiF_4
(d) MoS_2 (e) $SnCl_4$ (f) H_2Se
(g) P_4S_6

2.124 diselenium hexasulfide and diselenium tetrasulfide

Practice Exercises

3.1 $$\text{mol Al} = 3.47 \text{ g Al}\left(\frac{1 \text{ mol Al}}{26.98 \text{ g Al}}\right) = 0.129 \text{ mol Al}$$

3.2 $$\text{Uncertainty in moles} = \pm 0.002 \text{ g}\left(\frac{1 \text{ mol Si}}{28.0855 \text{ g Si}}\right) = \pm 7.12 \times 10^{-5} \text{ mol Si}$$

3.3 Find the mass of 5.64×10^{18} molecules of $Ca(NO_3)_2$ (MW = 164.09 g/mo)

$$\text{g} = 5.64 \times 10^{18}\left(\frac{1 \text{ mol Ca}(NO_3)_2}{6.022 \times 10^{23} \text{ molecules Ca}(NO_3)_2}\right)\left(\frac{164.09 \text{ g Ca}(NO_3)_2}{1 \text{ mol Ca}(NO_3)_2}\right) = 1.54 \times 10^{-3} \text{ g}$$

$\text{g} = 1.54 \times 10^{-3} \text{ g} = 0.00154 \text{ g}$
Many laboratory balances can measure 1 mg (0.001 g); therefore, it is possible to weigh 5.64×10^{18} molecules of $Ca(NO_3)_2$.

3.4 Formula mass of sucrose = (12 C)(12.011 g/mol) + (22 H)(1.0079 g/mol) + (11 O)(15.9994 g/mol) = 342.299 g/mol
0.002 g of uncertainty = ? mol of sucrose

$$\text{mol of sucrose} = 0.002 \text{ g}\left(\frac{1 \text{ mol sucrose}}{342.299 \text{ g}}\right) = 5.8 \times 10^{-6} \text{ mol sucrose}$$

$$\text{molecules of sucrose} = 5.8 \times 10^{-6} \text{ mol sucrose}\left(\frac{6.022 \times 10^{23} \text{ molecules sucrose}}{1 \text{ mol sucrose}}\right) =$$

3.5×10^{18} molecules of sucrose

3.5 Aluminum sulfate: $Al_2(SO_4)_3$, the aluminum is Al^{3+}

$$\text{mole } Al^{3+} = 0.0774 \text{ mol } SO_4^{2-}\left(\frac{2 \text{ mol } Al^{3+}}{3 \text{ mol } SO_4^{2-}}\right) = 0.0516 \text{ mol } Al^{3+}$$

3.6 $$\text{mol N} = (8.60 \text{ mol O})\left(\frac{2 \text{ mol N}}{5 \text{ mol O}}\right) = 3.44 \text{ mol N atoms}$$

3.7 $$\text{g Fe} = (25.6 \text{ g O})\left(\frac{1 \text{ mol O}}{16.0 \text{ g O}}\right)\left(\frac{2 \text{ mol Fe}}{3 \text{ mol O}}\right)\left(\frac{55.8 \text{ g Fe}}{1 \text{ mol Fe}}\right) = 59.5 \text{ g Fe}$$

3.8 $$\text{g Fe} = (15.0 \text{ g } Fe_2O_3)\left(\frac{1 \text{ mol } Fe_2O_3}{159.7 \text{ g } Fe_2O_3}\right)\left(\frac{2 \text{ mol Fe}}{1 \text{ mol } Fe_2O_3}\right)\left(\frac{55.85 \text{ g Fe}}{1 \text{ mol Fe}}\right) = 10.5 \text{ g Fe}$$

3.9 $$\text{g Fe} = (12.0 \text{ g O})\left(\frac{1 \text{ mol O}}{16.00 \text{ g O}}\right)\left(\frac{1 \text{ mol } Fe_2O_3}{3 \text{ mol O}}\right)\left(\frac{2 \text{ mol Fe}}{1 \text{ mol } Fe_2O_3}\right)\left(\frac{55.85 \text{ g Fe}}{1 \text{ mol Fe}}\right) = 27.9 \text{ g Fe}$$

3.10 $$\%\text{ H} = \left(\frac{\text{mass H}}{\text{total mass}}\right) \times 100\% = \left(\frac{0.0870 \text{ g H}}{0.6672 \text{ g total}}\right) \times 100\% = 13.04\%$$

$$\%\text{ C} = \left(\frac{\text{mass C}}{\text{total mass}}\right) \times 100\% = \left(\frac{0.3481 \text{ g H}}{0.6672 \text{ g total}}\right) \times 100\% = 52.17\%$$

It is likely that the compound contains another element since the percentages do not add up to 100%.

3.11 % N = 0.2012/0.5462 × 100% = 36.84% N
% O = 0.3450/0.5462 × 100% = 63.16% O
Since these two values constitute 100%, there are no other elements present.

3.12 We first determine the number of grams of each element that are present in one mol of sample:
2 mol N × 14.01 g/mol = 28.02 g N
4 mol O × 16.00 g/mol = 64.00 g O
The percentages by mass are then obtained using the formula mass of the compound (92.02 g):
% N = (28.02/92.02) × 100% = 30.45% N
% O = (64.00/92.02) × 100% = 69.55% O

3.13 N_2O: Formula mass = 44.02 g/mol
2 mol N × 14.01 g/mol = 28.02 g N % N = (28.02/44.02) × 100% = 63.65% N
1 mol O × 16.00 g/mol = 16.00 g O % O = (16.00/44.02) × 100% = 36.34% O
NO: Formula mass = 30.01 g/mol
1 mol N × 14.01 g/mol = 14.01 g N % N = (14.01/30.01) × 100% = 46.68% N
1 mol O × 16.00 g/mol = 16.00 g O % O = (16.00/30.01) × 100% = 53.32% O
NO_2: Formula mass = 46.01 g/mol
1 mol N × 14.01 g/mol = 14.01 g N % N = (14.01/46.01) × 100% = 30.45% N
2 mol O × 16.00 g/mol = 32.00 g O % O = (32.00/46.01) × 100% = 69.55% O
N_2O_3: Formula mass = 76.02 g/mol
2 mol N × 14.01 g/mol = 28.02 g N % N = (28.02/76.02) × 100% = 36.86% N
3 mol O × 16.00 g/mol = 48.00 g O % O = (48.00/76.02) × 100% = 63.14% O
N_2O_4: Formula mass = 92.02 g/mol
2 mol N × 14.01 g/mol = 28.02 g N % N = (28.02/92.02) × 100% = 30.45% N
4 mol O × 16.00 g/mol = 64.00 g O % O = (64.00/92.02) × 100% = 69.55% O
N_2O_5: Formula mass = 108.02 g/mol
2 mol N × 14.01 g/mol = 28.02 g N % N = (28.02/108.02) × 100% = 25.94% N
5 mol O × 16.00 g/mol = 80.00 g O % O = (80.00/108.02) × 100% = 74.06% O

The compound N_2O_3 corresponds to the data in Practice Exercise 3.11.

3.14 We first determine the number of mol of each element as follows:

$$\text{mol N} = (0.712 \text{ g N})\left(\frac{1 \text{ mol N}}{14.01 \text{ g N}}\right) = 0.0508 \text{ mol N}$$

We need to know the number of grams of O. Since there is a total of 1.525 g of compound and the only other element present is N, the mass of O = 1.525 g – 0.712 g = 0.813 g O.

$$\text{mol O} = (0.813 \text{ g O})\left(\frac{1 \text{ mol O}}{16.00 \text{ g O}}\right) = 0.0508 \text{ mol O}$$

Since these two mole amounts are the same, the empirical formula is NO.

3.15 First, find the number of moles of each element, then determine the empirical formula by comparing the ratio of the number of moles of each element.
Start with the number of moles of S:

$$\text{mol S} = 0.7625 \text{ g S}\left(\frac{1 \text{ mol S}}{32.066 \text{ g S}}\right) = 0.02378 \text{ mol S}$$

Then find the number of moles of O: since there are only two elements in the compound, S and O, the remaining mass is O
g O = 1.525 g compound – 0.7625 g S = 0.7625 g O

$$\text{mol O} = 0.7625 \text{ g O}\left(\frac{1 \text{ mol O}}{15.9994 \text{ g O}}\right) = 0.04766 \text{ mol O}$$

The empirical formula is
$S_{0.02378}O_{0.4766}$
The empirical formula must be in whole numbers, so divide by the smaller subscript:
$S_{\frac{0.02378}{0.02378}}O_{\frac{0.04766}{0.02378}}$ which becomes SO_2

3.16 $$\text{mol Al} = 5.68 \text{ tons Al}\left(\frac{2000 \text{ lb Al}}{1 \text{ ton Al}}\right)\left(\frac{454 \text{ g Al}}{1 \text{ lb Al}}\right)\left(\frac{1 \text{ mol Al}}{26.98 \text{ g Al}}\right) = 1.91 \times 10^5 \text{ mol Al}$$

$$\text{mol O} = 5.04 \text{ tons O}\left(\frac{2000 \text{ lb O}}{1 \text{ ton O}}\right)\left(\frac{454 \text{ g O}}{1 \text{ lb O}}\right)\left(\frac{1 \text{ mol O}}{16.00 \text{ g O}}\right) = 2.86 \times 10^5 \text{ mol O}$$

Empirical Formula: $Al_{1.91\times10^5}O_{2.86\times10^5}$
In whole numbers: $Al_{\frac{1.91\times10^5}{1.91\times10^5}}O_{\frac{2.86\times10^5}{1.91\times10^5}}$ which becomes $AlO_{1.5}$ and multiply the subscripts by 2: Al_2O_3

3.17 We first determine the number of mol of each element as follows:

$$\text{mol N} = (0.522 \text{ g N})\left(\frac{1 \text{ mol N}}{14.01 \text{ g N}}\right) = 0.0373 \text{ mol N}$$

We need to know the number of grams of O. Since there is a total of 2.012 g of compound and the only other element present is N, the mass of O = 2.012 g – 0.522 g = 1.490 g O.

$$\text{mol O} = (1.490 \text{ g O})\left(\frac{1 \text{ mol O}}{16.00 \text{ g O}}\right) = 0.0931 \text{ mol O}$$

Since these two mole amounts are the same, the empirical formula is $N_{0.0373}O_{0.0931}$; to have the empirical formula in whole numbers, first divide by the smaller number of moles: $N_{\frac{0.0373}{0.0373}}O_{\frac{0.0931}{0.0373}}$ which is $NO_{2.5}$, now to have whole numbers, multiply the subscripts by 2: N_2O_5.

3.18 It is convenient to assume that we have 100 g of the sample, so that the % by mass values may be taken directly to represent masses. Thus there is 32.4 g of Na, 22.6 g of S and (100.00 – 32.4 – 22.6) = 45.0 g of O. Now, convert these masses to a number of mol:

$$\text{mol Na} = (32.4 \text{ g Na})\left(\frac{1 \text{ mol Na}}{23.00 \text{ g Na}}\right) = 1.40 \text{ mol Na}\theta$$

$$\text{mol S} = (22.6 \text{ g S})\left(\frac{1 \text{ mol S}}{32.06 \text{ g S}}\right) = 0.705 \text{ mol S}$$

$$\text{mol O} = (45.0 \text{ g O})\left(\frac{1 \text{ mol O}}{16.00 \text{ g O}}\right) = 2.81 \text{ mol O}$$

Next, we divide each of these mol amounts by the smallest in order to deduce the simplest whole number ratio:
For Na: 1.40 mol/0.705 mol = 1.99
For S: 0.705 mol/0.705 mol = 1.00
For O: 2.81 mol/0.705 mol = 3.99
The empirical formula is Na_2SO_4.

3.19 It is convenient to assume that we have 100 g of the sample, so that the % by mass values may be taken directly to represent masses. Thus there is 81.79 g of C, 6.10 g of H and (100.00 – 81.79 – 6.10) = 12.11 g of O. Now, convert these masses to a number of mol:

$$\text{mol C} = (81.79 \text{ g C})\left(\frac{1 \text{ mol C}}{12.01 \text{ g C}}\right) = 6.81 \text{ mol C}$$

$$\text{mol H} = (6.10 \text{ g H})\left(\frac{1 \text{ mol H}}{1.008 \text{ g H}}\right) = 6.05 \text{ mol H}$$

$$\text{mol O} = (12.11 \text{ g O})\left(\frac{1 \text{ mol O}}{16.00 \text{ g O}}\right) = 0.757 \text{ mol O}$$

Next, we divide each of these mol amounts by the smallest in order to deduce the simplest whole number ratio:

For C: 6.81 mol/0.757 mol = 9.00
For H: 6.05 mol/0.757 mol = 7.99
For O: 0.757 mol/0.757 mol = 1.00

The empirical formula is C_9H_8O

3.20 Find the moles of S and C using the stoichiometric ratios, then find the empirical formula from the ratio of moles of S and C.
FM SO_2 = 64.06 g mol^{-1} FM CO_2 = 44.01 g mol^{-1}

$$\text{mol S} = 0.640 \text{ g } SO_2\left(\frac{1 \text{ mol } SO_2}{64.06 \text{ g } SO_2}\right)\left(\frac{1 \text{ mol S}}{1 \text{ mol } SO_2}\right) = 9.99 \times 10^{-3}$$

$$\text{mol C} = 0.220 \text{ g } CO_2\left(\frac{1 \text{ mol } CO_2}{44.01 \text{ g } CO_2}\right)\left(\frac{1 \text{ mol C}}{1 \text{ mol } CO_2}\right) = 5.00 \times 10^{-3}$$

Empirical Formula $C_{5.00\times10^{-3}}S_{9.99\times10^{-3}}$ divide both subscripts by 5.00×10^{-3} to get CS_2.

3.21 Since the entire amount of carbon that was present in the original sample appears among the products only as CO_2, we calculate the amount of carbon in the sample as follows:

$$\text{g C} = (7.406 \text{ g } CO_2)\left(\frac{1 \text{ mol } CO_2}{44.01 \text{ g } CO_2}\right)\left(\frac{1 \text{ mol C}}{1 \text{ mol } CO_2}\right)\left(\frac{12.01 \text{ g C}}{1 \text{ mol C}}\right) = 2.021 \text{ g C}$$

Similarly, the entire mass of hydrogen that was present in the original sample appears among the products only as H_2O. Thus the mass of hydrogen in the sample is:

$$\text{g H} = (3.027 \text{ g } H_2O)\left(\frac{1 \text{ mol } H_2O}{18.02 \text{ g } H_2O}\right)\left(\frac{2 \text{ mol H}}{1 \text{ mol } H_2O}\right)\left(\frac{1.008 \text{ g H}}{1 \text{ mol H}}\right) = 0.3386 \text{ g H}$$

The mass of oxygen in the original sample is determined by difference:

5.048 g – 2.021 g – 0.3386 g = 2.688 g O

Next, these mass amounts are converted to the corresponding mol amounts:

$$\text{mol C} = (2.021 \text{ g C})\left(\frac{1 \text{ mol C}}{12.01 \text{ g C}}\right) = 0.1683 \text{ mol C}$$

$$\text{mol H} = (0.3386 \text{ g H})\left(\frac{1 \text{ mol H}}{1.008 \text{ g H}}\right) = 0.3359 \text{ mol H}$$

$$\text{mol O} = (2.688 \text{ g O})\left(\frac{1 \text{ mol O}}{16.00 \text{ g O}}\right) = 0.1680 \text{ mol O}$$

The simplest formula is obtained by dividing each of these mol amounts by the smallest:
For C: 0.1683 mol/0.1680 mol= 1.002
for H: 0.3359 mol/0.1680 mol= 1.999
For O: 0.1680 mol/0.1680 mol = 1.000
These values give us the simplest formula directly, namely CH_2O.

3.22 To find the molecular formula, divide the molecular mass by the formula mass of the empirical formula, then multiply the subscripts of the empirical formula by that value.
Formula mass of CH_2Cl: 49.48 g mol^{-1}
Formula mass of CHCl: 48.47 g mol^{-1}

For CH_2Cl $\quad \frac{100}{49.48} = 2.02$ and $\frac{289}{49.48} = 5.84$

For CHCl: $\quad \frac{100}{48.47} = 2.06$ and $\frac{289}{48.47} = 5.96$

The CH_2Cl rounds better using the molecular mass of 100, therefore multiply the subscripts by 2 and the formula is $C_2H_4Cl_2$.
For CHCl, the molecular mass of 289 gives a multiple of 6, therefore the formula is $C_6H_6Cl_6$.

3.23 The formula mass of the empirical unit is 1 N + 2 H = 16.03. Since this is half of the molecular mass, the molecular formula is N_2H_4.

3.24 $AlCl_3(aq) + Na_3PO_4(aq) \rightarrow AlPO_4(s) + 3NaCl(aq)$

3.25 $3CaCl_2(aq) + 2K_3PO_4(aq) \rightarrow Ca_3(PO_4)_2(s) + 6KCl(aq)$

3.26 $$\text{mol } O_2 = (6.76 \text{ mol } SO_3)\left(\frac{1 \text{ mol } O_2}{2 \text{ mol } SO_3}\right) = 3.38 \text{ mol } O_2$$

3.27 $$\text{mol } H_2SO_4 = (0.366 \text{ mol NaOH})\left(\frac{1 \text{ mol } H_2SO_4}{2 \text{ mol NaOH}}\right) = 0.183 \text{ mol } H_2SO_4$$

3.28 $$\text{g } Al_2O_3 = (86.0 \text{ g Fe})\left(\frac{1 \text{ mol Fe}}{55.85 \text{ g Fe}}\right)\left(\frac{1 \text{ mol } Al_2O_3}{2 \text{ mol Fe}}\right)\left(\frac{102.0 \text{ g } Al_2O_3}{1 \text{ mol } Al_2O_3 \text{ mol } Al_2O_3}\right)$$
$$= 78.5 \text{ g } Al_2O_3$$

3.29 $$\text{g } CO_2 = (1.50 \times 10^2 \text{ g CaO})\left(\frac{1 \text{ mol CaO}}{56.08 \text{ g CaO}}\right)\left(\frac{1 \text{ mol } CO_2}{1 \text{ mol CaO}}\right)\left(\frac{44.01 \text{ g } CO_2}{1 \text{ mol } CO_2}\right) = 1.18 \times 10^2 \text{ g } CO_2$$

3.30 First determine the number of grams of $CaCO_3$ that would be required to react completely with the given amount of HCl:

$$\text{g } CaCO_3 = (125 \text{ g HCl})\left(\frac{1 \text{ mol HCl}}{36.461 \text{ g HCl}}\right)\left(\frac{1 \text{ mol } CaCO_3}{2 \text{ mol HCl}}\right)\left(\frac{100.088 \text{ g } CaCO_3}{1 \text{ mol } CaO_3}\right) = 171.57 \text{ g } CaCO_3$$

Since this is more than the amount that is available, we conclude that $CaCO_3$ is the limiting reactant. The rest of the calculation is therefore based on the available amount of $CaCO_3$:

$$g\ CO_2 = (125\ g\ CaCO_3)\left(\frac{1\ mol\ CaCO_3}{100.088\ g\ CaCO_3}\right)\left(\frac{1\ mol\ CO_2}{1\ mol\ CaCO_3}\right)\left(\frac{44.01\ g\ CO_2}{1\ mol\ CO_2}\right)$$

$$= 55.0\ g\ CO_2$$

For the number of grams of left over HCl, the excess reagent, find the amount of HCl used and then subtract that from the amount of HCl started with, 125 g.

$$g\ HCl\ used = (125\ g\ CaCO_3)\left(\frac{1\ mol\ CaCO_3}{100.088\ g\ CaCO_3}\right)\left(\frac{2\ mol\ HCl}{1\ mol\ CaCO_3}\right)\left(\frac{36.461\ g\ HCl}{1\ mol\ HCl}\right)$$

$$= 91.1\ g\ HCl$$

g HCl remaining = 125 g – 91.1 g = 34 g HCl remaining

3.31 First determine the number of grams of O_2 that would be required to react completely with the given amount of ammonia:

$$g\ O_2 = (30.00\ g\ NH_3)\left(\frac{1\ mol\ NH_3}{17.03\ g\ NH_3}\right)\left(\frac{5\ mol\ O_2}{4\ mol\ NH_3}\right)\left(\frac{32.00\ g\ O_2}{1\ mol\ O_2}\right)$$

$$= 70.46\ g\ O_2$$

Since this is more than the amount that is available, we conclude that oxygen is the limiting reactant. The rest of the calculation is therefore based on the available amount of oxygen:

$$g\ NO = (40.00\ g\ O_2)\left(\frac{1\ mol\ O_2}{32.00\ g\ O_2}\right)\left(\frac{4\ mol\ NO}{5\ mol\ O_2}\right)\left(\frac{30.01\ g\ NO}{1\ mol\ NO}\right)$$

$$= 30.01\ g\ NO$$

3.32 First determine the number of grams of salicylic acid, $HOOCC_6H_4OH$ that would be required to react completely with the given amount of acetic anhydride, $C_4H_6O_3$:

$$g\ HOOCC_6H_4OH = (15.6\ g\ C_4H_6O_3) \times$$

$$\left(\frac{1\ mol\ C_4H_6O_3}{102.09\ g\ C_4H_6O_3}\right)\left(\frac{2\ mol\ HOOCC_6H_4OH}{1\ mol\ C_4H_6O_3}\right)\left(\frac{138.12\ g\ HOOCC_6H_4OH}{1\ mol\ HOOCC_6H_4OH}\right)$$

$$= 42.2\ g\ HOOCC_6H_4OH$$

Since more salicylic acid is required than is available, it is the limiting reagent. Once 28.2 g of salicylic acid is reacted the reaction will stop, even though there are 15.6 g of acetic anhydride present. Therefore the salicylic acid is the limiting reactant. The theoretical yield of aspirin $HOOCC_6H_4O_2C_2H_3$ is therefore based on the amount of salicylic acid added. This is calculated below:

$$g\ HOOCC_6H_4O_2C_2H_3 = (28.2\ g\ HOOCC_6H_4OH) \times$$

$$\left(\frac{1\ mol\ HOOCC_6H_4OH}{138.12\ g\ HOOCC_6H_4OH}\right)\left(\frac{2\ mol\ HOOCC_6H_4O_2C_2H_3}{2\ mol\ HOOCC_6H_4OH}\right)\left(\frac{180.16\ g\ HOOCC_6H_4O_2C_2H_3}{1\ mol\ HOOCC_6H_4O_2C_2H_3}\right)$$

$$= 36.78\ g\ HOOCC_6H_4O_2C_2O_3$$

Now the percentage yield can be calculated from the amount of acetyl salicylic acid actually produced, 30.7 g:

$$\text{percent yield} = \left(\frac{\text{actual yield}}{\text{theoretical yield}}\right)\times 100\% = \left(\frac{30.7\ g\ HOOCC_6H_4O_2C_2H_3}{36.78\ g\ HOOCC_6H_4O_2C_2H_3}\right)\times 100\%$$

$$= 83.5\%$$

3.33 First determine the number of grams of C_2H_5OH that would be required to react completely with the given amount of sodium dichromate:

$$\text{g } C_2H_5OH = (90.0 \text{ g } Na_2Cr_2O_7)\left(\frac{1 \text{ mol } Na_2Cr_2O_7}{262.0 \text{ g } Na_2Cr_2O_7}\right)\left(\frac{3 \text{ mol } C_2H_5OH}{2 \text{ mol } Na_2Cr_2O_7}\right)\left(\frac{46.08 \text{ g } C_2H_5OH}{1 \text{ mol } C_2H_5OH}\right)$$

$$= 23.7 \text{ g } C_2H_5OH$$

Once this amount of C_2H_5OH is reacted the reaction will stop, even though there are 24.0 g C_2H_5OH present, because the $Na_2Cr_2O_7$ will be used up. Therefore $Na_2Cr_2O_7$ is the limiting reactant. The theoretical yield of acetic acid ($HC_2H_3O_2$) is therefore based on the amount of $Na_2Cr_2O_7$ added. This is calculated below:

$$\text{g } HC_2H_3O_2 = (90.0 \text{ g } Na_2Cr_2O_7)\left(\frac{1 \text{ mol } Na_2Cr_2O_7}{262.0 \text{ g } Na_2Cr_2O_7}\right)\left(\frac{3 \text{ mol } HC_2H_3O_2}{2 \text{ mol } Na_2Cr_2O_7}\right)\left(\frac{60.06 \text{ g } HC_2H_3O_2}{1 \text{ mol } HC_2H_3O_2}\right)$$

$$= 30.9 \text{ g } HC_2H_3O_2$$

Now the percentage yield can be calculated from the amount of acetic acid actually produced, 26.6 g:

$$\text{percent yield} = \left(\frac{\text{actual yield}}{\text{theoretical yield}}\right) \times 100 = \left(\frac{26.6 \text{ g } HC_2H_3O_2}{30.9 \text{ g } HC_2H_3O_2}\right) \times 100 = 86.1\%$$

Review Problems

3.25 1:2, 2 mol N to 4 mol O or in the smallest whole number ratio 1 mol N to 2 mol O

3.27 $$1.56 \times 10^{21} \text{ atoms Ta}\left(\frac{1 \text{ mol Ta}}{6.022 \times 10^{23} \text{ atoms Ta}}\right) = 2.59 \times 10^{-3} \text{ mole Ta}$$

3.29 (a) 6 atom C:11 atom H
(b) 12 mole C:11 mole O
(c) 12 atom H:11 atom O
(d) 12 mole H:11 mole O

3.31 $$\text{mol Bi} = (1.58 \text{ mol O})\left(\frac{2 \text{ mol Bi}}{3 \text{ mol O}}\right) = 1.05 \text{ mol Bi}$$

3.33 $$\text{mol Cr} = (2.16 \text{ mol } Cr_2O_3)\left(\frac{2 \text{ mol Cr}}{1 \text{ mol } Cr_2O_3}\right) = 4.32 \text{ mol Cr}$$

3.35 (a) $$\left(\frac{2 \text{ mol Al}}{3 \text{ mol S}}\right) \text{ or } \left(\frac{3 \text{ mol S}}{2 \text{ mol Al}}\right)$$

(b) $$\left(\frac{3 \text{ mol S}}{1 \text{ mol } Al_2(SO_4)_3}\right) \text{ or } \left(\frac{1 \text{ mol } Al_2(SO_4)_3}{3 \text{ mol S}}\right)$$

(c) $$\text{mol Al} = (0.900 \text{ mol S})\left(\frac{2 \text{ mol Al}}{3 \text{ mol S}}\right) = 0.600 \text{ mol Al}$$

(d) $$\text{mol S} = (1.16 \text{ mol } Al_2(SO_4)_3)\left(\frac{3 \text{ mol S}}{1 \text{ mol } Al_2(SO_4)_3}\right) = 3.48 \text{ mol S}$$

3.37 Based on the balanced equation:
$2\ NH_3(g) \rightarrow N_2(g) + 3H_2(g)$

From this equation the conversion factors can be written:

$$\left(\frac{1 \text{ mol } N_2}{2 \text{ mol } NH_3}\right) \text{ and } \left(\frac{3 \text{ mol } H_2}{2 \text{ mol } NH_3}\right)$$

To determine the moles produced, simply convert from starting moles to end moles:

$$\text{mole } N_2 = 0.145 \text{ mol } NH_3\left(\frac{1 \text{ mol } N_2}{2 \text{ mol } NH_3}\right) = 0.0725 \text{ mol } N_2$$

The moles of hydrogen are calculated similarly:

$$0.145 \text{ mol } NH_3\left(\frac{3 \text{ mol } H_2}{2 \text{ mol } NH_3}\right) = 0.218 \text{ mol } H_2$$

3.39 $\text{mol } UF_6 = (1.25 \text{ mol } CF_4)\left(\frac{4 \text{ mol F}}{1 \text{ mol } CF_4}\right)\left(\frac{1 \text{ mol } UF_6}{6 \text{ mol F}}\right) = 0.833 \text{ mol } CF_4$

3.41 $\text{atoms C} = (4.13 \text{ mol H})\left(\frac{1 \text{ mol } C_3H_8}{8 \text{ mol H}}\right)\left(\frac{6.022 \times 10^{23} \text{ molecules } C_3H_8}{1 \text{ mol } C_3H_8}\right)\left(\frac{3 \text{ atoms C}}{1 \text{ molecule } C_3H_8}\right)$

$= 9.33 \times 10^{23}$ atoms C

3.43 C, H and O atoms in glucose = 6 atoms C + 12 atoms H + 6 atoms O = 24 atoms

$$\text{atoms} = (0.260 \text{ mol glucose})\left(\frac{6.022 \times 10^{23} \text{ molecules glucose}}{1 \text{ mol glucose}}\right)\left(\frac{24 \text{ atoms}}{1 \text{ molecule glucose}}\right)$$

$= 3.76 \times 10^{24}$ atoms

3.45 $\text{mol C–12} = 6 \text{ g} \times \left(\frac{1 \text{ mol C-12}}{12.00 \text{ g C-12}}\right) = 0.5 \text{ mol C–12}$

$$\text{atoms C–12} = 0.5 \text{ mol}\left(\frac{6.022\times10^{23} \text{ atoms C-12}}{1 \text{ mol C-12}}\right) = 3.01 \times 10^{23} \text{ atoms C–12}$$

3.47 (a) $\text{g Fe} = (1.35 \text{ mol Fe})\left(\frac{55.85 \text{ g Fe}}{1 \text{ mole Fe}}\right) = 75.4 \text{ g Fe}$

(b) $\text{g O} = (24.5 \text{ mol O})\left(\frac{16.0 \text{ g O}}{1 \text{ mole O}}\right) = 392 \text{ g O}$

(c) $\text{g Ca} = (0.876)\left(\frac{40.08 \text{ g Ca}}{1 \text{ mole Ca}}\right) = 35.1 \text{ g Ca}$

3.49 $\text{g K} = 2.00 \times 10^{12} \text{ atoms K}\left(\frac{1 \text{ mol K}}{6.022\times10^{23} \text{ atoms K}}\right)\left(\frac{39.10 \text{ g K}}{1 \text{ mol K}}\right) = 1.30 \times 10^{-10} \text{ g K}$

3.51 $\text{mol Ni} = 17.7 \text{ g Ni}\left(\frac{1 \text{ mol Ni}}{58.69 \text{ g Ni}}\right) = 0.302 \text{ mol Ni}$

3.53 Note: all masses are in g/mole

(a) $NaHCO_3$ = 1Na + 1H + 1C + 3O
= (22.98977) + (1.00794) + (12.0107) + (3 × 15.9994)
= 84.00661 g/mole = 84.0066 g/mol

(b) $(NH_4)_2CO_3$ = 2N + 8H + C + 3O
= (2 × 14.0067) + (8 × 1.00794) + (12.0107) + (3 × 15.9994)
= 96.08582 g/mole = 96.0858 g/mol

(c) $CuSO_4{\cdot}5H_2O$ = 1Cu + 1S +9O + 10H
= 63.546 + 32.065 + (9 × 15.9994) + (10 × 1.00794)
= 249.685 g/mole

(d) $K_2Cr_2O_7$ = 2K + 2Cr + 7O
= (2 × 39.0983) + (2 × 51.9961) + (7 × 15.9994)
= 294.1846 g/mole

(e) $Al_2(SO_4)_3$ = 2Al + 3S + 12O
= (2 × 26.98154) + (3 × 32.065) + (12 × 15.9994)
= 342.15088 g/mole = 342.151 g/mol

3.55 (a) $$\text{g Ca}_3(\text{PO}_4)_2 = (1.25\text{ mol Ca}_3(\text{PO}_4)_2)\left(\frac{310.18\text{ g Ca}_3(\text{PO}_4)_2}{1\text{ mol Ca}_3(\text{PO}_4)_2}\right) = 388\text{ g Ca}_3(\text{PO}_4)_2$$

(b) $$\text{g Fe(NO}_3)_3 = (0.625\text{ mmol Fe(NO}_3)_3)\left(\frac{1\text{ mol Fe(NO}_3)_3}{1000\text{ mmol Fe(NO}_3)_3}\right)\left(\frac{241.86\text{ g Fe(NO}_3)_3}{1\text{ mol Fe(NO}_3)_3}\right) = 0.151\text{ g}$$
$Fe(NO_3)_3$

(c) $$\text{g C}_4\text{H}_{10} = (0.600\ \mu\text{mol C}_4\text{H}_{10})\left(\frac{1\text{ mol C}_4\text{H}_{10}}{1{\times}10^6\ \mu\text{mol C}_4\text{H}_{10}}\right)\left(\frac{58.12\text{ g C}_4\text{H}_{10}}{1\text{ mol C}_4\text{H}_{10}}\right) = 3.49 \times 10^{-5}\text{ g C}_4\text{H}_{10}$$

(d) $$\text{g (NH}_4)_2\text{CO}_3 = (1.45\text{ mol (NH}_4)_2\text{CO}_3)\left(\frac{96.09\text{ g (NH}_4)_2\text{CO}_3}{1\text{ mol (NH}_4)_2\text{CO}_3}\right) = 139\text{ g (NH}_4)_2\text{CO}_3$$

3.57 (a) $$\text{moles CaCO}_3 = (21.5\text{ g CaCO}_3)\left(\frac{1\text{ mole CaCO}_3}{100.09\text{ g CaCO}_3}\right) = 0.215\text{ moles CaCO}_3$$

(b) $$\text{moles NH}_3 = (1.56\text{ ng NH}_3)\left(\frac{1\text{ g NH}_3}{1{\times}10^9\text{ ng NH}_3}\right)\left(\frac{1\text{ mole NH}_3}{17.03\text{ g NH}_3}\right) = 9.16 \times 10^{-11}\text{ moles NH}_3$$

(c) $$\text{moles Sr(NO}_3)_2 = (16.8\text{ g Sr(NO}_3)_2)\left(\frac{1\text{ mole Sr(NO}_3)_2}{211.6\text{ g Sr(NO}_3)_2}\right) = 7.94 \times 10^{-2}\text{ moles Sr(NO}_3)_2$$

(d) $$\text{moles Na}_2\text{CrO}_4 = (6.98\ \mu\text{g Na}_2\text{CrO}_4)\left(\frac{1\text{ g Na}_2\text{CrO}_4}{10^6\ \mu\text{g Na}_2\text{CrO}_4}\right)\left(\frac{1\text{ mole Na}_2\text{CrO}_4}{162.0\text{ g Na}_2\text{CrO}_4}\right)$$
$$= 4.31 \times 10^{-8}\text{ moles Na}_2\text{CrO}_4$$

3.59 The formula CaC_2 indicates that there is 1 mole of Ca for every 2 moles of C. Therefore, if there are 0.150 moles of C there must be 0.0750 moles of Ca.

$$\text{g Ca} = (0.075\text{ mol Ca})\left(\frac{40.078\text{ g Ca}}{1\text{ mole Ca}}\right) = 3.01\text{ g Ca}$$

3.61 mol N = (0.650 mol $(NH_4)_2CO_3$)$\left(\frac{2\text{ moles N}}{1\text{ mole }(NH_4)_2CO_3}\right)$ = 1.30 mol N

g $(NH_4)_2CO_3$ = (0.650 mol $(NH_4)_2CO_3$)$\left(\frac{96.09\text{ g }(NH_4)_2CO_3}{1\text{ mole }(NH_4)_2CO_3}\right)$ = 62.5 g $(NH_4)_2CO_3$

3.63 kg fertilizer = (1 kg N)$\left(\frac{1000\text{ g N}}{1\text{ kg N}}\right)\left(\frac{1\text{ mol N}}{14.01\text{ g N}}\right)\left(\frac{1\text{ mol }(NH_4)_2CO_3}{2\text{ mol N}}\right)$

$$\times\left(\frac{96.09\text{ g }(NH_4)_2CO_3}{1\text{ mol }(NH_4)_2CO_3}\right)\left(\frac{1\text{ kg }(NH_4)_2CO_3}{1000\text{ g }(NH_4)_2CO_3}\right) = 3.43\text{ kg fertilizer}$$

3.65 Assume one mole total for each of the following.

(a) The molar mass of NaH_2PO_4 is 119.98 g/mol.

$$\%\text{ Na} = \frac{23.0\text{ g Na}}{119.98\text{ g }NaH_2PO_4} \times 100\% = 19.2\%$$

$$\%\text{ H} = \frac{2.02\text{ g H}}{119.98\text{ g }NaH_2PO_4} \times 100\% = 1.68\%$$

$$\%\text{ P} = \frac{31.0\text{ g P}}{119.98\text{ g }NaH_2PO_4} \times 100\% = 25.8\%$$

$$\%\text{ O} = \frac{64.0\text{ g O}}{119.98\text{ g }NaH_2PO_4} \times 100\ \% = 53.3\ \%$$

(b) The molar mass of $NH_4H_2PO_4$ is 115.05 g/mol.

$$\%\text{ N} = \frac{14.0\text{ g N}}{115.05\text{ g }NH_4H_2PO_4} \times 100\% = 12.2\%$$

$$\%\text{ H} = \frac{6.05\text{ g H}}{115.05\text{ g }NH_4H_2PO_4} \times 100\% = 5.26\%$$

$$\%\text{ P} = \frac{31.0\text{ g P}}{115.05\text{ g }NH_4H_2PO_4} \times 100\% = 26.9\%$$

$$\%\text{ O} = \frac{64.0\text{ g O}}{115.05\text{ g }NH_4H_2PO_4} \times 100\ \% = 55.6\ \%$$

(c) The molar mass of $(CH_3)_2CO$ is 58.08 g/mol

$$\%\text{ C} = \frac{36.0\text{ g C}}{58.08\text{ g }(CH_3)_2CO} \times 100\% = 62.0\%$$

$$\%\text{ H} = \frac{6.05\text{ g H}}{58.08\text{ g }(CH_3)_2CO} \times 100\% = 10.4\%$$

$$\%\text{ O} = \frac{16.0\text{ g O}}{58.08\text{ g }(CH_3)_2CO} \times 100\% = 27.6\%$$

(d) The molar mass of calcium sulfate dihydrate is 136.2 g/mol.

$$\% \text{ Ca} = \frac{40.1 \text{ g Ca}}{172.2 \text{ g CaSO}_4 \bullet 2\text{H}_2\text{O}} \times 100\% = 23.3\%$$

$$\% \text{ S} = \frac{32.1 \text{ g S}}{172.2 \text{ g CaSO}_4 \bullet 2\text{H}_2\text{O}} \times 100\% = 18.6\%$$

$$\% \text{ O} = \frac{96.0 \text{ g O}}{172.2 \text{ g CaSO}_4 \bullet 2\text{H}_2\text{O}} \times 100\% = 55.7\%$$

$$\% \text{ H} = \frac{4.03 \text{ g H}}{172.2 \text{ g CaSO}_4 \bullet 2\text{H}_2\text{O}} \times 100\ \% = 2.34\ \%$$

(e) The molar mass of $CaSO_4 \bullet 2H_2O$ is 172.2 g/mol.

$$\% \text{ Ca} = \frac{40.1 \text{ g Ca}}{172.2 \text{ g CaSO}_4 \bullet 2\text{H}_2\text{O}} \times 100\% = 23.3\%$$

$$\% \text{ S} = \frac{32.1 \text{ g S}}{172.2 \text{ g CaSO}_4 \bullet 2\text{H}_2\text{O}} \times 100\% = 18.6\%$$

$$\% \text{ O} = \frac{96.0 \text{ g O}}{172.2 \text{ g CaSO}_4 \bullet 2\text{H}_2\text{O}} \times 100\% = 55.7\%$$

$$\% \text{ H} = \frac{4.03 \text{ g H}}{172.2 \text{ g CaSO}_4 \bullet 2\text{H}_2\text{O}} \times 100\ \% = 2.34\ \%$$

3.67

$$\% \text{ O in morphine} = \frac{48.00 \text{ g O}}{285.36 \text{ g C}_{17}\text{H}_{19}\text{NO}_3} \times 100\% = 16.82\% \text{ O}$$

$$\% \text{ O in heroin} = \frac{80.00 \text{ g O}}{369.44 \text{ g C}_{21}\text{H}_{23}\text{NO}_5} \times 100\% = 21.65\% \text{ O}$$

Therefore heroin has a higher percentage oxygen.

3.69

$$\% \text{ Cl in Freon-12} = \frac{70.90 \text{ g Cl}}{120.92 \text{ g CCl}_2\text{F}_2} \times 100\% = 58.63\% \text{ Cl}$$

$$\% \text{ Cl in Freon 141b} = \frac{70.9 \text{ g Cl}}{116.95 \text{ g C}_2\text{H}_3\text{Cl}_2\text{F}} \times 100\% = 60.62\% \text{ Cl}$$

Therefore Freon 141b has a higher percentage chlorine.

3.71

$$\% \text{ P} = \frac{0.539 \text{ g P}}{2.35 \text{ g compound}} \times 100\% = 22.9\%$$

$$\% \text{ Cl} = 100\% - 22.9\% = 77.1\%$$

3.73 For $C_{17}H_{25}N$, the molar mass (17C + 25H + 1N) equals 243.43 g/mole, and the three theoretical values for % by weight are calculated as follows:

$$\% \text{ C} = \frac{204.2 \text{ g C}}{243.4 \text{ g C}_{17}\text{H}_{25}\text{N}} \times 100\% = 83.89\%$$

$$\% \text{ H} = \frac{25.20 \text{ g H}}{243.4 \text{ g C}_{17}\text{H}_{25}\text{N}} \times 100\% = 10.35\%$$

$$\% \text{ N} = \frac{14.01 \text{ g N}}{243.4 \text{ g C}_{17}\text{H}_{25}\text{N}} \times 100\% = 5.76\%$$

These data are consistent with the experimental values cited in the problem.

3.75 $$\text{g O} = (7.14 \times 10^{21} \text{ atoms N})\left(\frac{1 \text{ mol N}}{6.02 \times 10^{23} \text{ atoms N}}\right)\left(\frac{5 \text{ mol O}}{2 \text{ mol N}}\right)\left(\frac{16.0 \text{ g O}}{1 \text{ mol O}}\right) = 0.474 \text{ g O}$$

3.77 The molecular formula is some integer multiple of the empirical formula. This means that we can divide the molecular formula by the largest possible whole number that gives an integer ratio among the atoms in the empirical formula.
(a) SCl (b) CH_2O (c) NH_3 (d) AsO_3 (e) HO

3.79 We begin by realizing that the mass of oxygen in the compound may be determined by difference:
0.896 g total – (0.111 g Na + 0.477 g Tc) = 0.308 g O.
Next we can convert each mass of an element into the corresponding number of moles of that element as follows:

$$\text{mol Na} = (0.111 \text{ g Na})\left(\frac{1 \text{ mol Na}}{23.00 \text{ g Na}}\right) = 4.83 \times 10^{-3} \text{ mol Na}$$

$$\text{mol Tc} = (0.477 \text{ g Tc})\left(\frac{1 \text{ mol Tc}}{98.9 \text{ g Tc}}\right) = 4.82 \times 10^{-3} \text{ mol Tc}$$

$$\text{mol O} = (0.308 \text{ g O})\left(\frac{1 \text{ mol O}}{16.0 \text{ g O}}\right) = 1.93 \times 10^{-2} \text{ mol O}$$

Now we divide each of these numbers of moles by the smallest of the three numbers, in order to obtain the simplest mole ratio among the three elements in the compound:

for Na, 4.83×10^{-3} moles / 4.82×10^{-3} moles = 1.00
for Tc, 4.82×10^{-3} moles / 4.82×10^{-3} moles = 1.00
for O, 1.93×10^{-2} moles / 4.82×10^{-3} moles = 4.00

These relative mole amounts give us the empirical formula: $NaTcO_4$.

3.81 Assume a 100 g sample:

$$\text{mol C} = (14.5 \text{ g C})\left(\frac{1 \text{ mol C}}{12.01 \text{ g C}}\right) = 1.21 \text{ mol C}$$

$$\text{mol Cl} = (85.5 \text{ g Cl})\left(\frac{1 \text{ mol Cl}}{35.45 \text{ g Cl}}\right) = 2.41 \text{ mol Cl}$$

Now we divide each of these numbers of moles by the smallest of the three numbers, in order to obtain the simplest mole ratio among the three elements in the compound:

for C, 1.21 moles / 1.21 moles = 1.00
for Cl, 2.41 moles /1.21 moles = 2.000

These relative mole amounts give us the empirical formula CCl_2

3.83 Assume a 100 g sample:

$$\text{mol C} = (72.96 \text{ g C})\left(\frac{1 \text{ mol C}}{12.011 \text{ g C}}\right) = 6.074 \text{ mol C}$$

$$\text{mol H} = (5.40 \text{ g H})\left(\frac{1 \text{ mol H}}{1.008 \text{ g H}}\right) = 5.36 \text{ mol H}$$

To find the number of moles of O, first we have to find the number of grams of O:
100 g total = (72.96 g C) + (5.40 g H) + (x g O)
g O = 21.64 g O

$$\text{mol O} = (21.64 \text{ g O})\left(\frac{1 \text{ mol O}}{15.999 \text{ g O}}\right) = 1.353 \text{ mol O}$$

Now we divide each of these numbers of moles by the smallest of the three numbers, in order to obtain the simplest mole ratio among the three elements in the compound:

for C, 6.074 moles / 1.353 moles = 4.49
for H, 5.36 moles / 1.353 moles = 3.96
for O, 1.353 moles / 1.353 moles = 1.00
These relative mole amounts give us the empirical formula $C_{4.5}H_4O$
Since we cannot have decimals as subscripts, multiply all of the subscripts by 2 to get the formula: $C_9H_8O_2$

3.85 All of the carbon is converted to carbon dioxide so,

$$\text{g C} = (1.312 \text{ g CO}_2)\left(\frac{1 \text{ mol CO}_2}{44.01 \text{ g CO}_2}\right)\left(\frac{1 \text{ mol C}}{1 \text{ mol CO}_2}\right)\left(\frac{12.01 \text{ g C}}{1 \text{ mol C}}\right) = 0.358 \text{ g C}$$

$$\text{mol C} = (0.358 \text{ g C})\left(\frac{1 \text{ mol C}}{12.01 \text{ g C}}\right) = 2.98 \times 10^{-2} \text{ mol C}$$

All of the hydrogen is converted to H_2O, so

$$\text{g H} = (0.805 \text{ g H}_2\text{O})\left(\frac{1 \text{ mol H}_2\text{O}}{18.02 \text{ g H}_2\text{O}}\right)\left(\frac{2 \text{ mol H}}{1 \text{ mol H}_2\text{O}}\right)\left(\frac{1.008 \text{ g H}}{1 \text{ mol H}}\right) = 0.0901 \text{ g H}$$

$$\text{mol H} = (0.0901 \text{ g H})\left(\frac{1 \text{ mol H}}{1.008 \text{ g H}}\right) = 8.93 \times 10^{-2} \text{ mol H}$$

The amount of O in the compound is determined by subtracting the mass of C and the mass of H from the sample.

$$\text{g O} = 0.684 \text{ g} - 0.358 \text{ g} - 0.0901 \text{ g} = 0.236 \text{ g O}$$

$$\text{mol O} = (0.236 \text{ g O})\left(\frac{1 \text{ mol O}}{16.00 \text{ g O}}\right) = 1.48 \times 10^{-2} \text{ mol O}$$

The relative mole ratios are:
for C, 0.0298 moles / 0.0148 moles = 2.01
for H, 0.0893 moles/ 0.0148 moles = 6.03
for O, 0.0148 moles / 0.0148 moles = 1.00

The relative mole amounts give the empirical formula C_2H_6O

3.87 This type of combustion analysis takes advantage of the fact that the entire amount of carbon in the original sample appears as CO_2 among the products. Hence the mass of carbon in the original sample must be equal to the mass of carbon that is found in the CO_2.

$$\text{g C} = (19.73 \times 10^{-3} \text{ g CO}_2)\left(\frac{1 \text{ mole CO}_2}{44.01 \text{ g CO}_2}\right)\left(\frac{1 \text{ mole C}}{1 \text{ mole CO}_2}\right)\left(\frac{12.011 \text{ g C}}{1 \text{ mole C}}\right) = 5.385 \times 10^{-3} \text{ g C}$$

Similarly, the entire mass of hydrogen that was present in the original sample ends up in the products as H_2O:

$$\text{g H} = (6.391 \times 10^{-3} \text{ g H}_2\text{O})\left(\frac{1 \text{ mole H}_2\text{O}}{18.02 \text{ g H}_2\text{O}}\right)\left(\frac{2 \text{ mole H}}{1 \text{ mole H}_2\text{O}}\right)\left(\frac{1.008 \text{ g H}}{1 \text{ mole H}}\right) = 7.150 \times 10^{-4} \text{ g H}$$

The mass of oxygen is determined by subtracting the mass due to C and H from the total mass:

6.853 mg total – (5.385 mg C + 0.7150 mg H) = 0.753 mg O.

Now, convert these masses to a number of moles:

$$\text{mol C} = (5.385 \times 10^{-3}\ \text{g C})\left(\frac{1\ \text{mol C}}{12.011\ \text{g C}}\right) = 4.484 \times 10^{-4}\ \text{mol C}$$

$$\text{mol H} = (7.150 \times 10^{-4}\ \text{g H})\left(\frac{1\ \text{mol H}}{1.0079\ \text{g H}}\right) = 7.094 \times 10^{-4}\ \text{mol H}$$

$$\text{mol O} = (7.53 \times 10^{-4}\ \text{g O})\left(\frac{1\ \text{mol O}}{15.999\ \text{g O}}\right) = 4.71 \times 10^{-5}\ \text{mol H}$$

The relative mole amounts are:
for C, 4.483×10^{-4} mol / 4.71×10^{-5} mol = 9.52
for H, 7.094×10^{-4} mol / 4.71×10^{-5} mol = 15.1
for O, 4.71×10^{-5} mol / 4.71×10^{-5} mol = 1.00

The relative mole amounts are not whole numbers as we would like. However, we see that if we double the relative number of moles of each compound, there are approximately 19 moles of C, 30 moles of H and 2 moles of O. If we assume these numbers are correct, the empirical formula is $C_{19}H_{30}O_2$, for which the formula weight is 290 g/mole.

In most problems where we attempt to determine an empirical formula, the relative mole amounts should work out to give a "nice" set of values for the formula. Rarely will a problem be designed that gives very odd coefficients. With experience and practice, you will recognize when a set of values is reasonable.

3.89 (a) Formula mass = 135.1 g

$$\frac{270.4\ \text{g/mol}}{135.1\ \text{g/mol}} = 2.001$$

The molecular formula is $Na_2S_4O_6$

(b) Formula mass = 73.50 g

$$\frac{147.0\ \text{g/mol}}{73.50\ \text{g/mol}} = 2.000$$

The molecular formula is $C_6H_4Cl_2$

(c) Formula mass = 60.48 g

$$\frac{181.4\ \text{g/mol}}{60.48\ \text{g/mol}} = 2.999$$

The molecular formula is $C_6H_3Cl_3$

3.91 The formula mass for the compound $C_{19}H_{30}O_2$ is 290 g/mol. Thus, the empirical and molecular formulas are equivalent.

3.93 From the information provided, we can determine the mass of mercury as the difference between the total mass and the mass of bromine:

g Hg = 0.389 g compound – 0.111 g Br = 0.278 g Hg

To determine the empirical formula, first convert the two masses to a number of moles.

$$\text{mol Hg} = (0.278\ \text{g Hg})\left(\frac{1\ \text{mole Hg}}{200.59\ \text{g Hg}}\right) = 1.39 \times 10^{-3}\ \text{mol Hg}$$

$$\text{mol Br} = (0.111 \text{ g Br})\left(\frac{1 \text{ mole Br}}{79.904 \text{ g Br}}\right) = 1.39 \times 10^{-3} \text{ mol Br}$$

Now, we would divide each of these values by the smaller quantity to determine the simplest mole ratio between the two elements. By inspection, though, we can see there are the same number of moles of Hg and Br. Consequently, the simplest mole ratio is 1:1 and the empirical formula is HgBr.

To determine the molecular formula, recall that the ratio of the molecular mass to the empirical mass is equivalent to the ratio of the molecular formula to the empirical formula. Thus, we need to calculate an empirical mass:

(1 mole Hg)(200.59 g Hg/mole Hg) + (1 mole Br)(79.904 g Br/mole Br) = 280.49 g/mole HgBr.

The molecular mass, as reported in the problem is 561 g/mole. The ratio of these is:

$$\frac{561 \text{ g/mole}}{280.49 \text{ g/mole}} = 2.00$$

So, the molecular formula is two times the empirical formula or Hg_2Br_2.

3.95 First, determine the amount of oxygen in the sample by subtracting the masses of the other elements from the total mass:

0.6216 g – (0.1735 g C + 0.01455 g H + 0.2024 g N) = 0.2312 g O.

Now, convert these masses into a number of moles for each element:

$$\text{mol C} = (0.1735 \text{ g C})\left(\frac{1 \text{ mole C}}{12.011 \text{ g C}}\right) = 1.445 \times 10^{-2} \text{ mol C}$$

$$\text{mol H} = (0.01455 \text{ g H})\left(\frac{1 \text{ mole H}}{1.0079 \text{ g H}}\right) = 1.444 \times 10^{-2} \text{ mol H}$$

$$\text{mol N} = (0.2024 \text{ g N})\left(\frac{1 \text{ mole N}}{14.007 \text{ g N}}\right) = 1.444 \times 10^{-2} \text{ mol N}$$

$$\text{mol O} = (0.2312 \text{ g O})\left(\frac{1 \text{ mole O}}{15.999 \text{ g O}}\right) = 1.444 \times 10^{-2} \text{ mol O}$$

These are clearly all the same mole amounts, and we deduce that the empirical formula is CHNO, which has a formula weight of 43. It can be seen that the number 43 must be multiplied by the integer 3 in order to obtain the molar mass (3 × 43 = 129), and this means that the empirical formula should similarly be multiplied by 3 in order to arrive at the molecular formula, $C_3H_3N_3O_3$.

3.97 36 mol H

3.99 $4Fe(s) + 3O_2(g) \rightarrow 2Fe_2O_3(s)$

3.101
(a) $Ca(OH)_2 + 2HCl \rightarrow CaCl_2 + 2H_2O$
(b) $2AgNO_3 + CaCl_2 \rightarrow Ca(NO_3)_2 + 2AgCl$
(c) $Pb(NO_3)_2 + Na_2SO_4 \rightarrow PbSO_4 + 2NaNO_3$
(d) $2Fe_2O_3 + 3C \rightarrow 4Fe + 3CO_2$
(e) $2C_4H_{10} + 13O_2 \rightarrow 8CO_2 + 10H_2O$

3.103
(a) $Mg(OH)_2 + 2HBr \rightarrow MgBr_2 + 2H_2O$
(b) $2HCl + Ca(OH)_2 \rightarrow CaCl_2 + 2H_2O$
(c) $Al_2O_3 + 3H_2SO_4 \rightarrow Al_2(SO_4)_3 + 3H_2O$
(d) $2KHCO_3 + H_3PO_4 \rightarrow K_2HPO_4 + 2H_2O + 2CO_2$
(e) $C_9H_{20} + 14O_2 \rightarrow 9CO_2 + 10H_2O$

3.105 $2FeCl_3 + SnCl_2 \rightarrow 2FeCl_2 + SnCl_4$

3.107 (a) $\text{mol Na}_2\text{S}_2\text{O}_3 = (0.12\text{ mol Cl}_2)\left(\frac{1\text{ mole Na}_2\text{S}_2\text{O}_3}{4\text{ mole Cl}_2}\right) = 0.030\text{ mol Na}_2\text{S}_2\text{O}_3$

(b) $\text{mol HCl} = (0.12\text{ mol Cl}_2)\left(\frac{8\text{ mole HCl}}{4\text{ mole Cl}_2}\right) = 0.24\text{ mol HCl}$

(c) $\text{mol H}_2\text{O} = (0.12\text{ mol Cl}_2)\left(\frac{5\text{ mole H}_2\text{O}}{4\text{ mole Cl}_2}\right) = 0.15\text{ mol H}_2\text{O}$

(d) $\text{mol H}_2\text{O} = (0.24\text{ mol HCl})\left(\frac{5\text{ mole H}_2\text{O}}{8\text{ mole HCl}}\right) = 0.15\text{ mol H}_2\text{O}$

3.109 (a) $0.11\text{ mol Au(CN)}_2^-\left(\frac{1\text{ mol Zn}}{2\text{ mol Au(CN)}_2^-}\right)\left(\frac{65.39\text{ g Zn}}{1\text{ mol Zn}}\right) = 3.6\text{ g Zn}$

(b) $0.11\text{ mol Au(CN)}_2^-\left(\frac{2\text{ mol Au}}{2\text{ mol Au(CN)}_2^-}\right)\left(\frac{197.0\text{ g Au}}{1\text{ mol Au}}\right) = 22\text{ g Zn}$

(c) $0.11\text{ mol Zn}\left(\frac{2\text{ mol Au(CN)}_2^-}{1\text{ mol Zn}}\right)\left(\frac{249.0\text{ g Au(CN)}_2^-}{1\text{ mol Au(CN)}_2^-}\right) = 55\text{ g Au(CN)}_2^-$

3.111 (a) $4P + 5O_2 \rightarrow P_4O_{10}$

(b) $\text{g O}_2 = (6.85\text{ g P})\left(\frac{1\text{ mol P}}{30.97\text{ g P}}\right)\left(\frac{5\text{ mol O}_2}{4\text{ mol P}}\right)\left(\frac{32.0\text{ g O}_2}{1\text{ mol O}_2}\right) = 8.85\text{ g O}_2$

(c) $\text{g P}_4\text{O}_{10} = (8.00\text{ g O}_2)\left(\frac{1\text{ mol O}_2}{32.00\text{ g O}_2}\right)\left(\frac{1\text{ mol P}_4\text{O}_{10}}{5\text{ mol O}_2}\right)\left(\frac{283.9\text{ g P}_4\text{O}_{10}}{1\text{ mol P}_4\text{O}_{10}}\right) = 14.2\text{ g P}_2\text{O}_{10}$

(d) $\text{g P} = (7.46\text{ g P}_4\text{O}_{10})\left(\frac{1\text{ mol P}_4\text{O}_{10}}{283.9\text{ g P}_4\text{O}_{10}}\right)\left(\frac{4\text{ mol P}}{1\text{ mol P}_4\text{O}_{10}}\right)\left(\frac{30.97\text{ g P}}{1\text{ mol P}}\right) = 3.26\text{ g P}$

3.113 $\text{g HNO}_3 = (11.45\text{ g Cu})\left(\frac{1\text{ mol Cu}}{63.546\text{ g Cu}}\right)\left(\frac{8\text{ mol HNO}_3}{3\text{ mol Cu}}\right)\left(\frac{63.013\text{ g HNO}_3}{1\text{ mol HNO}_3}\right) = 30.28\text{ g HNO}_3$

3.115 $\text{kg O}_2 = 1.0\text{ kg H}_2\text{O}_2\left(\frac{1000\text{ g H}_2\text{O}_2}{1\text{ kg H}_2\text{O}_2}\right)\left(\frac{1\text{ mol H}_2\text{O}_2}{34.01\text{ g H}_2\text{O}_2}\right)\left(\frac{1\text{ mol O}_2}{2\text{ mol H}_2\text{O}_2}\right)\left(\frac{32.00\text{ g O}_2}{1\text{ mol O}_2}\right)\left(\frac{1\text{ kg O}_2}{1000\text{ g O}_2}\right)$

$= 0.47\text{ kg O}_2$

3.117 (a) First determine the amount of Fe_2O_3 that would be required to react completely with the given amount of Al:

$$\text{mol Fe}_2\text{O}_3 = (4.20\text{ mol Al})\left(\frac{1\text{ mol Fe}_2\text{O}_3}{2\text{ mol Al}}\right) = 2.10\text{ mol Fe}_2\text{O}_3$$

Since only 1.75 mol of Fe_2O_3 are supplied, it is the limiting reactant. This can be confirmed by calculating the amount of Al that would be required to react completely with all of the available Fe_2O_3:

$$\text{mol Al} = (1.75\text{ mol Fe}_2\text{O}_3)\left(\frac{2\text{ mol Al}}{1\text{ mol Fe}_2\text{O}_3}\right) = 3.50\text{ mol Al}$$

Since an excess (4.20 mol – 3.50 mol = 0.70 mol) of Al is present, Fe_2O_3 must be the limiting reactant, as determined above.

(b) $$\text{g Fe} = (1.75 \text{ mol Fe}_2\text{O}_3)\left(\frac{2 \text{ mol Fe}}{1 \text{ mol Fe}_2\text{O}_3}\right)\left(\frac{55.847 \text{ g Fe}}{1 \text{ mol Fe}}\right) = 195 \text{ g Fe}$$

3.119 $3AgNO_3 + FeCl_3 \rightarrow 3AgCl + Fe(NO_3)_3$
Calculate the amount of $FeCl_3$ that are required to react completely with all of the available silver nitrate:

$$\text{g Fe Cl}_3 = (18.0 \text{ g AgNO}_3)\left(\frac{1 \text{ mol AgNO}_3}{169.87 \text{ g AgNO}_3}\right)\left(\frac{1 \text{ mol FeCl}_3}{3 \text{ mol AgNO}_3}\right)\left(\frac{162.21 \text{ g FeCl}_3}{1 \text{ mol FeCl}_3}\right)$$
$$= 5.73 \text{ g FeCl}_3$$

Since more than this minimum amount is available, $FeCl_3$ is present in excess, and $AgNO_3$ must be the limiting reactant.

We know that only 5.73 g $FeCl_3$ will be used. Therefore, the amount left unused is:
32.4 g total – 5.73 g used = 26.7 g $FeCl_3$

3.121 First calculate the number of moles of water that are needed to react completely with the given amount of NO_2:

$$\text{g H}_2\text{O} = 0.0010 \text{ g NO}_2\left(\frac{1 \text{ mol NO}_2}{46.01 \text{ g NO}_2}\right)\left(\frac{1 \text{ mol H}_2\text{O}}{3 \text{ mol NO}_2}\right)\left(\frac{18.02 \text{ g H}_2\text{O}}{1 \text{ mol H}_2\text{O}}\right) = 1.3 \times 10^{-4} \text{ g H}_2\text{O}$$

Since this is less than the amount of water that is supplied, the limiting reactant must be NO_2. Therefore, to calculate the amount of HNO_3:

$$\text{g HNO}_3 = 0.0010 \text{ g NO}_2\left(\frac{1 \text{ mol NO}_2}{46.01 \text{ g NO}_2}\right)\left(\frac{2 \text{ mol HNO}_3}{3 \text{ mol NO}_2}\right)\left(\frac{63.02 \text{ g HNO}_3}{1 \text{ mol HNO}_3}\right) = 9.1 \times 10^{-4} \text{ g HNO}_3$$

3.123 First determine the theoretical yield:

$$\text{g BaSO}_4 = (75.00 \text{ g Ba(NO}_3)_2\left(\frac{1 \text{ mol Ba(NO}_3)_2}{261.34 \text{ g Ba(NO}_3)_2}\right)\left(\frac{1 \text{ mol BaSO}_4}{1 \text{ mol Ba(NO}_3)_2}\right)\left(\frac{233.39 \text{ g BaSO}_4}{1 \text{ mol BaSO}_4}\right)$$
$$= 66.98 \text{ g BaSO}_4$$

Then calculate a % yield:

$$\%\text{ yield} = \frac{\text{actual yield}}{\text{theoretical yield}} \times 100 = \frac{64.45 \text{ g}}{66.98 \text{ g}} \times 100 = 96.22\%$$

3.125 First, determine how much H_2SO_4 is needed to completely react with the $AlCl_3$

$$\text{g H}_2\text{SO}_4 = 25 \text{ g AlCl}_3\left(\frac{1 \text{ mol AlCl}_3}{133.34 \text{ g AlCl}_3}\right)\left(\frac{3 \text{ mol H}_2\text{SO}_4}{2 \text{ mol AlCl}_3}\right)\left(\frac{98.08 \text{ g H}_2\text{SO}_4}{1 \text{ mol H}_2\text{SO}_4}\right)$$
$$= 27.58 \text{ g H}_2\text{SO}_4$$

There is an excess of H_2SO_4 present.

Determine the theoretical yield:

$$\text{g Al}_2(\text{SO}_4)_3 = 25.00 \text{ g AlCl}_3\left(\frac{1 \text{ mol AlCl}_3}{133.33 \text{ g AlCl}_3}\right)\left(\frac{1 \text{ mol Al}_2(\text{SO}_4)_3}{2 \text{ mol AlCl}_3}\right)\left(\frac{342.17 \text{ g Al}_2(\text{SO}_4)_3}{1 \text{ mol Al}_2(\text{SO}_4)_3}\right)$$
$$= 32.08 \text{ g Al}_2(\text{SO}_4)_3$$

$$\%\text{ yield} = \frac{\text{actual yield}}{\text{theoretical yield}} \times 100 = \frac{28.46 \text{ g}}{32.08 \text{ g}} \times 100 = 88.72\%$$

3.127 If the yield for this reaction is only 71 % and we need to have 11.5 g of product, we will attempt to make 16 g of product. This is determined by dividing the actual yield by the percent yield. Recall that; $\% \text{ yield} = \dfrac{\text{actual yield}}{\text{theoretical yield}} \times 100$. If we rearrange this equation we can see that $\text{theoretical yield} = \dfrac{\text{actual yield}}{\% \text{ yield}} \times 100$. Substituting the values from this problem gives the 16 g of product mentioned above.

$$\text{g } C_7H_8 = 16 \text{ g } KC_7H_5O_2 \left(\frac{1 \text{ mol } KC_7H_5O_2}{160.21 \text{ g } KC_7H_5O_2}\right)\left(\frac{1 \text{ mol } C_7H_8}{1 \text{ mol } KC_7H_5O_2}\right)\left(\frac{92.14 \text{ g } C_7H_8}{1 \text{ mol } C_7H_8}\right) = 9.2 \text{ g } C_7H_8$$

Practice Exercises

4.1 (a) $FeCl_3(s) \rightarrow Fe^{3+}(aq) + 3Cl^-(aq)$
(b) $K_3PO_4(s) \rightarrow 3K^+(aq) + PO_4^{3-}(aq)$

4.2 (a) $MgCl_2(s) \rightarrow Mg^{2+}(aq) + 2Cl^-(aq)$
(b) $Al(NO_3)_3(s) \rightarrow Al^{3+}(aq) + 3NO_3^-(aq)$
(c) $Na_2CO_3(s) \rightarrow 2Na^+(aq) + CO_3^{2-}(aq)$

4.3 molecular: $(NH_4)_2SO_4(aq) + Ba(NO_3)_2(aq) \rightarrow BaSO_4(s) + 2NH_4NO_3(aq)$
ionic: $2NH_4^+(aq) + SO_4^{2-}(aq) + Ba^{2+}(aq) + 2NO_3^-(aq) \rightarrow BaSO_4(s) + 2NH_4^+(aq) + 2NO_3^-(aq)$
net ionic: $Ba^{2+}(aq) + SO_4^{2-}(aq) \rightarrow BaSO_4(s)$

4.4 molecular: $CdCl_2(aq) + Na_2S(aq) \rightarrow CdS(s) + 2NaCl(aq)$
ionic: $Cd^{2+}(aq) + 2Cl^-(aq) + 2Na^+(aq) + S^{2-}(aq) \rightarrow CdS(s) + 2Na^+(aq) + 2Cl^-(aq)$
net ionic: $Cd^{2+}(aq) + S^{2-}(aq) \rightarrow CdS(s)$

4.5 $HCHO_2(aq) + H_2O \rightarrow H_3O^+(aq) + CHO_2^-(aq)$

4.6 $H_3C_6H_5O_7(s) + H_2O \rightarrow H_3O^+(aq) + H_2C_6H_5O_7^-(aq)$
$H_2C_6H_5O_7^-(aq) + H_2O \rightarrow H_3O^+(aq) + HC_6H_5O_7^{2-}(aq)$
$HC_6H_5O_7^{2-}(aq) + H_2O \rightarrow H_3O^+(aq) + C_6H_5O_7^{3-}(aq)$

4.7 $(C_2H_5)_3N(aq) + H_2O \rightarrow (C_2H_5)_3NH^+(aq) + OH^-(aq)$

4.8 $HONH_2(aq) + H_2O \rightarrow HONH_3^+(aq) + OH^-(aq)$

4.9 $CH_3NH_2(aq) + H_2O \rightleftharpoons CH_3NH_3^+(aq) + OH^-(aq)$

4.10 $HNO_2(aq) + H_2O \rightleftharpoons H_3O^+(aq) + NO_2^-(aq)$

4.11 Sodium arsenate

4.12 Calcium formate

4.13 HF: Hydrofluoric acid, sodium salt = sodium fluoride (NaF)
HBr: Hydrobromic acid, sodium salt = sodium bromide (NaBr)

4.14 $NaHSO_3$, sodium hydrogen sulfite

4.15 $H_3PO_4(aq) + NaOH(aq) \rightarrow NaH_2PO_4(aq) + H_2O$ sodium dihydrogen phosphate
$NaH_2PO_4(aq) + NaOH(aq) \rightarrow Na_2HPO_4(aq) + H_2O$ sodium hydrogen phosphate
$Na_2HPO_4(aq) + NaOH(aq) \rightarrow Na_3PO_4(aq) + H_2O$ sodium phosphate

4.16 molecular: $Zn(NO_3)_2(aq) + Ca(C_2H_3O_2)_2(aq) \rightarrow Zn(C_2H_3O_2)_2(aq) + Ca(NO_3)_2(aq)$
ionic: $Zn^{2+}(aq) + 2NO_3^-(aq) + Ca^{2+}(aq) + 2C_2H_3O_2^-(aq) \rightarrow$
$Zn^{2+}(aq) + 2C_2H_3O_2^-(aq) + Ca^{2+}(aq) + 2NO_3^-(aq)$
net ionic: No reaction

4.17 (a) molecular: $AgNO_3(aq) + NH_4Cl(aq) \rightarrow AgCl(s) + NH_4NO_3(aq)$
ionic: $Ag^+(aq) + NO_3^-(aq) + NH_4^+(aq) + Cl^-(aq) \rightarrow AgCl(s) + NH_4^+(aq) + NO_3^-(aq)$
net ionic: $Ag^+(aq) + Cl^-(aq) \rightarrow AgCl(s)$

(b) molecular: $Na_2S(aq) + Pb(C_2H_3O_2)_2(aq) \rightarrow 2NaC_2H_3O_2(aq) + PbS(s)$
ionic: $2Na^+(aq) + S^{2-}(aq) + Pb^{2+}(aq) + 2C_2H_3O_2^-(aq) \rightarrow 2Na^+(aq) + 2C_2H_3O_2^-(aq) + PbS(s)$
net ionic: $S^{2-}(aq) + Pb^{2+}(aq) \rightarrow PbS(s)$

4.18 molecular: $2HNO_3(aq) + Ca(OH)_2(aq) \rightarrow Ca(NO_3)_2(aq) + 2H_2O$
ionic: $2H^+(aq) + 2NO_3^-(aq) + Ca^{2+}(aq) + 2OH^-(aq) \rightarrow Ca^{2+}(aq) + 2NO_3^-(aq) + 2H_2O$
net ionic: $H^+(aq) + OH^-(aq) \rightarrow H_2O$

4.19 (a) molecular: $HCl(aq) + KOH(aq) \rightarrow H_2O + KCl(aq)$
ionic: $H^+(aq) + Cl^-(aq) + K^+(aq) + OH^-(aq) \rightarrow H_2O + K^+(aq) + Cl^-(aq)$
net ionic: $H^+(aq) + OH^-(aq) \rightarrow H_2O$

(b) molecular: $HCHO_2(aq) + LiOH(aq) \rightarrow H_2O + LiCHO_2(aq)$
ionic: $HCHO_2(aq) + Li^+(aq) + OH^-(aq) \rightarrow H_2O + Li^+(aq) + CHO_2^-(aq)$
net ionic: $HCHO_2(aq) + OH^-(aq) \rightarrow H_2O + CHO_2^-(aq)$

(c) molecular: $N_2H_4(aq) + HCl(aq) \rightarrow N_2H_5Cl(aq)$
ionic: $N_2H_4(aq) + H^+(aq) + Cl^-(aq) \rightarrow N_2H_5^+(aq) + Cl^-(aq)$
net ionic: $N_2H_4(aq) + H^+(aq) \rightarrow N_2H_5^+(aq)$

4.20 molecular: $CH_3NH_2(aq) + HCHO_2(aq) \rightarrow CH_3NH_3CHO_2(aq)$
ionic: $CH_3NH_2(aq) + HCHO_2(aq) \rightarrow CH_3NH_3^+(aq) + CHO_2^-(aq)$
net ionic: $CH_3NH_2(aq) + HCHO_2(aq) \rightarrow CH_3NH_3^+(aq) + CHO_2^-(aq)$

4.21 molecular: $2HCHO_2(aq) + Co(OH)_2(s) \rightarrow Co(CHO_2)_2(aq) + 2H_2O$
ionic: $2HCHO_2(aq) + Co(OH)_2(s) \rightarrow 2CHO_2^-(aq) + Co^{2+}(aq) + 2H_2O$
net ionic: $2HCHO_2(aq) + Co(OH)_2(s) \rightarrow 2CHO_2^-(aq) + Co^{2+}(aq) + 2H_2O$

4.22 (a) Formic acid, a weak acid will form.
molecular: $KCHO_2(aq) + HCl(aq) \rightarrow KCl(aq) + HCHO_2(aq)$
ionic: $K^+(aq) + CHO_2^-(aq) + H^+(aq) + Cl^-(aq) \rightarrow K^+(aq) + Cl^-(aq) + HCHO_2(aq)$
net ionic: $CHO_2^-(aq) + H^+(aq) \rightarrow HCHO_2(aq)$

(b) Carbonic acid will form and it will further dissociate to water and carbon dioxide:
$CuCO_3(s) + 2H^+(aq) \rightarrow CO_2(g) + H_2O + Cu^{2+}(aq)$
molecular: $CuCO_3(s) + 2HC_2H_3O_2(aq) \rightarrow CO_2(g) + H_2O + Cu(C_2H_3O_2)_2(aq)$
ionic: $CuCO_3(s) + 2HC_2H_3O_2(aq) \rightarrow CO_2(g) + H_2O + Cu^{2+} + 2C_2H_3O_2^-(aq)$
net ionic: $CuCO_3(s) + 2HC_2H_3O_2(aq) \rightarrow CO_2(g) + H_2O + Cu^{2+} + 2C_2H_3O_2^-(aq)$

(c) NR

(d) Insoluble nickel hydroxide will precipitate.
$Ni^{2+}(aq) + 2OH^-(aq) \rightarrow Ni(OH)_2(s)$
molecular: $NiCl_2(aq) + 2NaOH(aq) \rightarrow Ni(OH)_2(s) + 2NaCl(aq)$
ionic: $Ni^{2+}(aq) + 2Cl^-(aq) + 2Na^+(aq) + 2OH^-(aq) \rightarrow Ni(OH)_2(s) + 2Na^+(aq) + 2Cl^-(aq)$
net ionic: $Ni^{2+}(aq) + 2OH^-(aq) \rightarrow Ni(OH)_2(s)$

4.23 $$\text{mol } HNO_3 = (16.9 \text{ g } HNO_3)\left(\frac{1 \text{ mol } HNO_3}{63.02 \text{ g } HNO_3}\right) = 0.268 \text{ mol } HNO_3$$

$$M\ HNO_3 = \left(\frac{0.268 \text{ mol } HNO_3}{175 \text{ ml solution}}\right)\left(\frac{1000 \text{ mL solution}}{1 \text{ L solution}}\right) = 1.53\ M\ HNO_3$$

The amount of HNO_3 does not change as the solution is diluted.

4.24 $$\text{mol NaCl in 1.223 g NaCl} = (1.223 \text{ g NaCl})\left(\frac{1 \text{ mol NaCl}}{58.443 \text{ g NaCl}}\right) = 0.02093 \text{ mol NaCl}$$

mol NaCl in 1.461 g NaCl = 0.02500 mol NaCl
Total mol NaCl = 0.02093 mol NaCl + 0.02500 mol NaCl = 0.04593 mol NaCl

$$\text{Molarity NaCl} = \left(\frac{0.04593 \text{ mol NaCl}}{250.0 \text{ mL solution}}\right)\left(\frac{1000 \text{ mL solution}}{1 \text{ L solution}}\right) = 0.1837\ M$$

4.25 $$\text{mol HCl} = 175 \text{ mL HCl solution}\left(\frac{1 \text{ L solution}}{1000 \text{ mL solution}}\right)\left(\frac{0.250 \text{ mol HCl}}{1 \text{ L solution}}\right) = 0.0438 \text{ mol HCl}$$

4.26 $$\text{mL } 0.250\ M \text{ HCl} = (1.30 \text{ g HCl})\left(\frac{1 \text{ mol HCl}}{36.46 \text{ g HCl}}\right)\left(\frac{1 \text{ L solution}}{0.250 \text{ mol HCl}}\right)\left(\frac{1000 \text{ mL solution}}{1 \text{ L solution}}\right)$$

$= 143$ mL of 0.250 M HCl

4.27 If we were working with a full liter of this solution, it would contain 0.2 mol of $Sr(NO_3)_2$. The molar mass of the salt is 211.62 g mol^{-1}, so 0.2 mol is slightly more than 40 g. However, we are working with just 50 mL, so the amount of $Sr(NO_3)_2$ needed is slightly more than a twentieth of 40 g, or 2 g. The answer, 2.11 g, is close to this, so it makes sense.

$$\text{g } Sr(NO_3)_2 = 50 \text{ mL}\left(\frac{1 \text{ L solution}}{1000 \text{ mL solution}}\right)\left(\frac{0.2 \text{ mol } Sr(NO_3)_2}{1 \text{ L solution}}\right)\left(\frac{211.62 \text{ g } Sr(NO_3)_2}{1 \text{ mol } Sr(NO_3)_2}\right) = 2.11 \text{ g } Sr(NO_3)_2$$

4.28 $$\text{g } AgNO_3 = (250 \text{ mL sol'n})\left(\frac{1 \text{ L sol'n}}{1000 \text{ mL sol'n}}\right)\left(\frac{0.0125 \text{ mol } AgNO_3}{1 \text{ L sol'n}}\right)\left(\frac{169.9 \text{ g } AgNO_3}{1 \text{ mol } AgNO_3}\right) = 0.531 \text{ g } AgNO_3$$

4.29 $$\text{mL} = (100.0 \text{ mL solution})\left(\frac{0.125 \text{ mol } H_2SO_4}{1 \text{ L solution}}\right)\frac{1}{(0.0500 \text{ M})} = 250.0 \text{ mL}$$

4.30 $$\text{mol HCl} = (150 \text{ mL solution})\left(\frac{1 \text{ L solution}}{1000 \text{ mL solution}}\right)\left(\frac{0.50 \text{ mol HCl}}{1 \text{ L solution}}\right) = 0.075 \text{ mol HCl}$$

$$\text{mL} = (0.075 \text{ mol HCl})\left(\frac{1 \text{ L solution}}{0.10 \text{ mol HCl}}\right)\left(\frac{1000 \text{ mL solution}}{1 \text{ L solution}}\right) = 750 \text{ mL}$$

To find the number of mL of water to add to the solution subtract the number of mL of the concentrated soltuion from the total volume:

750 mL solution – 150 mL = 600 mL

Add 600 mL of water.

4.31 $$\text{mol } H_3PO_4 = (45.0 \text{ mL KOH})\left(\frac{1 \text{ L solution}}{1000 \text{ mL solution}}\right)\left(\frac{0.100 \text{ mol KOH}}{1 \text{ L solution}}\right)\left(\frac{1 \text{ mol } H_3PO_4}{3 \text{ mol KOH}}\right)$$

$= 1.5 \times 10^{-3}$ mol H_3PO_4

$$\text{mL } H_3PO_4 = (1.5 \times 10^{-3} \text{ mol } H_3PO_4)\left(\frac{1 \text{ L solution}}{0.0475 \text{ mol } H_3PO_4}\right)\left(\frac{1000 \text{ mL solution}}{1 \text{ L solution}}\right) = 31.6 \text{ mL } H_3PO_4$$

4.32 $$\text{mL NaOH} = (15.4 \text{ mL } H_2SO_4)\left(\frac{1 \text{ L } H_2SO_4}{1000 \text{ mL } H_2SO_4}\right)\left(\frac{0.108 \text{ mol } H_2SO_4}{1 \text{ L } H_2SO_4}\right)$$

$$\times\left(\frac{2 \text{ mol NaOH}}{1 \text{ mol } H_2SO_4}\right)\left(\frac{1 \text{ L NaOH}}{0.124 \text{ mol NaOH}}\right)\left(\frac{1000 \text{ mL NaOH}}{1 \text{ L NaOH}}\right) = 26.8 \text{ mL NaOH}$$

4.33 $FeCl_3 \rightarrow Fe^{3+} + 3Cl^-$

$$M\,Fe^{3+} = \left(\frac{0.40\text{ mol FeCl}_3}{1\text{ L FeCl}_3\text{ soln}}\right)\left(\frac{1\text{ mol Fe}^{3+}}{1\text{ mol FeCl}_3}\right) = 0.40\,M\,Fe^{3+}$$

$$M\,Cl^- = \left(\frac{0.40\text{ mol FeCl}_3}{1\text{ L FeCl}_3\text{ soln}}\right)\left(\frac{3\text{ mol Cl}^-}{1\text{ mol FeCl}_3}\right) = 1.2\,M\,Cl^-$$

4.34 $$M\,Na^+ = \left(\frac{0.250\text{ mol PO}_4^{3-}}{1\text{ L Na}_3\text{PO}_4\text{ soln}}\right)\left(\frac{3\text{ mol Na}^+}{1\text{ mol PO}_4^{3-}}\right) = 0.750\,M\,Na^+$$

4.35 $$\text{mol CaCl}_2 = 18.4\text{ mL AgNO}_3\left(\frac{1\text{ L AgNO}_3}{1000\text{ mL AgNO}_3}\right)\left(\frac{0.100\text{ mol AgNO}_3}{1\text{ L AgNO}_3}\right)\left(\frac{1\text{ mol Ag}^+}{1\text{ mol AgNO}_3}\right)$$

$$\times\left(\frac{1\text{ mol Cl}^-}{1\text{ mol Ag}^+}\right)\left(\frac{1\text{ mol CaCl}_2}{2\text{ mol Cl}^-}\right) = 9.20\times10^{-4}\text{ mol CaCl}_2$$

$$M\,CaCl_2 = \left(\frac{9.20\times10^{-4}\text{ mol CaCl}_2}{20.5\text{ mL CaCl}_2}\right)\left(\frac{1000\text{ mL CaCl}_2}{1\text{ L CaCl}_2}\right) = 0.0449\,M\,CaCl_2$$

4.36 The balanced net ionic equation is: $Fe^{2+}(aq) + 2OH^-(aq) \rightarrow Fe(OH)_2(s)$.
First determine the number of moles of Fe^{2+} present.

$$\text{mol Fe}^{2+} = (60.0\text{ mL Fe}^{2+})\left(\frac{0.250\text{ mol FeCl}_2}{1000\text{ mL solution}}\right)\left(\frac{1\text{ mol Fe}^{2+}}{1\text{ mol FeCl}_2}\right) = 1.50\times10^{-2}\text{ mol Fe}^{2+}$$

Now, determine the amount of KOH needed to react with the Fe^{2+}.

$$\text{mL KOH} = (1.50\times10^{-2}\text{ mol Fe}^{2+})\left(\frac{2\text{ mol OH}^-}{1\text{ mol Fe}^{2+}}\right)\left(\frac{1\text{ mol KOH}}{1\text{ mol OH}^-}\right)\left(\frac{1000\text{ mL solution}}{0.500\text{ mol KOH}}\right) = 60.0\text{ mL KOH}$$

4.37 $$\text{g Na}_2\text{SO}_4 = 28.40\text{ mL BaCl}_2\left(\frac{0.150\text{ mol BaCl}_2}{1000\text{ mL BaCl}_2}\right)\left(\frac{1\text{ mol Ba}^{2+}}{1\text{ mol BaCl}_2}\right)\left(\frac{1\text{ mol SO}_4^{2-}}{1\text{ mol Ba}^{2+}}\right)$$

$$\times\left(\frac{1\text{ mol Na}_2\text{SO}_4}{1\text{ mol SO}_4^{2-}}\right)\left(\frac{142.05\text{ g Na}_2\text{SO}_4}{1\text{ mol Na}_2\text{SO}_4}\right) = 0.605\text{ g Na}_2\text{SO}_4$$

4.38 (a) $$\text{mol Ca}^{2+} = (0.736\text{ g CaSO}_4)\left(\frac{1\text{ mol CaSO}_4}{136.14\text{ g CaSO}_4}\right)\left(\frac{1\text{ mol Ca}^{2+}}{1\text{ mol CaSO}_4}\right) = 5.41\times10^{-3}\text{ mol Ca}^{2+}$$

(b) Since all of the Ca^{2+} is precipitated as $CaSO_4$, there were originally 5.41×10^{-3} moles of Ca^{2+} in the sample.

(c) All of the Ca^{2+} comes from $CaCl_2$, so there were 5.41×10^{-3} moles of $CaCl_2$ in the sample.

(d) $$\text{g CaCl}_2 = (5.41\times10^{-3}\text{ mol CaCl}_2)\left(\frac{110.98\text{ g CaCl}_2}{1\text{ mol CaCl}_2}\right) = 0.600\text{ g CaCl}_2$$

(e) $$\%\text{ CaCl}_2 = \frac{0.600\text{ g CaCl}_2}{2.000\text{ g sample}}\times100\% = 30.0\%\text{ CaCl}_2$$

4.39 Balanced equation:
$H_2SO_4(aq) + 2NaOH(aq) \rightarrow Na_2SO_4(aq) + 2H_2O$

$$\text{mol } H_sSO_4 = (36.42 \text{ mL NaOH})\left(\frac{1 \text{ L NaOH soln}}{1000 \text{ mL NaOH soln}}\right)\left(\frac{0.147 \text{ mol NaOH}}{1 \text{ L NaOH soln}}\right)\left(\frac{1 \text{ mol } H_2SO_4}{2 \text{ mol NaOH}}\right)$$

$= 2.68 \times 10^{-3}$ mol H_2SO_4

$$M\ H_2SO_4 = \left(\frac{2.68\times10^{-3} \text{ mol } H_2SO_4}{15.00 \text{ mL } H_2SO_4}\right)\left(\frac{1000 \text{ mL } H_2SO_4}{1 \text{ L } H_2SO_4}\right) = 0.178\ M\ H_2SO_4$$

4.40 $$\text{mol HCl} = (11.00 \text{ mL})\left(\frac{1 \text{ L KOH soln}}{1000 \text{ mL NaOH soln}}\right)\left(\frac{0.0100 \text{ mol KOH soln}}{1 \text{ L KOH soln}}\right)\left(\frac{1 \text{ mol HCl}}{1 \text{ mol KOH}}\right) = 1.1 \times 10^{-4} \text{ mol HCl}$$

$$M\ \text{HCl} = \left(\frac{1.1\times10^{-4} \text{ mol HCl}}{5.00 \text{ mL HCl soln}}\right)\left(\frac{1000 \text{ mL HCl soln}}{1 \text{ L HCl soln}}\right) = 0.0220\ M \text{ HCl}$$

$$\text{g HCl} = (5.00 \text{ mL HCl})\left(\frac{0.0220 \text{ mol HCl}}{1000 \text{ mL HCl}}\right)\left(\frac{36.5 \text{ g HCl}}{1 \text{ mol HCl}}\right) = 4.02 \times 10^{-3} \text{ g HCl}$$

$$\text{weight \%} = \frac{4.02 \times 10^{-3} \text{ g}}{5.00 \text{ g}} \times 100\% = 0.0803\%$$

Review Problems

4.49 (a) ionic: $2NH_4^+(aq) + CO_3^{2-}(aq) + Mg^{2+}(aq) + 2Cl^-(aq) \rightarrow 2NH_4^+(aq) + 2Cl^-(aq) + MgCO_3(s)$
net: $Mg^{2+}(aq) + CO_3^{2-}(aq) \rightarrow MgCO_3(s)$

(b) ionic: $Cu^{2+}(aq) + 2Cl^-(aq) + 2Na^+(aq) + 2OH^-(aq) \rightarrow Cu(OH)_2(s) + 2Na^+(aq) + 2Cl^-(aq)$
net: $Cu^{2+}(aq) + 2OH^-(aq) \rightarrow Cu(OH)_2(s)$

(c) ionic: $3Fe^{2+}(aq) + 3SO_4^{2-}(aq) + 6Na^+(aq) + 2PO_4^{3-}(aq) \rightarrow Fe_3(PO_4)_2(s) + 6Na^+(aq) + 3SO_4^{2-}(aq)$
net: $3Fe^{2+}(aq) + 2PO_4^{3-}(aq) \rightarrow Fe_3(PO_4)_2(s)$

(d) ionic: $2Ag^+(aq) + 2C_2H_3O_2^-(aq) + Ni^{2+}(aq) + 2Cl^-(aq) \rightarrow 2AgCl(s) + Ni^{2+}(aq) + 2C_2H_3O_2^-(aq)$
net: $Ag^+(aq) + Cl^-(aq) \rightarrow AgCl(s)$

4.51 This is an ionization reaction: $HClO_4(l) + H_2O \rightarrow H_3O^+(aq) + ClO_4^-(aq)$

4.53 $N_2H_4(aq) + H_2O \rightleftharpoons N_2H_5^+(aq) + OH^-(aq)$

4.55 $HNO_2(aq) + H_2O \rightleftharpoons H_3O^+(aq) + NO_2^-(aq)$

4.57 $H_2CO_3(aq) + H_2O \rightleftharpoons H_3O^+(aq) + HCO_3^-(aq)$
$HCO_3^-(aq) + H_2O \rightleftharpoons H_3O^+(aq) + CO_3^{2-}(aq)$

4.59 (a) ionic: $3Fe^{2+}(aq) + 3SO_4^{2-}(aq) + 6K^+(aq) + 2PO_4^{3-}(aq) \rightarrow Fe_3(PO_4)_2(s) + 6K^+(aq) + 3SO_4^{2-}(aq)$
net: $3Fe^{2+}(aq) + 2PO_4^{3-}(aq) \rightarrow Fe_3(PO_4)_2(s)$

(b) ionic: $3Ag^+(aq) + 3C_2H_3O_2^-(aq) + Al^{3+}(aq) + 3Cl^-(aq) \rightarrow 3AgCl(s) + Al^{3+}(aq) + 3C_2H_3O_2^-(aq)$
net: $Ag^+(aq) + Cl^-(aq) \rightarrow AgCl(s)$

4.61 molecular: $Na_2S(aq) + Cu(NO_3)_2(aq) \rightarrow CuS(s) + 2NaNO_3(aq)$
ionic: $2Na^+(aq) + S^{2-}(aq) + Cu^{2+}(aq) + 2NO_3^-(aq) \rightarrow CuS(s) + 2Na^+(aq) + 2NO_3^-(aq)$
net: $Cu^{2+}(aq) + S^{2-}(aq) \rightarrow CuS(s)$

4.63 The soluble ones are (a), (b), and (d).

4.65 (a) molecular: $Ca(OH)_2(aq) + 2HNO_3(aq) \rightarrow Ca(NO_3)_2(aq) + 2H_2O$
ionic: $Ca^{2+}(aq) + 2OH^-(aq) + 2H^+(aq) + 2NO_3^-(aq) \rightarrow Ca^{2+}(aq) + 2NO_3^-(aq) + 2H_2O$
net: $H^+(aq) + OH^-(aq) \rightarrow H_2O$

(b) molecular: $Al_2O_3(s) + 6HCl(aq) \rightarrow 2AlCl_3(aq) + 3H_2O$
ionic: $Al_2O_3(s) + 6H^+(aq) + 6Cl^-(aq) \rightarrow 2Al^{3+}(aq) + 6Cl^-(aq) + 3H_2O$
net: $Al_2O_3(s) + 6H^+(aq) \rightarrow 2Al^{3+}(aq) + 3H_2O$

(c) molecular: $Zn(OH)_2(s) + H_2SO_4(aq) \rightarrow ZnSO_4(aq) + 2H_2O$
ionic: $Zn(OH)_2(s) + 2H^+(aq) + SO_4^{2-}(aq) \rightarrow Zn^{2+}(aq) + SO_4^{2-}(aq) + 2H_2O$
net: $Zn(OH)_2(s) + 2H^+(aq) \rightarrow Zn^{2+}(aq) + 2H_2O$

4.67 The electrical conductivity would decrease regularly, until one solution had neutralized the other, forming a nonelectrolyte:

$Ba^{2+}(aq) + 2OH^-(aq) + 2H^+(aq) + SO_4^{2-}(aq) \rightarrow BaSO_4(s) + 2H_2O$

Once the point of neutralization had been reached, the addition of excess sulfuric acid would cause the conductivity to increase, because sulfuric acid is a strong electrolyte itself.
Normally, sulfuric acid in an aqueous solution would be written as $H^+(aq)$ and $HSO_4^-(aq)$ since the second ionization is weak, but in this case, the $Ba^{2+}(aq)$ displaces the hydronium ion from the bisulfate ion, therefore it is written as the sulfate ion.

4.69 (a) $2H^+(aq) + CO_3^{2-}(aq) \rightarrow H_2O + CO_2(g)$
(b) $NH_4^+(aq) + OH^-(aq) \rightarrow NH_3(g) + H_2O$

4.71 These reactions have the following "driving forces":
(a) formation of insoluble $Cr(OH)_3$
(b) formation of water, a weak electrolyte

4.73 (a) molecular: $3HNO_3(aq) + Cr(OH)_3(s) \rightarrow Cr(NO_3)_3(aq) + 3H_2O$
ionic: $3H^+(aq) + 3NO_3^-(aq) + Cr(OH)_3(s) \rightarrow Cr^{3+}(aq) + 3NO_3^-(aq) + 3H_2O$
net: $3H^+(aq) + Cr(OH)_3(s) \rightarrow Cr^{3+}(aq) + 3H_2O$

(b) molecular: $HClO_4(aq) + NaOH(aq) \rightarrow NaClO_4(aq) + H_2O$
ionic: $H^+(aq) + ClO_4^-(aq) + Na^+(aq) + OH^-(aq) \rightarrow Na^+(aq) + ClO_4^-(aq) + H_2O$
net: $H^+(aq) + OH^-(aq) \rightarrow H_2O$

(c) molecular: $Cu(OH)_2(s) + 2HC_2H_3O_2(aq) \rightarrow Cu(C_2H_3O_2)_2(aq) + 2H_2O$
ionic: $Cu(OH)_2(s) + 2H^+ + 2C_2H_3O_2^-(aq) \rightarrow Cu^{2+}(aq) + 2C_2H_3O_2^-(aq) + 2H_2O$
net: $Cu(OH)_2(s) + 2H^+(aq) \rightarrow Cu^{2+}(aq) + 2H_2O$

(d) molecular: $ZnO(s) + H_2SO_4(aq) \rightarrow ZnSO_4(aq) + H_2O$
ionic: $ZnO(s) + 2H^+(aq) + SO_4^{2-}(aq) \rightarrow Zn^{2+}(aq) + SO_4^{2-}(aq) + H_2O$
net: $ZnO(s) + 2H^+(aq) \rightarrow Zn^{2+}(aq) + H_2O$

4.75 (a) molecular: $Na_2SO_3(aq) + Ba(NO_3)_2(aq) \rightarrow BaSO_3(s) + 2NaNO_3(aq)$
ionic: $2Na^+(aq) + SO_3^{2-}(aq) + Ba^{2+}(aq) + 2NO_3^-(aq) \rightarrow BaSO_3(s) + 2Na^+(aq) + 2NO_3^-(aq)$
net: $Ba^{2+}(aq) + SO_3^{2-}(aq) \rightarrow BaSO_3(s)$

(b) molecular: $2HCHO_2(aq) + K_2CO_3(aq) \rightarrow CO_2(g) + H_2O + 2KCHO_2(aq)$
ionic: $2H^+(aq) + 2CHO_2^-(aq) + 2K^+(aq) + CO_3^{2-}(aq) \rightarrow CO_2(g) + H_2O + 2K^+(aq) + 2CHO_2^-(aq)$
net: $2H^+(aq) + CO_3^{2-}(aq) \rightarrow CO_2(g) + H_2O$

(c) molecular: $2NH_4Br(aq) + Pb(C_2H_3O_2)_2(aq) \rightarrow 2NH_4C_2H_3O_2(aq) + PbBr_2(s)$
ionic: $2NH_4^+(aq) + 2Br^-(aq) + Pb^{2+}(aq) + 2C_2H_3O_2^-(aq) \rightarrow 2NH_4^+(aq) + 2C_2H_3O_2^-(aq) + PbBr_2(s)$
net: $Pb^{2+}(aq) + 2Br^-(aq) \rightarrow PbBr_2(s)$

(d) molecular: $2NH_4ClO_4(aq) + Cu(NO_3)_2(aq) \rightarrow Cu(ClO_4)_2(aq) + 2NH_4NO_3(aq)$
ionic: $2NH_4^+(aq) + 2ClO_4^-(aq) + Cu^{2+}(aq) + 2NO_3^-(aq) \rightarrow$
$Cu^{2+}(aq) + 2ClO_4^-(aq) + 2NO_3^-(aq) + 2NH_4^+(aq)$
net: N.R.

4.77 There are numerous possible answers. One of many possible sets of answers would be:
(a) $NaHCO_3(aq) + HCl(aq) \rightarrow NaCl(aq) + CO_2(g) + H_2O$
(b) $FeCl_2(aq) + 2NaOH(aq) \rightarrow Fe(OH)_2(s) + 2NaCl(aq)$
(c) $Ba(NO_3)_2(aq) + K_2SO_3(aq) \rightarrow BaSO_3(s) + 2KNO_3(aq)$
(d) $2AgNO_3(aq) + Na_2S(aq) \rightarrow Ag_2S(s) + 2NaNO_3(aq)$
(e) $ZnO(s) + 2HCl(aq) \rightarrow ZnCl_2(aq) + H_2O$

4.79 (a) $NaOH \rightarrow Na^+ + OH^-$

$$\text{mol NaOH} = (4.00\text{ g NaOH})\left(\frac{1\text{ mol NaOH}}{40.00\text{ g NaOH}}\right) = 0.100\text{ mol NaOH}$$

$$M\text{ NaOH solution} = \left(\frac{0.100\text{ mol NaOH}}{100.0\text{ mL NaOH}}\right)\left(\frac{1000\text{ mL NaOH}}{1\text{ L NaOH}}\right) = 1.00\ M\text{ NaOH}$$

(b) $CaCl_2 \rightarrow Ca^{2+} + 2Cl^-$

$$\text{mol CaCl}_2 = (16.0\text{ g CaCl}_2)\left(\frac{1\text{ mol CaCl}_2}{110.98\text{ g CaCl}_2}\right) = 0.144\text{ mol CaCl}_2$$

$$M\text{ CaCl}_2\text{ solution} = \left(\frac{0.144\text{ mol CaCl}_2}{250.0\text{ mL CaCl}_2}\right)\left(\frac{1000\text{ mL CaCl}_2}{1\text{ L CaCl}_2}\right) = 0.577\ M\text{ CaCl}_2$$

4.81 $$\text{mL NaC}_2\text{H}_3\text{O}_2 = (14.3\text{ g NaC}_2\text{H}_3\text{O}_2)\left(\frac{1\text{ mol NaC}_2\text{H}_3\text{O}_2}{82.03\text{ g NaC}_2\text{H}_3\text{O}_2}\right)\left(\frac{1\text{ L NaC}_2\text{H}_3\text{O}_2}{0.265\text{ mol NaC}_2\text{H}_3\text{O}_2}\right)$$

$$\left(\frac{1000\text{ mL NaC}_2\text{H}_3\text{O}_2}{1\text{ L NaC}_2\text{H}_3\text{O}_2}\right) = 658\text{ mL NaC}_2\text{H}_3\text{O}_2$$

4.83 (a) $$\text{g NaCl} = 125\text{ mL soln}\left(\frac{1\text{ L}}{1000\text{ mL}}\right)\left(\frac{0.200\text{ mol NaCl}}{1\text{ L soln}}\right)\left(\frac{58.44\text{ g NaCl}}{1\text{ mol NaCl}}\right) = 1.46\text{ g NaCl}$$

(b) $$\text{g C}_2\text{H}_{12}\text{O}_6 = 250\text{ mL soln}\left(\frac{1\text{ L}}{1000\text{ mL}}\right)\left(\frac{0.360\text{ mol C}_6\text{H}_{12}\text{O}_6}{1\text{ L soln}}\right)\left(\frac{180.2\text{ g C}_6\text{H}_{12}\text{O}_6}{1\text{ mol C}_6\text{H}_{12}\text{O}_6}\right)$$

$$= 16.2\text{ g C}_2\text{H}_{12}\text{O}_6$$

(c) $$\text{g H}_2\text{SO}_4 = 250\text{ mL soln}\left(\frac{1\text{ L}}{1000\text{ mL}}\right)\left(\frac{0.250\text{ mol H}_2\text{SO}_4}{1\text{ L soln}}\right)\left(\frac{98.08\text{ g H}_2\text{SO}_4}{1\text{ mol H}_2\text{SO}_4}\right) = 6.13\text{ g H}_2\text{SO}_4$$

4.85 $$\text{mol of H}_2\text{SO}_4 = 25.0\text{ mL H}_2\text{SO}_4\left(\frac{1\text{ L soln}}{1000\text{ mL soln}}\right)\left(\frac{0.56\text{ mol H}_2\text{SO}_4}{1\text{ L soln}}\right) = 0.014\text{ mol H}_2\text{SO}_4$$

$$M\text{ of final solution} = \left(\frac{0.014\text{ mol H}_2\text{SO}_4}{125\text{ mL H}_2\text{SO}_4}\right)\left(\frac{1000\text{ mL H}_2\text{SO}_4}{1\text{ L H}_2\text{SO}_4}\right) = 0.11\ M\text{ H}_2\text{SO}_4$$

4.87 $M_1V_1 = M_2V_2$ $\quad V_2 = \left(\frac{M_1V_1}{M_2}\right)$

$$V_2 = \frac{(18.0 \text{ M } H_2SO_4)(25.0 \text{ mL})}{1.50 \text{ M } H_2SO_4} = 300 \text{ mL } H_2SO_4$$

The 25.0 mL of H_2SO_4 must be diluted to 300 mL.

4.89 $M_1V_1 = M_2V_2$ $\quad V_2 = \left(\frac{M_1V_1}{M_2}\right)$

$$V_2 = \frac{(2.50 \text{ M KOH})(150.0 \text{ mL})}{1.0 \text{ M KOH}} = 375 \text{ mL KOH}$$

The 150.0 mL of KOH must be diluted to 375 mL. The volume of water to be added is:
375 mL of V_2 – 150 mL of V_1 = 225 mL water

4.91 (a) $CaCl_2 \rightarrow Ca^{2+} + 2Cl^-$

mol $CaCl_2$ = 0.45 mol/L × 0.0323 L = 0.0145 mol $CaCl_2$

$$0.0145 \text{ mol } CaCl_2\left(\frac{1 \text{ mol } Ca^{2+}}{1 \text{ mol } CaCl_2}\right) = 0.0145 \text{ mol } Ca^{2+}$$

$$0.0145 \text{ mol } CaCl_2\left(\frac{2 \text{ mol } Cl^-}{1 \text{ mol } CaCl_2}\right) = 0.0290 \text{ mol } Cl^-$$

(b) $AlCl_3 \rightarrow Al^{3+} + 3Cl^-$

mol $AlCl_3$ = 0.40 mol/L × 0.0500 L = 0.020 mol $AlCl_3$

$$0.020 \text{ mol } AlCl_3\left(\frac{1 \text{ mol } Al^{3+}}{1 \text{ mol } AlCl_3}\right) = 0.020 \text{ mol } Al^{3+}$$

$$0.020 \text{ mol } AlCl_3\left(\frac{3 \text{ mol } Cl^-}{1 \text{ mol } AlCl_3}\right) = 0.060 \text{ mol } Cl^-$$

4.93 (a) $Cr(NO_3)_2 \rightarrow Cr^{2+} + 2NO_3^-$

$$\text{M } Cr^{2+} = \left(\frac{0.25 \text{ mol } Cr(NO_3)_2}{1 \text{ L } Cr(NO_3)_2 \text{ soln}}\right)\left(\frac{1 \text{ mol } Cr^{2+}}{1 \text{ mol } Cr(NO_3)_2}\right) = 0.25 \text{ M } Cr^{2+}$$

$$\text{M } NO_3^- = \left(\frac{0.25 \text{ mol } Cr(NO_3)_2}{1 \text{ L } Cr(NO_3)_2 \text{ soln}}\right)\left(\frac{2 \text{ mol } NO_3^-}{1 \text{ mol } Cr(NO_3)_2}\right) = 0.50 \text{ M } NO_3^-$$

(b) $CuSO_4 \rightarrow Cu^{2+} + SO_4^{2-}$

$$\text{M } Cu^{2+} = \left(\frac{0.10 \text{ mol } CuSO_4}{1 \text{ L } CuSO_4 \text{ soln}}\right)\left(\frac{1 \text{ mol } Cu^{2+}}{1 \text{ mol } CuSO_4}\right) = 0.10 \text{ M } Cu^{2+}$$

$$\text{M } SO_4^{2-} = \left(\frac{0.10 \text{ mol } CuSO_4}{1 \text{ L } CuSO_4 \text{ soln}}\right)\left(\frac{1 \text{ mol } SO_4^{2-}}{1 \text{ mol } CuSO_4}\right) = 0.10 \text{ M } SO_4^{2-}$$

(c) $Na_3PO_4 \rightarrow 3Na^+ + PO_4^{3-}$

$$\text{M } Na^+ = \left(\frac{0.16 \text{ mol } Na_3PO_4}{1 \text{ L } Na_3PO_4 \text{ soln}}\right)\left(\frac{3 \text{ mol } Na^+}{1 \text{ mol } Na_3PO_4}\right) = 0.48 \text{ M } Na^+$$

$$\mathrm{M\ PO_4^{3-}} = \left(\frac{0.16\ \mathrm{mol\ Na_3PO_4}}{1\ \mathrm{L\ Na_3PO_4\ soln}}\right)\left(\frac{1\ \mathrm{mol\ PO_4^{3-}}}{1\ \mathrm{mol\ Na_3PO_4}}\right) = 0.16\ \mathrm{M\ PO_4^{3-}}$$

(d) $Al_2(SO_4)_3 \rightarrow 2Al^{3+} + 3SO_4^{2-}$

$$\mathrm{M\ Al^{3+}} = \left(\frac{0.075\ \mathrm{mol\ Al_2(SO_4)_3}}{1\ \mathrm{L\ Al_2(SO_4)_3\ soln}}\right)\left(\frac{2\ \mathrm{mol\ Al^{3+}}}{1\ \mathrm{mol\ Al_2(SO_4)_3}}\right) = 0.15\ \mathrm{M\ Al^{3+}}$$

$$\mathrm{M\ SO_4^{2-}} = \left(\frac{0.075\ \mathrm{mol\ Al_2(SO_4)_3}}{1\ \mathrm{L\ Al_2(SO_4)_3\ soln}}\right)\left(\frac{3\ \mathrm{mol\ SO_4^{2-}}}{1\ \mathrm{mol\ Al_2(SO_4)_3}}\right) = 0.22\ \mathrm{M\ SO_4^{2-}}$$

4.95 $\mathrm{g\ Al_2(SO_4)_3} = (50.0\ \mathrm{mL\ soln})\left(\frac{0.12\ \mathrm{mol\ Al^{3+}}}{1000\ \mathrm{mL\ soln}}\right)\left(\frac{1\ \mathrm{mol\ Al_2(SO_4)_3}}{2\ \mathrm{mol\ Al^{3+}}}\right)\left(\frac{342.14\ \mathrm{g\ Al_2(SO_4)_3}}{1\ \mathrm{mol\ Al_2(SO_4)_3}}\right)$

$= 1.0\ \mathrm{g\ Al_2(SO_4)_3}$

4.97 $\mathrm{mL\ NiCl_2\ soln} = 20.0\ \mathrm{mL\ soln}\left(\frac{0.15\ \mathrm{mol\ Na_2CO_3}}{1000\ \mathrm{mL\ soln}}\right)\left(\frac{1\ \mathrm{mol\ NiCl_2}}{1\ \mathrm{mol\ Na_2CO_3}}\right)\left(\frac{1000\ \mathrm{mL\ soln}}{0.25\ \mathrm{mol\ NiCl_2}}\right)$

$= 12.0\ \mathrm{mL\ NiCl_2\ soln}$

$$\mathrm{g\ NiCO_3} = 12.0\ \mathrm{mL\ NiCl_2\ soln}\left(\frac{0.25\ \mathrm{mol\ NiCl_2}}{1000\ \mathrm{mL\ soln}}\right)\left(\frac{1\ \mathrm{mol\ NiCO_3}}{1\ \mathrm{mol\ NiCl_2}}\right)\left(\frac{118.7\ \mathrm{g\ NiCO_3}}{1\ \mathrm{mol\ NiCO_3}}\right) = 0.36\ \mathrm{g\ NiCO_3}$$

4.99 $$M\ \mathrm{KOH} = \frac{(20.78\ \mathrm{mL\ HCl\ soln})\left(\frac{1\ \mathrm{L\ HCl\ soln}}{1000\ \mathrm{mL\ HCl\ soln}}\right)\left(\frac{0.116\ \mathrm{mol\ HCl}}{1\ \mathrm{L\ HCl}}\right)\left(\frac{1\ \mathrm{mol\ KOH}}{1\ \mathrm{mol\ HCl}}\right)}{21.34\ \mathrm{mL\ KOH}\left(\frac{1\ \mathrm{L\ KOH}}{1000\ \mathrm{mL\ KOH}}\right)}$$

$= 0.113\ M\ \mathrm{KOH}$

$KOH(aq) + HCl(aq) \rightarrow KCl(aq) + H_2O$

4.101 $Al_2(SO_4)_3(aq) + 3Ba(OH)_2(aq) \rightarrow 2Al(OH)_3(s) + 3BaSO_4(s)$

$$\mathrm{g\ Al_2(SO_4)_3} = 85.0\ \mathrm{mL\ soln}\left(\frac{0.0500\ \mathrm{mol\ Ba(OH)_2}}{1000\ \mathrm{mL\ soln}}\right)\left(\frac{1\ \mathrm{mol\ Al_2(SO_4)_3}}{3\ \mathrm{mol\ Ba(OH)_2}}\right)\left(\frac{342.2\ \mathrm{g\ Al_2(SO_4)_3}}{1\ \mathrm{mol\ Al_2(SO_4)_3}}\right)$$

$= 0.485\ \mathrm{g\ Al_2(SO_4)_3}$

4.103 $\mathrm{mL\ FeCl_3\ soln} = 20.0\ \mathrm{mL\ AgNO_3}\left(\frac{0.0450\ \mathrm{mol\ AgNO_3}}{1000\ \mathrm{mL\ soln}}\right)\left(\frac{1\ \mathrm{mol\ Ag^+}}{1\ \mathrm{mol\ AgNO_3}}\right)\left(\frac{1\ \mathrm{mol\ Cl^-}}{1\ \mathrm{mol\ Ag^+}}\right)$

$\times\left(\frac{1\ \mathrm{mol\ FeCl_3}}{3\ \mathrm{mol\ Cl^-}}\right)\left(\frac{1000\ \mathrm{mL\ soln}}{0.150\ \mathrm{mol\ FeCl_3}}\right) = 2.00\ \mathrm{mL\ FeCl_3\ soln}$

$$\mathrm{g\ AgCl} = (20.0\ \mathrm{mL\ AgNO_3})\left(\frac{1\ \mathrm{L\ AgNO_3}}{1000\ \mathrm{mL\ AgNO_3}}\right)\left(\frac{0.0450\ \mathrm{mol\ AgNO_3}}{1\ \mathrm{L\ AgNO_3}}\right)\left(\frac{3\ \mathrm{mol\ AgCl}}{3\ \mathrm{mol\ AgNO_3}}\right)\left(\frac{143.32\ \mathrm{AgCl}}{1\ \mathrm{mol\ AgCl}}\right)$$

$= 0.129\ \mathrm{g\ AgCl}$

4.105 $Ag^+ + Cl^- \rightarrow AgCl(s)$

$$\text{mL AlCl}_3 = (20.0\text{ mL AgC}_2\text{H}_3\text{O}_2)\left(\frac{0.500\text{ mol AgC}_2\text{H}_3\text{O}_2}{1000\text{ mL AgC}_2\text{H}_3\text{O}_2}\right)\left(\frac{1\text{ mol Ag}^+}{1\text{ mol AgC}_2\text{H}_3\text{O}_2}\right)\left(\frac{1\text{ mol Cl}^-}{1\text{ mol Ag}^+}\right)$$

$$\times\left(\frac{1\text{ mol AlCl}_3}{3\text{ mol Cl}^-}\right)\left(\frac{1000\text{ mL AlCl}_3}{0.250\text{ moles AlCl}_3}\right) = 13.3\text{ mL AlCl}_3$$

4.107 $Fe_2O_3 + 6HCl \rightarrow 2FeCl_3 + 3H_2O$
$0.0250\text{ L HCl} \times 0.500\text{ mol/L} = 1.25 \times 10^{-2}\text{ mol HCl}$

$$\text{mol Fe}^{3+} = (1.25 \times 10^{-2}\text{ mol HCl})\left(\frac{1\text{ mol Fe}_2\text{O}_3}{6\text{ mol HCl}}\right)\left(\frac{2\text{ mol Fe}^{3+}}{1\text{ mol Fe}_2\text{O}_3}\right) = 4.17 \times 10^{-3}\text{ mol Fe}^{3+}$$

$$M\text{ Fe}^{3+} = \frac{4.17 \times 10^{-3}\text{ mol Fe}^{3+}}{0.0250\text{ L soln}} = 0.167\ M\text{ Fe}^{3+}$$

$$\text{g Fe}_2\text{O}_3 = (4.17 \times 10^{-3}\text{ mol Fe}^{3+})\left(\frac{1\text{ mol Fe}_2\text{O}_3}{2\text{ mol Fe}^{3+}}\right)\left(\frac{159.69\text{ g Fe}_2\text{O}_3}{1\text{ mol Fe}_2\text{O}_3}\right) = 0.333\text{ g Fe}_2\text{O}_3$$

Therefore, the mass of Fe_2O_3 that remains unreacted is:
$(4.00\text{ g} - 0.333\text{ g}) = 3.67\text{ g}$

4.109 The equation for the reaction indicates that the two materials react in equimolar amounts, i.e. the stoichiometry is 1 to 1:

$$\text{AgNO}_3(aq) + \text{NaCl}(aq) \rightarrow \text{AgCl}(s) + \text{NaNO}_3(aq)$$

(a) Because this reaction is 1:1, we can see by inspection that the $AgNO_3$ is the limiting reagent. We know this because the concentration of the $AgNO_3$ is lower than the NaCl. Since we start with equal volumes, there are fewer moles of the $AgNO_3$.

$$\text{mol AgCl} = (25.0\text{ mL AgNO}_3\text{ soln})\left(\frac{0.320\text{ mol AgNO}_3}{1000\text{ mL AgNO}_3\text{ soln}}\right)\left(\frac{1\text{ mol AgCl}}{1\text{ mol AgNO}_3}\right)$$

$$= 8.00 \times 10^{-3}\text{ mol AgCl}$$

(b) Assuming that AgCl is essentially insoluble, the concentration of silver ion can be said to be zero since all of the $AgNO_3$ reacted. The number of moles of chloride ion would be reduced by the precipitation of 8.00×10^{-3} mol AgCl, such that the final number of moles of chloride ion would be:

$$0.0250\text{ L} \times 0.440\text{ mol/L} - 8.00 \times 10^{-3}\text{ mol} = 3.0 \times 10^{-3}\text{ mol Cl}^-$$

The final concentration of Cl^- is, therefore:

$$3.0 \times 10^{-3}\text{ mol} \div 0.0500\text{ L} = 0.060\ M\text{ Cl}^-$$

All of the original number of moles of NO_3^- and of Na^+ would still be present in solution, and their concentrations would be:

For NO_3^-:

$$M\text{ NO}_3^- = \frac{(25.0\text{ mL AgNO}_3\text{ soln})\left(\frac{0.320\text{ mol AgNO}_3}{1000\text{ mL AgNO}_3\text{ soln}}\right)\left(\frac{1\text{ mol NO}_3^-}{1\text{ mol AgNO}_3}\right)}{(50.0\text{ mL soln})\left(\frac{1\text{ L soln}}{1000\text{ mL soln}}\right)} = 0.160\ M\text{ NO}_3^-$$

For Na^+:

$$M\ Na^+ = \frac{(25.0 \text{ mL NaCl soln})\left(\frac{0.440 \text{ mol NaCl}}{1000 \text{ mL NaCl soln}}\right)\left(\frac{1 \text{ mol Na}^+}{1 \text{ mol NaCl}}\right)}{(50.0 \text{ mL soln})\left(\frac{1 \text{ L soln}}{1000 \text{ mL soln}}\right)} = 0.220\ M\ Na^+$$

4.111 First, calculate the number of moles HCl based on the titration according to the following equation:

$NaOH(aq) + HCl(aq) \rightarrow NaCl(aq) + H_2O$

$$\text{mol HCl} = (23.25 \text{ mL NaOH})\left(\frac{0.105 \text{ mol NaOH}}{1000 \text{ mL NaOH}}\right)\left(\frac{1 \text{ mol HCl}}{1 \text{ mol NaOH}}\right) = 2.44 \times 10^{-3} \text{ mol HCl}$$

Next, determine the concentration of the HCl solution:
2.44×10^{-3} mol $\div$ 0.02145 L = 0.114 M HCl

4.113 Since lactic acid is monoprotic, it reacts with sodium hydroxide on a one to one mole basis:

$$\text{mol } HC_3H_5O_3 = (17.25 \text{ mL NaOH})\left(\frac{0.155 \text{ mol NaOH}}{1000 \text{ mL NaOH}}\right)\left(\frac{1 \text{ mol } HC_3H_5O_3}{1 \text{ mol NaOH}}\right) = 2.67 \times 10^{-3} \text{ mol } HC_3H_5O_3$$

4.115 $MgSO_4(aq) + BaCl_2(aq) \rightarrow BaSO_4(s) + Mg^{2+}(aq) + 2Cl^-(aq)$

$$\text{mol } BaSO_4 = (1.174 \text{ g } BaSO_4)\left(\frac{1 \text{ mol } BaSO_4}{233.39 \text{ g } BaSO_4}\right) = 5.030 \times 10^{-3} \text{ mol } BaSO_4$$

There are as many moles of $MgSO_4$ as of $BaSO_4$, namely 5.030×10^{-3} mol $MgSO_4$. We know this because we can see that there is one mole of SO_4^{2-} in both the Ba and Mg compounds. Since both of these elements are in the same family, the reaction to produce the barium salt from the magnesium salt must be 1:1.

First determine the mass of $MgSO_4$ that was present:

5.030×10^{-3} mol $MgSO_4 \times 120.37$ g/mol = 0.6055 g $MgSO_4$

and subtract this from the total mass of the sample to find the mass of water in the original sample:

1.24 g – 0.6055 g = 0.63 g H_2O

We need to know the number of moles of water that are involved:

0.63 g $\div$ 18.0 g/mol = 3.5×10^{-2} mol H_2O

and the relative mole amounts of water and $MgSO_4$:
For $MgSO_4$, 5.030×10^{-3} moles/5.030×10^{-3} moles = 1.000
For water, 3.5×10^{-2} mol/5.030×10^{-3} mol = 7.0
Hence the formula is $MgSO_4{\cdot}7H_2O$

4.117 $$\text{g Pb} = (29.22 \text{ mL})\left(\frac{1 \text{ L } Na_2SO_4}{1000 \text{ mL } Na_2SO_4}\right)\left(\frac{0.122 \text{ mol } Na_2SO_4}{1 \text{ L } Na_2SO_4}\right)\left(\frac{1 \text{ mol } PbSO_4}{1 \text{ mol } Na_2SO_4}\right)\left(\frac{1 \text{ mol } Pb^{2+}}{1 \text{ mol } PbSO_4}\right)$$

$$\times\left(\frac{207.2 \text{ g } Pb^{2+}}{1 \text{ mol } Pb^{2+}}\right) = 0.7386 \text{ g Pb}$$

The percentage of Pb in the sample can be calculated as

$$\left(\frac{0.7386 \text{ g Pb}}{1.526 \text{ g sample}}\right) \times 100\% = 48.40\% \text{ Pb in the sample.}$$

4.119 The HCl that is added will react with the Na_2CO_3. Therefore, the amount of Na_2CO_3 in the mixture may be determined from the amount of HCl needed to react with it. We determine this amount by measuring the amount of HCl added in excess. First determine the number of moles HCl added:

$$\text{mol HCl} = (50.0 \text{ mL HCl})\left(\frac{0.240 \text{ mol HCl}}{1000 \text{ mL HCl}}\right) = 1.20 \times 10^{-2} \text{ mol HCl}$$

Next determine how many moles of HCl remain after the reaction:

$$\text{mol HCl} = (22.90 \text{ mL NaOH})\left(\frac{0.100 \text{ mol NaOH}}{1000 \text{ mL NaOH}}\right)\left(\frac{1 \text{ mol HCl}}{1 \text{ mol NaOH}}\right) = 2.29 \times 10^{-3} \text{ mol HCl}$$

Therefore, we know that;
1.20×10^{-2} moles HCl – 2.29×10^{-3} moles HCl = 9.71×10^{-3} moles HCl reacted with the Na_2CO_3 present in the mixture. The balanced equation for the reaction is:

$$2\text{ HCl} + Na_2CO_3 \rightarrow 2NaCl + H_2O + CO_2$$

$$\text{g } Na_2CO_3 = (9.71 \times 10^{-3} \text{ mol HCl})\left(\frac{1 \text{ mol } Na_2CO_3}{2 \text{ mol HCl}}\right)\left(\frac{105.99 \text{ g } Na_2CO_3}{1 \text{ mol } Na_2CO_3}\right) = 0.515 \text{ g } Na_2CO_3$$

$$\text{g NaCl} = 1.243 \text{ g} - 0.515 \text{ g} = 0.728 \text{ g}$$

$$\% \text{ NaCl} = \frac{0.728 \text{ g}}{1.243 \text{ g}} \times 100\% = 58.6\%$$

Practice Exercises

5.1 $2Na(s) + O_2(g) \rightarrow Na_2O_2(s)$
Oxygen is reduced since it gains electrons.
Sodium is oxidized since it loses electrons.

5.2 $2Al(s) + 3Cl_2(g) \rightarrow 2AlCl_3(aq)$
Aluminum is oxidized and is, therefore, the reducing agent.
Chlorine is reduced and is, therefore, the oxidizing agent.

5.3 ClO_2^-: O –2 Cl +3

5.4 a) Ni +2; Cl –1
b) Mg +2; Ti +4; O –2
c) K +1; Cr +6; O –2
d) H +1; P +5, O –2
e) V +3; C 0; H +1; O –2

5.5 There is a total charge of +8, divided over three atoms, so the average charge is +8/3.

5.6 First the oxidation numbers of all atoms must be found.

$$KClO_3 + 6HNO_2 \rightarrow KCl + HNO_3$$

Reactants:	Products:
K = +1	K = +1
Cl = +5	Cl = –1
O = –2	
H = +1	H = +1
N = +3	N = +5
O = –2	O = –2

The oxidation numbers for K and Na do not change. However, the oxidation numbers for all chlorines atoms drop. The oxidation numbers for nitrogen increase.

Therefore, $KClO_3$ is reduced and HNO_2 is oxidized.
This means $KClO_3$ is the oxidizing agent and HNO_2 is the reducing agent.
This reaction is the redox reaction. In the other reaction, the oxidation numbers of the atoms do not change.

5.7 First the oxidation numbers of all atoms must be found.

$$Cl_2 + 2NaClO_2 \rightarrow 2ClO_2 + 2NaCl$$

Reactants:	Products:
Cl = 0	Cl = +4
	O = –2
Na = +1	Na = +1
Cl = +3	Cl = –1
O = –2	

The oxidation numbers for O and Na do not change. However, the oxidation numbers for all chlorine atoms change. There is no simple way to tell which chlorines are reduced and which are oxidized in this reaction.

One analysis would have the Cl in Cl_2 end up as the Cl in NaCl, while the Cl in $NaClO_2$ ends up as the Cl in ClO_2. In this case Cl_2 is reduced and is the oxidizing agent, while $NaClO_2$ is oxidized and is the reducing agent.

5.8 If H_2O_2 acts as an oxidizing agent, it gets reduced itself in the process. Examining the oxidation numbers:

H_2O_2	$H = +1, O = -1$
H_2O	$H = +1, O = -2$
O_2	$O = 0$

If H_2O_2 is reduced it must form water, since the oxidation number of oxygen drops from –1 to –2 in the formation of water (a reduction).

The product is therefore water.

5.9 $Al(s) + Cu^{2+}(aq) \rightarrow Al^{3+}(aq) + Cu(s)$

First, we break the reaction above into half-reactions:

$$Al(s) \rightarrow Al^{3+}(aq)$$
$$Cu^{2+}(aq) \rightarrow Cu(s)$$

Each half-reaction is already balanced with respect to atoms, so next we add electrons to balance the charges on both sides of the equations:

$$Al(s) \rightarrow Al^{3+}(aq) + 3e^-$$
$$2e^- + Cu^{2+}(aq) \rightarrow Cu(s)$$

Next, we multiply both equations so that the electrons gained equals the electrons lost,

$$2(Al(s) \rightarrow Al^{3+}(aq) + 3e^-)$$
$$3(2e^- + Cu^{2+}(aq) \rightarrow Cu(s))$$

which gives us:

$$2Al(s) \rightarrow 2Al^{3+}(aq) + 6e^-$$
$$6e^- + 3Cu^{2+}(aq) \rightarrow 3Cu(s)$$

Now, by adding the half-reactions back together, we have our balanced equation:

$$2Al(s) + 3Cu^{2+}(aq) \rightarrow 2Al^{3+}(aq) + 3Cu(s)$$

5.10 $TcO_4^- + Sn^{2+} \rightarrow Tc^{4+} + Sn^{4+}$

First, we break the reaction above into half-reactions:

$$TcO_4^- \rightarrow Tc^{4+}$$
$$Sn^{2+} \rightarrow Sn^{4+}$$

Each half-reaction is already balanced with respect to atoms other than O and H, so next we balance the O atoms by using water:

$$TcO_4^- \rightarrow Tc^{4+} + 4H_2O$$
$$Sn^{2+} \rightarrow Sn^{4+}$$

Now we balance H by using H^+:

$$8H^+ + TcO_4^- \rightarrow Tc^{4+} + 4H_2O$$
$$Sn^{2+} \rightarrow Sn^{4+}$$

Next, we add electrons to balance the charges on both sides of the equations:

$$3e^- + 8H^+ + TcO_4^- \rightarrow Tc^{4+} + 4H_2O$$
$$Sn^{2+} \rightarrow Sn^{4+} + 2e^-$$

We multiply the equations so that the electrons gained equals the electrons lost,

$$2(3e^- + 8H^+ + TcO_4^- \rightarrow Tc^{4+} + 4H_2O)$$
$$3(Sn^{2+} \rightarrow Sn^{4+} + 2e^-)$$

which gives us:

$$6e^- + 16H^+ + 2TcO_4^- \rightarrow 2Tc^{4+} + 8H_2O$$
$$3Sn^{2+} \rightarrow 3Sn^{4+} + 6e^-$$

Now, by adding the half-reactions back together, we have our balanced equation:

$$3Sn^{2+} + 16H^+ + 2TcO_4^- \rightarrow 2Tc^{4+} + 8H_2O + 3Sn^{4+}$$

5.11 $(Cu \rightarrow Cu^{2+} + 2e^-) \times 4$
$2NO_3^- + 10H^+ + 8e^- \rightarrow N_2O + 5H_2O$
$4Cu + 2NO_3^- + 10H^+ \rightarrow 4Cu^{2+} + N_2O + 5H_2O$

5.12 $2H_2O + SO_2 \rightarrow SO_4^{2-} + 4H^+ + 2e^-$
$4OH^- + 2H_2O + SO_2 \rightarrow SO_4^{2-} + 4H^+ + 2e^- + 4OH^-$
$4OH^- + 2H_2O + SO_2 \rightarrow SO_4^{2-} + 2e^- + 4H_2O$
$4OH^- + SO_2 \rightarrow SO_4^{2-} + 2e^- + 2H_2O$

5.13 $(MnO_4^- + 4H^+ + 3e^- \rightarrow MnO_2 + 2H_2O) \times 2$
$(C_2O_4^{2-} + 2H_2O \rightarrow 2CO_3^{2-} + 4H^+ + 2e^-) \times 3$
$2MnO_4^- + 3C_2O_4^{2-} + 2H_2O \rightarrow 2MnO_2 + 6CO_3^{2-} + 4H^+$
Adding $4OH^-$ to both sides of the above equation we get:

$$2MnO_4^- + 3C_2O_4^{2-} + 2H_2O + 4OH^- \rightarrow 2MnO_2 + 6CO_3^{2-} + 4H_2O$$

which simplifies to give:

$$2MnO_4^- + 3C_2O_4^{2-} + 4OH^- \rightarrow 2MnO_2 + 6CO_3^{2-} + 2H_2O$$

5.14 $Zn + H^+ \rightarrow Zn^{2+} + H_2$
Divide the reaction into two half reactions and balance the number of atoms
$Zn \rightarrow Zn^{2+}$
$2H^+ \rightarrow H_2$
Balance the charges with electrons
$Zn \rightarrow Zn^{2+} + 2e^-$
$2H^+ + 2e^- \rightarrow H_2$

5.15 $HNO_3 + Mg \rightarrow NH_4^+ + Mg^{2+}$
Divide the reaction into two half reactions and balance the number of atoms
$HNO_3 \rightarrow NH_4^+$
$Mg \rightarrow Mg^{2+}$
$HNO_3 \rightarrow NH_4^+ + 3H_2O$
$Mg \rightarrow Mg^{2+}$
$HNO_3 + 9H^+ \rightarrow NH_4^+ + 3H_2O$
$Mg \rightarrow Mg^{2+}$
Balance the charges with electrons
$8e^- + HNO_3 + 9H^+ \rightarrow NH_4^+ + 3H_2O$
$Mg \rightarrow Mg^{2+} + 2e^-$
Multiply by a factor to make the number of electrons the same
$8e^- + HNO_3 + 9H^+ \rightarrow NH_4^+ + 3H_2O$
$(Mg \rightarrow Mg^{2+} + 2e^-) \times 4$
Add together
$HNO_3 + 9H^+ + 4Mg \rightarrow NH_4^+ + 4Mg^{2+} + 3H_2O$

5.16 a) molecular: $Mg(s) + 2HCl(aq) \rightarrow MgCl_2(aq) + H_2(g)$
ionic: $Mg(s) + 2H^+(aq) + 2Cl^-(aq) \rightarrow Mg^{2+}(aq) + 2Cl^-(aq) + H_2(g)$
net ionic: $Mg(s) + 2H^+(aq) \rightarrow Mg^{2+}(aq) + H_2(g)$

b) molecular: $2Al(s) + 6HCl(aq) \rightarrow 2AlCl_3(aq) + 3H_2(g)$
ionic: $2Al(s) + 6H^+(aq) + 6Cl^-(aq) \rightarrow 2Al^{3+}(aq) + 6Cl^-(aq) + 3H_2(g)$
net ionic: $2Al(s) + 6H^+(aq) \rightarrow 2Al^{3+}(aq) + 3H_2(g)$

5.17 $Cu^{2+}(aq) + Mg(s) \rightarrow Cu(s) + Mg^{2+}(aq)$

5.18 a) $2Al(s) + 3Cu^{2+}(aq) \rightarrow 2Al^{3+}(aq) + 3Cu(s)$
b) $Ag(s) + Mg^{2+}(aq) \rightarrow$ No reaction

5.19 $2C_{20}H_{42}(s) + 21O_2(g) \rightarrow 40C(s) + 42H_2O(g)$

5.20 $2C_4H_{10}(\ell) + 13O_2(g) \rightarrow 8CO_2(g) + 10H_2O(g)$

5.21 $C_2H_5OH(\ell) + 3O_2(g) \rightarrow 2CO_2(g) + 3H_2O(g)$

5.22 $2Sr(s) + O_2(g) \rightarrow 2SrO(s)$

5.23 $4Fe(s) + 3O_2(g) \rightarrow 2Fe_2O_3(s)$

5.24 First we need a balanced equation:
$C_2O_4^{2-} \rightarrow CO_2$
$MnO_4^- \rightarrow Mn^{2+}$
Balance the atoms
$C_2O_4^{2-} \rightarrow 2CO_2$
$8H^+ + MnO_4^- \rightarrow Mn^{2+} + 4H_2O$
Balance the charges
$C_2O_4^{2-} \rightarrow 2CO_2 + 2e^-$
$5e^- + 8H^+ + MnO_4^- \rightarrow Mn^{2+} + 4H_2O$
Add the reactions together
$5C_2O_4^{2-} + 16H^+ + 2MnO_4^- \rightarrow 10CO_2 + 2Mn^{2+} + 8H_2O$

$$\text{mol } C_2O_4^{2-} = (18.30 \text{ mL KMnO}_4)\left(\frac{1 \text{ L KMnO}_4}{1000 \text{ mL KMnO}_4}\right)\left(\frac{0.02000 \text{ mol KMnO}_4}{1 \text{ L KMnO}_4}\right)\left(\frac{5 \text{ mol } C_2O_4^{-2}}{2 \text{ mol KMnO}_4}\right) =$$
$$0.0009150 \text{ mol } C_2O_4^{2-}$$

$$M\, H_2C_2O_4 = \left(\frac{0.0009150 \text{ mol } C_2O_4^{2-}}{15.00 \text{ mL } C_2O_4^{2-}}\right)\left(\frac{1000 \text{ mL } C_2O_4^{2-}}{1 \text{ L } C_2O_4^{2-}}\right)\left(\frac{1 \text{ mol } H_2C_2O_4}{1 \text{ mol } C_2O_4^{2-}}\right) = 0.06100\, M\, H_2C_2O_4$$

5.25 First we need a balanced equation:
$Cl_2 + 2e^- \rightarrow 2Cl^-$
$S_2O_3^{2-} + 5H_2O \rightarrow 2SO_4^{2-} + 10H^+ + 8e^-$
$4Cl_2 + S_2O_3^{2-} + 5H_2O \rightarrow 8Cl^- + 2SO_4^{2-} + 10H^+$

$$\text{g } Na_2S_2O_3 = (4.25 \text{ g } Cl_2)\left(\frac{1 \text{ mol } Cl_2}{70.906 \text{ g } Cl_2}\right)\left(\frac{1 \text{ mol } Na_2S_2O_3}{4 \text{ mol } Cl_2}\right)\left(\frac{158.132 \text{ g } Na_2S_2O_3}{1 \text{ mol } Na_2S_2O_3}\right) = 2.37 \text{ g } Na_2S_2O_3$$

5.26 a) $(Sn^{2+} \rightarrow Sn^{4+} + 2e^-) \times 5$
$(MnO_4^- + 8H^+ + 5e^- \rightarrow Mn^{2+} + 4H_2O) \times 2$
$5Sn^{2+} + 2MnO_4^- + 16H^+ \rightarrow 5Sn^{4+} + 2Mn^{2+} + 8H_2O$

b)
$$\text{g Sn} = (8.08 \text{ mL KMnO}_4 \text{ soln})\left(\frac{0.0500 \text{ mol KMnO}_4}{1000 \text{ mL KMnO}_4}\right)\left(\frac{1 \text{ mol } MnO_4^-}{1 \text{ mol KMnO}_4}\right)\left(\frac{5 \text{ mol } Sn^{2+}}{2 \text{ mol } MnO_4^-}\right)$$
$$\times \left(\frac{1 \text{ mol Sn}}{1 \text{ mol } Sn^{2+}}\right)\left(\frac{118.71 \text{ g Sn}}{1 \text{ mol Sn}}\right) = 0.120 \text{ g Sn}$$

c)
$$\% \text{ Sn} = \frac{0.120 \text{ g Sn}}{0.300 \text{ g sample}} \times 100\% = 40.0\% \text{ Sn}$$

d) g SnO_2 = (8.08 mL $KMnO_4$ soln)

$$\left(\frac{0.0500 \text{ mol } KMnO_4}{1000 \text{ mL } KMnO_4}\right)\left(\frac{1 \text{ mol } MnO_4^-}{1 \text{ mol } KMnO_4}\right)\left(\frac{5 \text{ mol } Sn^{2+}}{2 \text{ mol } MnO_4^-}\right)$$

$$\times\left(\frac{1 \text{ mol } SnO_2}{1 \text{ mol } Sn^{2+}}\right)\left(\frac{150.71 \text{ g } SnO_2}{1 \text{ mol } SnO_2}\right) = 0.152 \text{ g } SnO_2$$

$$\% \ SnO_2 = \frac{0.152 \text{ g } SnO_2}{0.300 \text{ g sample}} \times 100\% = 50.7\% \ SnO_2$$

Review Problems

5.25 The sum of the oxidation numbers should be zero:
(a) S^{2-}: –2
(b) SO_2: S +4, O –2
(c) P_4: P 0
(d) PH_3: P –3, H +1

5.27 The sum of the oxidation numbers should be zero:
(a) O: –2, Na: +1, Cl: +1
(b) O: –2, Na: +1, Cl: +3
(c) O: –2, Na: +1, Cl: +5
(d) O: –2, Na: +1, Cl: +7

5.29 The sum of the oxidation numbers should be zero:
(a) S: –2, Pb: +2
(b) Cl: –1, Ti: +4
(c) Cs +1, O –1/2 (The Cs can only have an oxidation number of +1 or 0.)
(d) F –1, O +1

5.31 $Cl_2(aq) + H_2O \rightleftarrows H^+(aq) + Cl^-(aq) + HOCl(aq)$
In the forward direction: The oxidation number of the chlorine atoms decreases from 0 to –1. Therefore **Cl_2 is reduced**. However, in HOCl, chlorine has an oxidation number of +1, so **Cl_2 also oxidized!** (One atom is reduced, the other is oxidized.)

In the reverse direction: The Cl^- ion begins with an oxidation number of –1 and ends with an oxidation number of 0. Therefore the Cl^- ion is oxidized: This means **Cl^- is the reducing agent**. Since the oxidation number of H^+ does not change, **HOCl must be the oxidizing agent.**

5.33 (a) substance reduced (and oxidizing agent): HNO_3
substance oxidized (and reducing agent): H_3AsO_3
(b) substance reduced (and oxidizing agent): HOCl
substance oxidized (and reducing agent): NaI
(c) substance reduced (and oxidizing agent): $KMnO_4$
substance oxidized (and reducing agent): $H_2C_2O_4$
(d) substance reduced (and oxidizing agent): H_2SO_4
substance oxidized (and reducing agent): Al

5.35 (a)
$2S_2O_3^{2-} \rightarrow S_4O_6^{2-} + 2e^-$
$OCl^- + 2H^+ + 2e^- \rightarrow Cl^- + H_2O$
$OCl^- + 2S_2O_3^{2-} + 2H^+ \rightarrow S_4O_6^{2-} + Cl^- + H_2O$

(b) $(NO_3^- + 2H^+ + e^- \rightarrow NO_2 + H_2O) \times 2$
$Cu \rightarrow Cu^{2+} + 2e^-$
$2NO_3^- + Cu + 4H^+ \rightarrow 2NO_2 + Cu^{2+} + 2H_2O$

(c) $IO_3^- + 6H^+ + 6e^- \rightarrow I^- + 3H_2O$
$(H_2O + AsO_3^{3-} \rightarrow AsO_4^{3-} + 2H^+ + 2e^-) \times 3$
$IO_3^- + 3AsO_3^{3-} + 6H^+ + 3H_2O \rightarrow I^- + 3AsO_4^{3-} + 3H_2O + 6H^+$
which simplifies to give:
$3AsO_3^{3-} + IO_3^- \rightarrow I^- + 3AsO_4^{3-}$

(d) $SO_4^{2-} + 4H^+ + 2e^- \rightarrow SO_2 + 2H_2O$
$Zn \rightarrow Zn^{2+} + 2e^-$
$Zn + SO_4^{2-} + 4H^+ \rightarrow Zn^{2+} + SO_2 + 2H_2O$

(e) $NO_3^- + 10H^+ + 8e^- \rightarrow NH_4^+ + 3H_2O$
$(Zn \rightarrow Zn^{2+} + 2e^-) \times 4$
$NO_3^- + 4Zn + 10H^+ \rightarrow 4Zn^{2+} + NH_4^+ + 3H_2O$

(f) $2Cr^{3+} + 7H_2O \rightarrow Cr_2O_7^{2-} + 14H^+ + 6e^-$
$(BiO_3^- + 6H^+ + 2e^- \rightarrow Bi^{3+} + 3H_2O) \times 3$
$2Cr^{3+} + 3BiO_3^- + 18H^+ + 7H_2O \rightarrow Cr_2O_7^{2-} + 14H^+ + 3Bi^{3+} + 9H_2O$
which simplifies to give:
$2Cr^{3+} + 3BiO_3^- + 4H^+ \rightarrow Cr_2O_7^{2-} + 3Bi^{3+} + 2H_2O$

(g) $I_2 + 6H_2O \rightarrow 2IO_3^- + 12H^+ + 10e^-$
$(OCl^- + 2H^+ + 2e^- \rightarrow Cl^- + H_2O) \times 5$
$I_2 + 5OCl^- + H_2O \rightarrow 2IO_3^- + 5Cl^- + 2H^+$

(h) $(Mn^{2+} + 4H_2O \rightarrow MnO_4^- + 8H^+ + 5e^-) \times 2$
$(BiO_3^- + 6H^+ + 2e^- \rightarrow Bi^{3+} + 3H_2O) \times 5$
$2Mn^{2+} + 5BiO_3^- + 30H^+ + 8H_2O \rightarrow 2MnO_4^- + 5Bi^{3+} + 16H^+ + 15H_2O$
which simplifies to:
$2Mn^{2+} + 5BiO_3^- + 14H^+ \rightarrow 2MnO_4^- + 5Bi^{3+} + 7H_2O$

(i) $(H_3AsO_3 + H_2O \rightarrow H_3AsO_4 + 2H^+ + 2e^-) \times 3$
$Cr_2O_7^{2-} + 14H^+ + 6e^- \rightarrow 2Cr^{3+} + 7H_2O$
$3H_3AsO_3 + Cr_2O_7^{2-} + 3H_2O + 14H^+ \rightarrow 3H_3AsO_4 + 2Cr^{3+} + 6H^+ + 7H_2O$
which simplifies to give:
$3H_3AsO_3 + Cr_2O_7^{2-} + 8H^+ \rightarrow 3H_3AsO_4 + 2Cr^{3+} + 4H_2O$

(j) $2I^- \rightarrow I_2 + 2e^-$
$HSO_4^- + 3H^+ + 2e^- \rightarrow SO_2 + 2H_2O$
$2I^- + HSO_4^- + 3H^+ \rightarrow I_2 + SO_2 + 2H_2O$

5.37 For redox reactions in basic solution, we proceed to balance the half reactions as if they were in acid solution, and then add enough OH^- to each side of the resulting equation in order to neutralize (titrate) all of the H^+. This gives a corresponding amount of water ($H^+ + OH^- \rightarrow H_2O$) on one side of the equation, and an excess of OH^- on the other side of the equation, as befits a reaction in basic solution.

(a) $(CrO_4^{2-} + 4H^+ + 3e^- \rightarrow CrO_2^- + 2H_2O) \times 2$
$(S^{2-} \rightarrow S + 2e^-) \times 3$
$2CrO_4^{2-} + 3S^{2-} + 8H^+ \rightarrow 2CrO_2^- + 2S + 4H_2O$
Adding $8OH^-$ to both sides of the above equation we obtain:
$2CrO_4^{2-} + 3S^{2-} + 8H_2O \rightarrow 2CrO_2^- + 8OH^- + 3S + 4H_2O$
which simplifies to:
$2CrO_4^{2-} + 3S^{2-} + 4H_2O \rightarrow 2CrO_2^- + 3S + 8OH^-$

(b) $(C_2O_4^{2-} \rightarrow 2CO_2 + 2e^-) \times 3$
$(MnO_4^- + 4H^+ + 3e^- \rightarrow MnO_2 + 2H_2O) \times 2$
$3C_2O_4^{2-} + 2MnO_4^- + 8H^+ \rightarrow 6CO_2 + 2MnO_2 + 4H_2O$
Adding $8OH^-$ to both sides of the above equation we get:
$3C_2O_4^{2-} + 2MnO_4^- + 8H_2O \rightarrow 6CO_2 + 2MnO_2 + 4H_2O + 8OH^-$
which simplifies to give:
$3C_2O_4^{2-} + 2MnO_4^- + 4H_2O \rightarrow 6CO_2 + 2MnO_2 + 8OH^-$

(c) $(ClO_3^- + 6H^+ + 6e^- \rightarrow Cl^- + 3H_2O) \times 4$
$(N_2H_4 + 2H_2O \rightarrow 2NO + 8H^+ + 8e^-) \times 3$
$4ClO_3^- + 3N_2H_4 + 24H^+ + 6H_2O \rightarrow 4Cl^- + 6NO + 12H_2O + 24H^+$
which needs no OH^-, because it simplifies directly to:
$4ClO_3^- + 3N_2H_4 \rightarrow 4Cl^- + 6NO + 6H_2O$

(d) $NiO_2 + 2H^+ + 2e^- \rightarrow Ni(OH)_2$
$2Mn(OH)_2 \rightarrow Mn_2O_3 + H_2O + 2H^+ + 2e^-$
$NiO_2 + 2Mn(OH)_2 \rightarrow Ni(OH)_2 + Mn_2O_3 + H_2O$

(e) $(SO_3^{2-} + H_2O \rightarrow SO_4^{2-} + 2H^+ + 2e^-) \times 3$
$(MnO_4^- + 4H^+ + 3e^- \rightarrow MnO_2 + 2H_2O) \times 2$
$3SO_3^{2-} + 3H_2O + 8H^+ + 2MnO_4^- \rightarrow 3SO_4^{2-} + 6H^+ + 2MnO_2 + 4H_2O$
Adding $8OH^-$ to both sides of the equation we obtain:
$3SO_3^{2-} + 11H_2O + 2MnO_4^- \rightarrow 3SO_4^{2-} + 10H_2O + 2MnO_2 + 2OH^-$
which simplifies to:
$3SO_3^{2-} + 2MnO_4^- + H_2O \rightarrow 3SO_4^{2-} + 2MnO_2 + 2OH^-$

5.39 $(OCl^- + 2H^+ + 2e^- \rightarrow Cl^- + H_2O) \times 4$
$S_2O_3^{2-} + 5H_2O \rightarrow 2SO_4^{2-} + 10H^+ + 8e^-$
$4OCl^- + S_2O_3^{2-} + 5H_2O + 8H^+ \rightarrow 4Cl^- + 2SO_4^{2-} + 10H^+ + 4H_2O$
which simplifies to:
$4OCl^- + S_2O_3^{2-} + H_2O \rightarrow 4Cl^- + 2SO_4^{2-} + 2H^+$

5.41 $O_3 + 6H^+ + 6e^- \rightarrow 3H_2O$
$Br^- + 3H_2O \rightarrow BrO_3^- + 6H^+ + 6e^-$
$O_3 + Br^- + 3H_2O + 6H^+ \rightarrow BrO_3^- + 3H_2O + 6H^+$
which simplifies to:
$O_3 + Br^- \rightarrow BrO_3^-$

5.43 (a) m: $Mn(s) + 2HCl(aq) \rightarrow MnCl_2(aq) + H_2(g)$
I: $Mn(s) + 2H^+(aq) + 2Cl^-(aq) \rightarrow Mn^{2+}(aq) + 2Cl^-(aq) + H_2(g)$
NI: $Mn(s) + 2H^+(aq) \rightarrow Mn^{2+}(aq) + H_2(g)$

(b) m: $Cd(s) + 2HCl(aq) \rightarrow CdCl_2(aq) + H_2(g)$
I: $Cd(s) + 2H^+(aq) + 2Cl^-(aq) \rightarrow Cd^{2+}(aq) + Cl^-(aq) + H_2(g)$
NI: $Cd(s) + 2H^+(aq) \rightarrow Cd^{2+}(aq) + H_2(g)$

(c) m: $Sn(s) + 2HCl(aq) \rightarrow SnCl_2(aq) + H_2(g)$
I: $Sn(s) + 2H^+(aq) + 2Cl^-(aq) \rightarrow Sn^{2+}(aq) + 2Cl^-(aq) + H_2(g)$
NI: $Sn(s) + 2H^+(aq) \rightarrow Sn^{2+}(aq) + H_2(g)$

5.45 (a) $3Ag(s) + 4HNO_3(aq) \rightarrow 3AgNO_3(aq) + 2H_2O + NO(g)$
(b) $Ag(s) + 2HNO_3(aq) \rightarrow AgNO_3(aq) + H_2O + NO_2(aq)$

5.47 In each case, the reaction should proceed to give the less reactive of the two metals, together with the ion of the more reactive of the two metals. The reactivity is taken from the reactivity series table 6.2.

(a) N.R.

(b) $2Cr(s) + 3Pb^{2+}(aq) \rightarrow 2Cr^{3+}(aq) + 3Pb(s)$

(c) $2Ag^{+}(aq) + Fe(s) \rightarrow 2Ag(s) + Fe^{2+}(aq)$

(d) $3Ag(s) + Au^{3+}(aq) \rightarrow Au(s) + 3Ag^{+}(aq)$

5.49 Increasing ease of oxidation: Pt, Ru, Tl, Pu

5.51 The equation given shows that Cd is more active than Ru. Coupled with the information in Review Problem 5.49, we also see that Cd is more active than Tl^{+}. This means that in a mixture of Cd and Tl, Cd will be oxidized and Tl^{+} will be reduced:

$$Cd(s) + 2TlCl(aq) \rightarrow CdCl_2(aq) + 2Tl(s)$$

(The Tl(s) and the $Cd(NO_3)_2(aq)$ will not react.)

5.53 (a) $2C_6H_6(\ell) + 15O_2(g) \rightarrow 12CO_2(g) + 6H_2O(g)$

(b) $C_3H_8(g) + 5O_2(g) \rightarrow 3CO_2(g) + 4H_2O(g)$

(c) $C_{21}H_{44}(s) + 32O_2(g) \rightarrow 21CO_2(g) + 22H_2O(g)$

5.55 (a) $2C_6H_6(\ell) + 9O_2(g) \rightarrow 12CO(g) + 6H_2O(g)$

$2C_3H_8(g) + 7O_2(g) \rightarrow 6CO(g) + 8H_2O(g)$

$2C_{21}H_{44}(s) + 43O_2(g) \rightarrow 42CO(g) + 44H_2O(g)$

(b) $2C_6H_6(\ell) + 3O_2(g) \rightarrow 12C(s) + 6H_2O(g)$

$C_3H_8(g) + 2O_2(g) \rightarrow 3(s) + 4H_2O(g)$

$C_{21}H_{44}(s) + 11O_2(g) \rightarrow 21C(s) + 22H_2O(g)$

5.57 $2CH_3OH(\ell) + 3O_2(g) \rightarrow 2CO_2(g) + 4H_2O(g)$

5.59 $2(CH_3)_2S(g) + 9O_2(g) \rightarrow 4CO_2(g) + 6H_2O(g) + 2SO_2(g)$

5.61 (a) $2Zn(s) + O_2(g) \rightarrow 2ZnO(s)$

(b) $4Al(s) + 3O_2(g) \rightarrow 2Al_2O_3(s)$

(c) $2Mg(s) + O_2(g) \rightarrow 2MgO(s)$

(d) $4Fe(s) + 3O_2(g) \rightarrow 2Fe_2O_3(s)$

5.63 (a) $IO_3^- + 6H^+ + 6e^- \rightarrow I^- + 3H_2O$

$[SO_3^{2-} + H_2O \rightarrow SO_4^{2-} + 2H^+ + 2e^-] \times 3$

$IO_3^- + 3SO_3^{2-} + 6H^+ + 3H_2O \rightarrow I^- + 3SO_4^{2-} + 3H_2O + 6H^+$

Which simplifies to:

$IO_3^- + 3SO_3^{2-} \rightarrow I^- + 3SO_4^{2-}$

(b) $$\text{g } Na_2SO_3 = (5.00 \text{ g } NaIO_3)\left(\frac{1 \text{ mol } NaIO_3}{197.9 \text{ g } NaIO_3}\right)\left(\frac{3 \text{ mol } Na_2SO_3}{1 \text{ mol } NaIO_3}\right)\left(\frac{126.0 \text{ g } Na_2SO_3}{1 \text{ mol } Na_2SO_3}\right)$$

$$= 9.55 \text{ g } Na_2SO_3$$

5.65 $Cu + 2Ag^+ \rightarrow Cu^{2+} + Ag$

$$\text{g Cu} = (12.0 \text{ g Ag})\left(\frac{1 \text{ mol Ag}}{107.868 \text{ g Ag}}\right)\left(\frac{1 \text{ mol Cu}}{2 \text{ mol Ag}}\right)\left(\frac{63.54 \text{ g Cu}}{1 \text{ mol Cu}}\right) = 3.53 \text{ g Cu}$$

5.67 (a)

$[MnO_4^- + 8H^+ + 5e^- \rightarrow Mn^{2+} + 4H_2O] \times 2$
$[Sn^{2+} \rightarrow Sn^{4+} + 2e^-] \times 5$
$2MnO_4^- + 5Sn^{2+} + 16H^+ \rightarrow 2Mn^{2+} + 5Sn^{4+} + 8H_2O$

(b)

$$\text{mL KMnO}_4 = (40.0\text{ mL SnCl}_2)\left(\frac{0.250\text{ mol SnCl}_2}{1000\text{ mL SnCl}_2}\right)\left(\frac{1\text{ mol Sn}^{2+}}{1\text{ mol SnCl}_2}\right)\left(\frac{2\text{ mol MnO}_4^-}{5\text{ mol Sn}^{2+}}\right)$$

$$\times\left(\frac{1\text{ mol KMnO}_4}{1\text{ mol MnO}_4^-}\right)\left(\frac{1000\text{ mL KMnO}_4}{0.230\text{ mol KMnO}_4}\right) = 17.4\text{ mL KMnO}_4$$

5.69 (a)

$$\text{mol of I}_3^- = 0.0421\text{ g NaIO}_3\left(\frac{1\text{ mol NaIO}_3}{197.89\text{ g NaIO}_3}\right)\left(\frac{1\text{ mol IO}_3^-}{1\text{ mol NaIO}_3}\right)\left(\frac{3\text{ mol I}_3^-}{1\text{ mol IO}_3^-}\right)$$

$$= 6.38 \times 10^{-4}\text{ mol I}_3^-$$

$$\text{Molarity of I}_3^- = \left(\frac{6.38\times10^{-4}\text{ mol I}_3^-}{100\text{ mL}}\right)\left(\frac{1000\text{ mL}}{1\text{ L}}\right) = 6.38 \times 10^{-3}\ M\text{ I}_3^-$$

(b)

$$\text{g SO}_2 = 2.47\text{ mL I}_3^-\left(\frac{1\text{ L I}_3^-}{1000\text{ mL I}_3^-}\right)\left(\frac{6.38\times10^{-3}\text{ mol I}_3^-}{1\text{ L I}_3^-}\right)$$

$$\left(\frac{1\text{ mol SO}_2}{1\text{ mol I}_3^-}\right)\left(\frac{64.07\text{ g SO}_2}{1\text{ mol SO}_2}\right) = 1.01 \times 10^{-3}\text{ g SO}_2$$

(c) The density of the wine was 0.96 g/mL and the SO_2 concentration was 1.01×10^{-3} g SO_2/mL

$$\text{concentration SO}_2 = \frac{7.58\times10^{-4}\text{ g SO}_2}{50\text{ mL}} = 1.01 \times 10^{-3}\text{ g SO}_2\text{/mL}$$

In 1 mL of solution there are 0.96 g of wine and 1.01×10^{-3} g SO_2
Therefore the percentage of SO_2 in the wine is

$$\frac{1.01\times10^{-3}\text{ g SO}_2}{0.96\text{ g wine}} \times 100\% = 2.10 \times 10^{-3}\,\%$$

(d)

$$\text{ppm SO}_2 = \frac{2.10\times10^{-3}\text{ g SO}_2}{0.96\text{ g wine}} \times 10^6\text{ ppm} = 21\text{ ppm}$$

5.71 (a)

$$\text{mol Cu}^{2+} = (29.96\text{ mL S}_2\text{O}_3^{2-})\left(\frac{0.02100\text{ mol S}_2\text{O}_3^{2-}}{1000\text{ mL S}_2\text{O}_3^{2-}}\right)$$

$$\times\left(\frac{1\text{ mol I}_3^-}{2\text{ mol S}_2\text{O}_3^{2-}}\right)\left(\frac{2\text{ mol Cu}^{2+}}{1\text{ mol I}_3^-}\right) = 6.292 \times 10^{-4}\text{ mol Cu}^{2+} = 6.929 \times 10^{-2}\text{ mol Cu}^{2+}$$

$$\text{g Cu} = (6.292 \times 10^{-4}\text{ mol Cu}) \times (63.546\text{ g Cu/mol Cu})$$
$$= 3.998 \times 10^{-2}\text{ g Cu}$$

$$\%\text{ Cu} = (3.998 \times 10^{-2}\text{ g Cu}/0.4225\text{ g sample}) \times 100 = 9.463\%$$

(b)

$$\text{g CuCO}_3 = (6.292 \times 10^{-4}\text{ mol Cu})\left(\frac{1\text{ mol CuCO}_3}{1\text{ mol Cu}}\right)\left(\frac{123.56\text{ g CuCO}_3}{1\text{ mol CuCO}_3}\right) = 0.07774\text{ g CuCO}_3$$

$$\%\text{ CuCO}_3 = \left(\frac{0.07774\text{ g CuCO}_3}{0.4225\text{ g sample}}\right) \times 100\% = 18.40\%$$

5.73 (a) $$g\ H_2O_2 = (17.60\ mL\ KMnO_4)\left(\frac{0.02000\ mol\ KMnO_4}{1000\ mL\ KMnO_4}\right)\left(\frac{1\ mol\ MnO_4^-}{1\ mol\ KMnO_4}\right)$$

$$\times\left(\frac{5\ mol\ H_2O_2}{2\ mol\ MnO_4^-}\right)\left(\frac{34.02\ g\ H_2O_2}{1\ mol\ H_2O_2}\right) = 0.02994\ g\ H_2O_2$$

(b) $(0.02994\ g/1.000\ g) \times 100 = 2.994\%\ H_2O_2$

5.75 (a) $2CrO_4^{2-} + 3SO_3^{2-} + H_2O \rightarrow 2CrO_2^- + 3SO_4^{2-} + 2OH^-$

(b) $$mol\ CrO_4^{2-} = (3.18\ g\ Na_2SO_3)\left(\frac{1\ mol\ Na_2SO_3}{126.04\ g\ Na_2SO_3}\right)$$

$$\times\left(\frac{1\ mol\ SO_3^{2-}}{1\ mol\ Na_2SO_3}\right)\left(\frac{2\ mol\ CrO_4^{2-}}{3\ mol\ SO_3^{2-}}\right) = 1.68 \times 10^{-2}\ mol\ CrO_4^{2-}$$

Since there is one mole of Cr in each mole of CrO_4^{2-}, then the above number of moles of CrO_4^{2-} is also equal to the number of moles of Cr that were present:

$0.0168\ mol\ Cr \times 52.00\ g/mol = 0.875\ g$ Cr in the original alloy.

(c) $(0.875\ g/3.450\ g) \times 100 = 25.4\%$ Cr

5.77 (a) $$mol\ C_2O_4^{2-} = (21.62\ mL\ KMnO_4)\left(\frac{0.1000\ mol\ KMnO_4}{1000\ mL\ KMnO_4}\right)\left(\frac{5\ mol\ C_2O_4^{2-}}{2\ mol\ KMnO_4}\right)$$

$$= 5.405 \times 10^{-3}\ mol\ C_2O_4^{2-}$$

(b) The stoichiometry for calcium is as follows:

$1\ mol\ C_2O_4^{2-} = 1\ mol\ Ca^{2+} = 1\ mol\ CaCl_2$

Thus the number of grams of $CaCl_2$ is given simply by:

$5.405 \times 10^{-3}\ mol\ CaCl_2 \times 110.98\ g/mol = 0.5999\ g\ CaCl_2$

(c) $(0.5999\ g/2.463\ g) \times 100 = 24.35\%\ CaCl_2$

Practice Exercises

6.1 $\Delta T_{water} = 30.0\ °C - 20.0\ °C = 10.0\ °C$

$$\text{q gained by water} = (10.0\ °C)(250\ g\ H_2O)\left(\frac{4.184\ J}{g\ °C}\right) = 10{,}460\ J$$

q lost by ball bearing = –q gained by water = –10,460 J

$C = q/\Delta T$

$C = -10{,}460\ J/(30.0\ °C - 220\ °C) = 55.1\ J/°C$

6.2 The amount of heat transferred into the water is:

$$J = (250\ g\ H_2O)(4.184\ J\ g^{-1}\ °C^{-1})(30.0\ °C - 25.0\ °C) = 5230\ J$$

$$kJ = (5230\ J)\left(\frac{1\ kJ}{1000\ J}\right) = 5.23\ kJ$$

$$cal = (5230\ J)\left(\frac{1\ cal}{4.184\ J}\right) = 1250\ cal$$

$$kcal = (1250\ cal)\left(\frac{1\ kcal}{1000\ cal}\right) = 1.25\ kcal$$

6.3 $$\text{mol } CH_3OH = (2.85\ g\ CH_3OH)\left(\frac{1\ mol\ CH_3OH}{32.04\ g\ CH_3OH}\right) = 0.0890\ mol\ CH_3OH$$

$$\text{kJ heat released} = (0.0890\ mol\ CH_3OH)\left(\frac{-715\ kJ}{1\ mol\ CH_3OH}\right) = -63.6\ kJ$$

$$\text{Heat capacity} = \left(\frac{63.6\ kJ}{29.19\ °C - 24.05\ °C}\right) = 12.4\ kJ\ °C^{-1}$$

6.4 $$\text{Heat absorbed by calorimeter} = (25.51\ °C - 20.00\ °C)\left(\frac{8.930\ kJ}{1\ °C}\right) = 49.2\ kJ$$

$$\text{mol C} = (1.50\ g\ C)\left(\frac{1\ mol\ C}{12.01\ g\ C}\right) = 0.125\ mol\ C$$

ΔE = energy/mol = 49.2 kJ/0.125 mol C = 394 kJ/mol C

6.5 Since the mole ratio of NaOH to HCl is 1:1 the number of moles of NaOH equals the number of moles of HCl, therefore the amount of heat needed to neutralize HCl equals the amount of heat needed to react NaOH, or –58 kJ mol^{-1} NaOH

6.6 q = specific heat × mass × temperature change
= 4.184 J/g °C × (175 g + 4.90 g) × (14.9 °C – 10.0 °C)
= 3.7×10^3 J = 3.7 kJ of heat released by the process.

This should then be converted to a value representing kJ per mole of reactant, remembering that the sign of ΔH is to be negative, since the process releases heat energy to surroundings. The number of moles of sulfuric acid is:

$$\text{mol } H_2SO_4 = (4.90\ g\ H_2SO_4)\left(\frac{1\ mol\ H_2SO_4}{98.06\ g\ H_2SO_4}\right) = 5.00 \times 10^{-2}\ mol\ H_2SO_4$$

and the enthalpy change in kJ/mole is given by:

3.7 kJ ÷ 0.0500 moles = 74 kJ/mole

6.7 $\frac{1}{4}CH_4(g) + \frac{1}{2}O_2(g) \rightarrow \frac{1}{4}CO_2(g) + \frac{1}{2}H_2O(l)$ $\Delta H = -222.6$ kJ

6.8 We can proceed by multiplying both the equation and the thermochemical value of Example 6.6 by 2.5:

$H_2(g) + \frac{1}{2}O_2(g) \rightarrow H_2O(l)$ $\Delta H = -285.9$ kJ

$2.500\ H_2(g) + \frac{2.500}{2}O_2(g) \rightarrow 2.500\ H_2O(l)$ $\Delta H = -714.75$ kJ

6.9

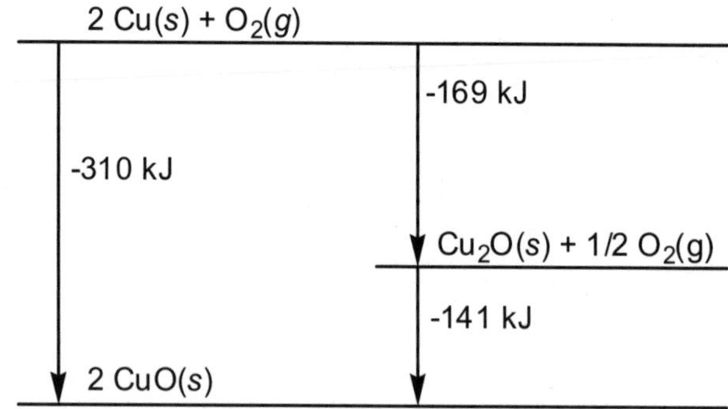

The reaction is exothermic.

6.10

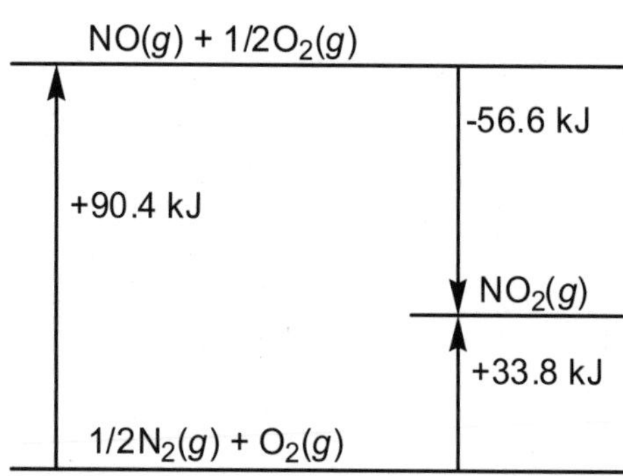

The reaction is endothermic.

6.11 $H_2(g) + \frac{1}{2}O_2(g) \rightarrow H_2O$ $\Delta H = -285.9$ kJ

Reverse the reaction: $H_2O \rightarrow H_2(g) + \frac{1}{2}O_2(g)$ $\Delta H = 285.9$ kJ

Multiply the reaction by 3: $3H_2O \rightarrow 3H_2(g) + \frac{3}{2}O_2(g)$ $\Delta H = +857.7$ kJ

6.12 For this problem: divide the second reaction by two:

$N_2O(g) + \frac{3}{2}O_2(g) \rightarrow 2NO_2(g)$ $\Delta H = -14.0$ kJ

Reverse the first reaction

$2NO_2(g) \rightarrow 2NO(g) + O_2(g)$ $\Delta H = 113.2$ kJ

Add the reactions together:

$$N_2O(g) + \tfrac{3}{2}\,O_2(g) \rightarrow 2NO_2(g) \qquad \Delta H = -14.0 \text{ kJ}$$

$$2NO_2(g) \rightarrow 2NO(g) + O_2(g) \qquad \Delta H = 113.2 \text{ kJ}$$

$$N_2O(g) + \tfrac{1}{2}\,O_2(g) \rightarrow 2NO(g) \qquad \Delta H = +99.2 \text{ kJ}$$

6.13 This problem requires that we add the reverse of the second equation (remembering to change the sign of the associated ΔH value) to the first equation:

$$C_2H_4(g) + 3O_2(g) \rightarrow 2CO_2(g) + 2H_2O(l), \qquad \Delta H^\circ = -1411.1 \text{ kJ}$$

$$2CO_2(g) + 3H_2O(l) \rightarrow C_2H_5OH(l) + 3O_2(g), \qquad \Delta H^\circ = +1367.1 \text{ kJ}$$

which gives the following net equation and value for ΔH°:

$$C_2H_4(g) + H_2O(l) \rightarrow C_2H_5OH(l) \qquad \Delta H^\circ = -44.0 \text{ kJ}$$

6.14 $$\text{kJ} = (12.5 \text{ g } C_3H_6O)\left(\frac{1 \text{ moles } C_3H_6O}{58.077 \text{ g } C_3H_6O}\right)\left(\frac{1790.4 \text{ kJ/mol}}{1 \text{ moles } C_8H_{18}}\right) = 385 \text{ kJ}$$

6.15 $$\text{kJ} = (480 \text{ mol } C_8H_{18})\left(\frac{5450.5 \text{ kJ/mol}}{1 \text{ moles } C_8H_{18}}\right) = 2.62 \times 10^6 \text{ kJ}$$

6.16 $\tfrac{1}{2}\,N_2(g) + 2H_2(g) + \tfrac{1}{2}\,Cl_2(g) \rightarrow NH_4Cl(s) \qquad \Delta H_f^\circ = -315.4 \text{ kJ}$

6.17 $Na(s) + \tfrac{1}{2}\,H_2(g) + C(s) + \tfrac{3}{2}\,O_2(g) \rightarrow NaHCO_3(s), \qquad \Delta H_f^\circ = -947.7 \text{ kJ/mol}$

6.18 $\Delta H^\circ = \text{sum } \Delta H_f^\circ[\text{products}] - \text{sum } \Delta H_f^\circ[\text{reactants}]$

$\Delta H^\circ = \{\Delta H_f^\circ[CaSO_4(s)] + 2\,\Delta H_f^\circ[HCl(g)]\} - \{\Delta H_f^\circ[CaCl_2(s)] + \Delta H_f^\circ\,[H_2SO_4(l)]\}$

$\Delta H^\circ = \{[1 \text{ mol} \times (-1432.7 \text{ kJ/mol})] + [2 \text{ mol} \times (-92.30 \text{ kJ/mol})]\} -$
$\{[1 \text{ mol} \times (-795.0 \text{ kJ/mol})] + [2 \text{ mol} \times (-811.32 \text{ kJ/mol})]\}$

$\Delta H^\circ = +800.34 \text{ kJ}$

6.19 $S(s) + \tfrac{3}{2}\,O_2(g) \rightarrow SO_3(g) \;\; \Delta H_f^\circ = -395.2 \text{ kJ/mol}$

$S(s) + O_2(g) \rightarrow SO_2(g) \;\; \Delta H_f^\circ = -296.9 \text{ kJ/mol}$

Reverse the first reaction and add the two reactions together to get

$SO_3(g) \rightarrow SO_2(g) + \tfrac{1}{2}\,O_2(g) \;\; \Delta H_f^\circ = +98.3 \text{ kJ}$

$\Delta H^\circ = \text{sum } \Delta H_f^\circ[\text{products}] - \text{sum } \Delta H_f^\circ[\text{reactants}]$

$\Delta H^\circ = \{\Delta H_f^\circ[SO_2(g)] + \tfrac{1}{2}\,\Delta H_f^\circ[O_2(g)]\} - \Delta H_f^\circ[SO_3(s)]$

$\Delta H^\circ = \{[1 \text{ mol} \times (-296.9 \text{ kJ/mol})] + [\tfrac{1}{2} \text{ mol} \times 0 \text{ kJ/mol}]\} - [1 \text{ mol} \times (-395.2 \text{ kJ/mol})]$

$\Delta H^\circ = +98.3 \text{ kJ}$

The answers for the enthalpy of reaction are the same using either method.

6.20 a) $\Delta H^\circ = \text{sum } \Delta H_f^\circ[\text{products}] - \text{sum } \Delta H_f^\circ[\text{reactants}]$

$= 2\,\Delta H_f^\circ[NO_2(g)] - \{2\,\Delta H_f^\circ[NO(g)] + \Delta H_f^\circ[O_2(g)]\}$

$= 2 \text{ mol} \times 33.8 \text{ kJ/mol} - [2 \text{ mol} \times 90.37 \text{ kJ/mol} + 1 \text{ mol} \times 0 \text{ kJ/mol}]$

$= -113.1 \text{ kJ}$

b) $\Delta H° = \{ \Delta H_f^\circ[H_2O(l)] + \Delta H_f^\circ[NaCl(s)]\} - \{ \Delta H_f^\circ[NaOH(s)] + \Delta H_f^\circ[HCl(g)]\}$
$= [(-285.9 \text{ kJ/mol}) + (-411.0 \text{ kJ/mol})] - [(-426.8 \text{ kJ/mol}) + (-92.30 \text{ kJ/mol})]$
$= -177.8 \text{ kJ}$

Review Problems

6.41 $\Delta E = q + w = 28 \text{ J} - 45 \text{ J} = -17 \text{ J}$

6.43 Here, ΔE must = 0 in order for there to be no change in energy for the cycle.
$\Delta E = q + w$
$0 = q + (-100 \text{ J})$
$q = +100 \text{ J}$

6.45 $\Delta T = 15.0\ °C - 25.0\ °C = -10.0\ °C$

$$J = (1.75 \text{ mol } H_2O) \times \left(\frac{18.02 \text{ g } H_2O}{1 \text{ mol } H_2O}\right) \times 4.184 \text{ J g}^{-1}\ °C^{-1} \times -10.0\ °C = -1320 \text{ J}$$

$$\text{cal} = (-1320 \text{ J}) \times \left(\frac{1 \text{ cal}}{4.184 \text{ J}}\right) = -315 \text{ cal}$$

6.47 (a) $J = 4.184 \text{ J g}^{-1}\ °C^{-1} \times 100 \text{ g} \times 4.0\ °C = 1.67 \times 10^3 \text{ J}$
(b) $1.67 \times 10^3 \text{ J}$
(c) $1.67 \times 10^3 \text{ J}/(100 - 28.0)°C = 23.2 \text{ J}\ °C^{-1}$
(d) $23.2 \text{ J}\ °C^{-1} \div 5.00 \text{ g} = 4.64 \text{ J g}^{-1}\ °C^{-1}$

6.49 $$\frac{J}{\text{mol } °C} = \left(\frac{0.4498 \text{ J}}{\text{g } °C}\right)\left(\frac{55.847 \text{ g Fe}}{1 \text{ mol Fe}}\right) = 25.12 \ {}^{J}\!/_{\text{mol } °C}$$

6.51 $4.18 \text{ J g}^{-1}\ °C^{-1} \times (4.54 \times 10^3 \text{ g}) \times (58.65 - 60.25)\ °C = -3.04 \times 10^4 \text{ J} = -30.4 \text{ kJ}$

6.53 $HNO_3(aq) + KOH(aq) \rightarrow KNO_3(aq) + H_2O(l)$
Keep in mind that the total mass must be considered in this calculation, and that both liquids, once mixed, undergo the same temperature increase:
heat $= (4.18 \text{ J/g } °C) \times (55.0 \text{ g} + 55.0 \text{ g}) \times (31.8\ °C - 23.5\ °C)$
$= 3.8 \times 10^3$ J of heat energy released

Next determine the number of moles of reactant involved in the reaction:
$0.0550 \text{ L} \times 1.3 \text{ mol/L} = 0.072$ mol of acid and of base.

Thus the enthalpy change is: $$\frac{kJ}{mol} = \frac{(3.8 \times 10^3 \text{ J})\left(\frac{1 \text{ kJ}}{1000 \text{ J}}\right)}{(0.072 \text{ mol})} = -53 \ {}^{kJ}\!/_{mol}$$

6.55 (a) $C_3H_8(g) + 5O_2(g) \rightarrow 3CO_2(g) + 4H_2O(l)$
(b) $J = (97.1 \text{ kJ/}°C)(27.282\ °C - 25.000\ °C) = 222 \text{ kJ} = 2.22 \times 10^5 \text{ J}$
(c) $\Delta H° = -222$ kJ/mol

6.57 (a) Multiply the given equation by the fraction 2/3.
$2CO(g) + O_2(g) \rightarrow 2CO_2(g)$, $\Delta H° = -566$ kJ
(b) To determine ΔH for 1 mol, simply multiply the original ΔH by 1/3; –283 kJ/mol.

6.59 $\text{kJ} = (6.54 \text{ g Mg})\left(\frac{-1203 \text{ kJ}}{2 \text{ mol Mg}}\right)\left(\frac{1 \text{ mol Mg}}{24.305 \text{ g Mg}}\right) = -162 \text{ kJ}$

162 kJ of heat are evolved

6.61

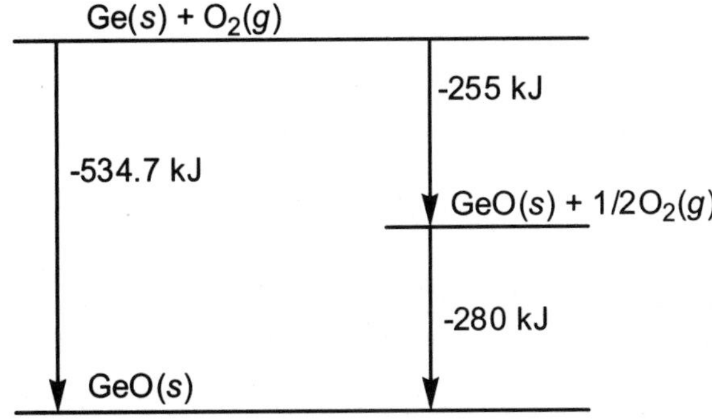

The enthalpy change for the reaction $GeO(s) + \frac{1}{2}O_2(g) \rightarrow GeO_2(s)$ is –280 kJ as seen in the figure above.

6.63 Since NO_2 does not appear in the desired overall reaction, the two steps are to be manipulated in such a manner so as to remove it by cancellation. Add the second equation to the inverse of the first, remembering to change the sign of the first equation, since it is to be reversed:

$2NO_2(g) \rightarrow N_2O_4(g)$, $\Delta H° = -57.93$ kJ
$2NO(g) + O_2(g) \rightarrow 2NO_2(g)$, $\Delta H° = -113.14$ kJ

Adding, we have:

$2NO(g) + O_2(g) \rightarrow N_2O_4(g)$, $\Delta H° = -171.07$ kJ

6.65 If we label the four known thermochemical equations consecutively, 1, 2, 3, and 4, then the sum is made in the following way: Divide equation #3 by two, and reverse all of the other equations (#1, #2, and #4), while also dividing each by two:

$\frac{1}{2}Na_2O(s) + HCl(g) \rightarrow \frac{1}{2}H_2O(l) + NaCl(s)$, $\Delta H° = -253.66$ kJ
$NaNO_2(s) \rightarrow \frac{1}{2}Na_2O(s) + \frac{1}{2}NO_2(g) + \frac{1}{2}NO(g)$, $\Delta H° = +213.57$ kJ
$\frac{1}{2}NO(g) + \frac{1}{2}NO_2(g) \rightarrow \frac{1}{2}N_2O(g) + \frac{1}{2}O_2(g)$, $\Delta H° = -21.34$ kJ
$\frac{1}{2}H_2O(l) + \frac{1}{2}O_2(g) + \frac{1}{2}N_2O(g) \rightarrow HNO_2(l)$, $\Delta H° = -17.18$ kJ

Adding gives:

$HCl(g) + NaNO_2(s) \rightarrow HNO_2(l) + NaCl(s)$, $\Delta H° = -78.61$ kJ

6.67 Multiply all of the equations by $\frac{1}{2}$ and add them together.

$\frac{1}{2}CaO(s) + \frac{1}{2}Cl_2(g) \rightarrow \frac{1}{2}CaOCl_2(s)$ $\Delta H° = \frac{1}{2}(-110.9\text{kJ})$
$\frac{1}{2}H_2O(l) + \frac{1}{2}CaOCl_2(s) + NaBr(s) \rightarrow NaCl(s) + \frac{1}{2}Ca(OH)_2(s) + \frac{1}{2}Br_2(l)$ $\Delta H° = \frac{1}{2}(-60.2 \text{ kJ})$
$\frac{1}{2}Ca(OH)_2(s) \rightarrow \frac{1}{2}CaO(s) + \frac{1}{2}H_2O(l)$ $\Delta H° = \frac{1}{2}(+65.1 \text{ kJ})$

$\frac{1}{2}Cl_2(g) + NaBr(s) \rightarrow NaCl(s) + \frac{1}{2}Br_2(l)$ $\Delta H° = \frac{1}{2}(-106 \text{ kJ}) = -53 \text{ kJ}$

6.69 We need to eliminate the NO_2 from the two equations. To do this, multiply the first reaction by 3 and the second reaction by two and add them together.

$$12NH_3(g) + 21O_2(g) \rightarrow 12NO_2(g) + 18H_2O(g) \qquad \Delta H^\circ = 3(-1132\ \text{kJ})$$
$$12NO_2(g) + 16NH_3(g) \rightarrow 14N_2(g) + 24H_2O(g) \qquad \Delta H^\circ = 2(-2740\ \text{kJ})$$
$$28NH_3(g) + 21O_2(g) \rightarrow 14N_2(g) + 42H_2O(g) \qquad \Delta H^\circ = -8876\ \text{kJ}$$

Now divide this equation by 7 to get

$$4NH_3(g) + 3O_2(g) \rightarrow 2N_2(g) + 6H_2O(g) \qquad \Delta H^\circ = 1/7(-8876\ \text{kJ}) = -1268\ \text{kJ}$$

6.71 (a) $2C(\textit{graphite}) + 2H_2(g) + O_2(g) \rightarrow HC_2H_3O_2(l),\qquad \Delta H_f^\circ = -487.0\ \text{kJ}$

(b) $2C(\textit{graphite}) + \frac{1}{2}O_2(g) + 3H_2(g) \rightarrow C_2H_5OH(l) \qquad \Delta H_f^\circ = -277.63\ \text{kJ}$

(c) $Ca(s) + \frac{1}{8}S_8(s) + 3O_2(g) + 2H_2(g) \rightarrow CaSO_4{\cdot}2H_2O(s) \quad \Delta H_f^\circ = -2021.1\ \text{kJ}$

6.73 (a) $\Delta H_f^\circ = \Delta H_f^\circ[O_2(g)] + 2\,\Delta H_f^\circ[H_2O(l)] - 2\,\Delta H_f^\circ\ \Delta H_f^\circ[H_2O_2(l)]$

$\Delta H_f^\circ = 0\ \text{kJ/mol} + [2\ \text{mol} \times (-285.9\ \text{kJ/mol})] - [2 \times (-187.6\ \text{kJ/mol})]$

$= -196.6\ \text{kJ}$

(b) $\Delta H_f^\circ = \Delta H_f^\circ[H_2O(l)] + \Delta H_f^\circ[NaCl(s)] - \Delta H_f^\circ[HCl(g)] - \Delta H_f^\circ[NaOH(s)]$

$= [1\ \text{mol} \times (-285.9\ \text{kJ/mol})] + [1\ \text{mol} \times (-411.0\ \text{kJ/mol})]$

$- [1\ \text{mol} \times (-92.30\ \text{kJ/mol})] - [1\ \text{mol} \times (-426.8\ \text{kJ/mol})]$

$= -177.8\ \text{kJ}$

6.75 $C_{12}H_{22}O_{11}(s) + 12O_2(g) \rightarrow 12CO_2(g) + 11H_2O(l) \qquad \Delta H^\circ_{\text{combustion}} = -5.65 \times 10^3\ \text{kJ/mol}$

$\Delta H^\circ_{\text{combustion}} = \Sigma\Delta H_f^\circ(\text{products}) - \Sigma\Delta H_f^\circ(\text{reactants})$

$=[12\ \text{mol}\ CO_2 \times \Delta H_f^\circ(CO_2(g)) + 11\ \text{mol}\ H_2O \times \Delta H_f^\circ(H_2O(l))]$

$- [1\ \text{mol}\ C_{12}H_{22}O_{11} \times \Delta H_f^\circ(C_{12}H_{22}O_{11}(s)) + 12\ \text{mol}\ O_2 \times \Delta H_f^\circ(O_2(g))]$

Rearranging and realizing the $\Delta H_f^\circ\ O_2(g) = 0$ we get

$\Delta H_f^\circ(C_{12}H_{22}O_{11}(s)) = 12\Delta H_f^\circ(CO_2(g)) + 11\Delta H_f^\circ(H_2O(l)) - \Delta H^\circ_{\text{combustion}}$

$= 12(-393\text{kJ}) + 11(-285.9\ \text{kJ}) - (-5.65 \times 10^3\ \text{kJ}) = -2.21 \times 10^3\ \text{kJ}$

Practice Exercises

7.1 $\lambda = (588 \text{ nm})\left(\dfrac{1\times10^{-9}\text{m}}{1 \text{ nm}}\right) = 5.88 \times 10^{-7}\text{ m}$

$= \dfrac{c}{\lambda} = \dfrac{3.00\times10^{8}\text{m/s}}{5.88\times10^{-7}\text{m}} = 5.10 \times 10^{14}\text{ s}^{-1} = 5.10 \times 10^{14}\text{ Hz}$

7.2 $\lambda = (10.0\ \mu\text{m})\left(\dfrac{1\times10^{-6}\text{m}}{1\ \mu\text{m}}\right) = 1.00 \times 10^{-5}\text{ m}$

$\nu = \dfrac{c}{\lambda} = \dfrac{3.00\times10^{8}\text{m/s}}{1.00\times10^{-5}\text{m}} = 3.00 \times 10^{13}\text{ s}^{-1} = 3.00 \times 10^{13}\text{ Hz}$

7.3 $\lambda = \dfrac{c}{\nu} = \dfrac{2.9979 \times 10^{8}\text{ m s}^{-1}}{104.3 \times 10^{6}\text{ s}^{-1}} = 2.874\text{ m}$

7.4 $\dfrac{1}{\lambda} = 109{,}678\text{ cm}^{-1} \times \left(\dfrac{1}{4^2} - \dfrac{1}{6^2}\right) = 109{,}678\text{ cm}^{-1} \times (0.0625 - 0.02778)$

$\dfrac{1}{\lambda} = 3.808 \times 10^{3}\text{ cm}^{-1}$

$\lambda = 2.63 \times 10^{-4}\text{ cm} = 2.63\ \mu\text{m}$

7.5 $\dfrac{1}{\lambda} = 109{,}678\text{ cm}^{-1} \times \left(\dfrac{1}{2^2} - \dfrac{1}{3^2}\right) = 109{,}678\text{ cm}^{-1} \times (0.2500 - 0.1111)$

$\dfrac{1}{\lambda} = 1.523 \times 10^{4}\text{ cm}^{-1}$

$\lambda = 6.566 \times 10^{-5}$ cm = 656.6 nm, which is red.

7.6 Period 1: 1 subshell
Period 2: 2 subshells
Period 3: 2 subshells
Period 4: 3 subshells
Period 5: 3 subshells
Period 6: 4 subshells

7.7 When $n = 3$, $\ell = 0, 1, 2$. Thus we have s, p and d subshells.

When $n = 4$, $\ell = 0, 1, 2, 3$. Thus we have s, p, d and f subshells.

7.8 (a) Mg: $1s^22s^22p^63s^2$
(b) Ge: $1s^22s^22p^63s^23p^63d^{10}4s^24p^2$
(c) Cd: $1s^22s^22p^63s^23p^63d^{10}4s^24p^64d^{10}5s^2$
(d) Gd: $1s^22s^22p^63s^23p^63d^{10}4s^24p^64d^{10}4f^75s^25p^65d^16s^2$

7.9 The electron configuration of an element follows the periodic table. The electrons are filled in the order of the periodic table and the energy levels are determined by the row the element is in and the subshell is given by the column, the first two columns are the *s*-block, the last six columns are the *p*-block, the *d*-block has ten columns, and the *f*-block has 14 columns.

7.10 (a) O: $1s^22s^22p^4$
S: $1s^22s^22p^63s^23p^4$
Se: $1s^22s^22p^63s^23p^63d^{10}4s^24p^4$

(b) P: $1s^22s^22p^63s^23p^3$
N: $1s^22s^22p^3$
Sb: $1s^22s^22p^63s^23p^63d^{10}4s^24p^64d^{10}5s^25p^3$
The elements have the same number of electrons in the valence shell, and the only differences between the valence shells are the energy levels.

7.11 (a) Na:
1s ↑↓ 2s ↑↓ 2p ↑↓ ↑↓ ↑↓ 3s ↑ 3p __ __ __ 4s __ 3d __ __ __ __ __

(b) S
1s ↑↓ 2s ↑↓ 2p ↑↓ ↑↓ ↑↓ 3s ↑↓ 3p ↑↓ ↑ ↑ 4s __ 3d __ __ __ __ __

(c) Fe
1s ↑↓ 2s ↑↓ 2p ↑↓ ↑↓ ↑↓ 3s ↑↓ 3p ↑↓ ↑↓ ↑↓ 4s ↑↓ 3d ↑↓ ↑ ↑ ↑ ↑

7.12 (a) Mg
1s ↑↓ 2s ↑↓ 2p ↑↓ ↑↓ ↑↓ 3s ↑↓ 3p __ __ __ 4s __ 3d __ __ __ __ __ 4s __ 4p __ __ __
0 unpaired electrons

(b) Ge
1s ↑↓ 2s ↑↓ 2p ↑↓ ↑↓ ↑↓ 3s ↑↓ 3p ↑↓ ↑↓ ↑↓ 4s ↑↓ 3d ↑↓ ↑↓ ↑↓ ↑↓ ↑↓ 4s ↑↓ 4p ↑ ↑ __
2 unpaired electrons

(c) Cd
1s ↑↓ 2s ↑↓ 2p ↑↓ ↑↓ ↑↓ 3s ↑↓ 3p ↑↓ ↑↓ ↑↓ 4s ↑↓ 3d ↑↓ ↑↓ ↑↓ ↑↓ ↑↓ 4s ↑↓ 4p ↑↓ ↑↓ ↑↓
5s ↑↓ 4d ↑↓ ↑↓ ↑↓ ↑↓ ↑↓
0 unpaired electrons

(d) Gd
1s ↑↓ 2s ↑↓ 2p ↑↓ ↑↓ ↑↓ 3s ↑↓ 3p ↑↓ ↑↓ ↑↓ 4s ↑↓ 3d ↑↓ ↑↓ ↑↓ ↑↓ ↑↓ 4p ↑↓ ↑↓ ↑↓
3s ↑↓ 3p ↑↓ ↑↓ ↑↓ 4s ↑↓ 3d ↑↓ ↑↓ ↑↓ ↑↓ ↑↓ 4p ↑↓ ↑↓ ↑↓ 5s ↑↓ 4d ↑↓ ↑↓ ↑↓ ↑↓ ↑↓ 5p ↑↓ ↑↓ ↑↓
6s ↑↓ 5d ↑ __ __ __ __ 4f ↑ ↑ ↑ ↑ ↑ ↑ ↑
8 unpaired electrons

7.13 Yes, oxygen has eight electrons, but it is paramagnetic since it has two unpaired electrons in the 2*p* orbitals.

7.14 (a) P: [Ne]$3s^23p^3$
[Ne] 3s ↑↓ 3p ↑ ↑ ↑
(3 unpaired electrons)

(b) Sn: $[Kr]4d^{10}5s^25p^2$

[Kr] ⇅ ⇅ ⇅ ⇅ ⇅ ⇅ ↑ ↑ _

4d 5s 5p (2 unpaired electrons)

7.15 Copper, zinc, silver, cadmium, gold, and mercury are examples of elements with valence shells with more than eight electrons.

7.16 (a) Se: $4s^24p^4$ (b) Sn: $5s^25p^2$ (c) I: $5s^25p^5$

7.17 (a) Sn (b) Ga (c) Cr (d) S^{2-}

7.18 (a) P (b) Fe^{3+} (c) Fe (d) Cl^-

7.19 (a) Be (b) C

7.20 (a) C^{2+} (b) Mg^{2+}

Review Problems

The number used for the speed of light, c, depends on the number of significant figures. For one to three significant figures, the value for c is 3.00×10^8 m/s, for four significant figures, the value for c is 2.998×10^8 m/s.

7.73 $$\nu = \frac{c}{\lambda} = \frac{3.00 \times 10^8 \text{ m/s}}{430 \times 10^{-9} \text{ m}} = 6.98 \times 10^{14} \text{ s}^{-1} = 6.98 \times 10^{14} \text{ Hz}$$

7.75 $$\nu = \frac{c}{\lambda} = \frac{3.00 \times 10^8 \text{ m/s}}{6.85 \times 10^{-6} \text{ m}} = 4.38 \times 10^{13} \text{ s}^{-1} = 4.38 \times 10^{13} \text{ Hz}$$

7.77 $295 \text{ nm} = 295 \times 10^{-9}$ m

$$\nu = \frac{c}{\lambda} = \frac{3.00 \times 10^8 \text{ m/s}}{295 \times 10^{-9} \text{ m}} = 1.02 \times 10^{15} \text{ s}^{-1} = 1.02 \times 10^{15} \text{ Hz}$$

7.79 $101.1 \text{ MHz} = 101.1 \times 10^6 \text{ Hz} = 101.1 \times 10^6 \text{ s}^{-1}$

$$\lambda = \frac{c}{\nu} = \frac{3.00 \times 10^8 \text{ m/s}}{101.1 \times 10^6 \text{ s}^{-1}} = 2.97 \text{ m}$$

7.81 $$\lambda = \frac{c}{\nu} = \frac{3.00 \times 10^8 \text{ m/s}}{60 \text{ s}^{-1}} = 5.0 \times 10^6 \text{ m} = 5.0 \times 10^3 \text{ km}$$

7.83 $E = h\nu = 6.63 \times 10^{-34} \text{ J s} \times (4.0 \times 10^{14} \text{ s}^{-1}) = 2.7 \times 10^{-19}$ J

$$\frac{\text{J}}{\text{mol}} = \left(\frac{2.7 \times 10^{-19} \text{ J}}{1 \text{ photon}}\right)\left(\frac{6.022 \times 10^{23} \text{ photons}}{1 \text{ mol}}\right) = 1.6 \times 10^5 \text{ J mol}^{-1}$$

8.85 (a) violet (see Figure 7.7)

(b) $\nu = c/\lambda = 2.998 \times 10^8 \text{ m s}^{-1} / 410.3 \times 10^{-9} \text{ m} = 7.307 \times 10^{14} \text{ s}^{-1}$

(c) $E = h\nu = (6.626 \times 10^{-34} \text{ J s}) \times (7.307 \times 10^{14} \text{ s}^{-1}) = 4.842 \times 10^{-19}$ J

7.87 $\frac{1}{\lambda} = 109{,}678 \text{ cm}^{-1} \times \left(\frac{1}{3^2} - \frac{1}{6^2}\right) = 109{,}678 \text{ cm}^{-1} \times (0.1111 - 0.02778) = 9.140 \times 10^3 \text{ cm}^{-1}$

$\lambda = 1.094 \times 10^{-4}$ cm = 1094 nm

We would not expect to see the light since it is not in the visible region.

7.89 $\frac{1}{\lambda} = 109{,}678 \text{ cm}^{-1} \times \left(\frac{1}{4^2} - \frac{1}{10^2}\right) = 5.758 \times 10^3 \text{ cm}^{-1}$

$\lambda = 1.737 \times 10^{-6}$ m, this is in the infrared region

7.91 (a) p (b) f

7.93 (a) $n = 3$, $\ell = 0$ (b) $n = 5$, $\ell = 2$

7.95 0, 1, 2, 3, 4, 5

7.97 (a) $m_\ell = 1, 0,$ or -1 (b) $m_\ell = 3, 2, 1, 0, -1, -2,$ or -3

7.99 When $m_\ell = -4$ the minimum value of ℓ is 4 and the minimum value of n is 5.

7.101

n	ℓ	m_ℓ	m_s
2	1	−1	+1/2
2	1	−1	−1/2
2	1	0	+1/2
2	1	0	−1/2
2	1	+1	+1/2
2	1	+1	−1/2

7.103 21 electrons have $\ell = 1$, 20 electrons have $\ell = 2$

7.105
(a) S $1s^22s^22p^63s^23p^4$
(b) K $1s^22s^22p^63s^23p^64s^1$
(c) Ti $1s^22s^22p^63s^23p^63d^24s^2$
(d) Sn $1s^22s^22p^63s^23p^64s^23d^{10}4p^64d^{10}5s^25p^2$

7.107
(a) Mn is $[Ar]4s^23d^5$, ∴ five unpaired electrons, paramagnetic
(b) As is $[Ar]\ 3d^{10}4s^24p^3$, ∴ three unpaired electrons, paramagnetic
(c) S is $[Ne]3s^23p^4$, ∴ two unpaired electrons, paramagnetic
(d) Sr is $[Kr]5s^2$, ∴ zero unpaired electrons, not paramagnetic
(b) Ar is $1s^22s^22p^63s^23p^6$, ∴ zero unpaired electrons, not paramagnetic

7.109
(a) Mg is $1s^22s^22p^63s^2$, zero unpaired electrons
(b) P is $1s^22s^22p^63s^23p^3$, three unpaired electrons
(c) V is $1s^22s^22p^63s^23p^63d^34s^2$, three unpaired electrons

7.111
(a) Ni $[Ar]3d^84s^2$
(b) Cs $[Xe]6s^1$
(c) Ge $[Ar]\ 3d^{10}4s^24p^2$
(d) Br $[Ar]\ 3d^{10}4s^24p^5$
(e) Bi $[Xe]\ 4f^{14}5d^{10}6s^26p^3$

7.113 (a) Mg

↑↓ ↑↓ ↑↓ ↑↓ ↑↓ ↑↓ ___ ___ ___ ___ ___ ___ ___ ___ ___

1s 2s 2p 3s 3p 4s 3d

(b) Ti

↑↓ ↑↓ ↑↓ ↑↓ ↑↓ ↑↓ ↑↓ ↑↓ ↑↓ ↑↓ ↑ ↑ ___ ___ ___

1s 2s 2p 3s 3p 4s 3d

7.115 (a) Ni

[Ar] ↑↓ ↑↓ ↑↓ ↑↓ ↑ ↑ ___ ___ ___

4s 3d 4p

(b) Cs

[Xe] ↑ ___ ___ ___ ___ ___ ___ ___ ___ ___ ___ ___ ___ ___ ___ ___

6s 5d 4f 6p

(c) Ge

[Ar] ↑↓ ↑↓ ↑↓ ↑↓ ↑↓ ↑↓ ↑ ↑ ___

4s 3d 4p

(d) Br

[Ar] ↑↓ ↑↓ ↑↓ ↑↓ ↑↓ ↑↓ ↑↓ ↑↓ ↑

4s 3d 4p

7.117 The value corresponds to the row in which the element resides:
(a) 5 (b) 4 (c) 4 (d) 6

7.119 (a) Na $3s^1$ (b) Al $3s^2 3p^1$ (c) Ge $4s^2 4p^2$ (d) P $3s^2 3p^3$

7.121 (a) Na

↑

3s

(b) Al

↑↓ ↑ ___ ___

3s 3p

(c) Ge

↑↓ ↑ ↑ ___

4s 4p

(d) P

↑↓ ↑ ↑ ↑

3s 3p

7.123 (a) 1 (b) 6 (c) 7

7.125 (a) Mg (b) Bi

7.127 Sb is a bit larger than Sn, according to Figure 7.29. (Based upon trends, Sn would be predicted to be larger, but this is one area in which an exception to the trend exists. See Figure 7.29.)

7.129 Cations are generally smaller than the corresponding atom, and anions are generally larger than the corresponding atom:
(a) Na (b) Co^{2+} (c) Cl^-

7.131 (a) N (b) S (c) Cl

7.133 (a) Br (b) As

7.135 Mg

Practice Exercises

8.1 There is one electron missing, and it should go into the 5*s* orbital, and the 5*p* orbital should be empty.
$1s^1 2s^2 2p^6 3s^2 3p^6 3d^{10} 4s^2 4p^6 4d^{10} 5s^2$

8.2 Cr: $[Ar]3d^5 4s^1$

(a) Cr^{2+}: $[Ar]3d^4$ The 4*s* electrons is lost and one 3*d* electrons are lost.
(b) Cr^{3+}: $[Ar]3d^3$ The 4*s* electron and two 3*d* electrons are lost.
(c) Cr^{6+}: [Ar] The 4*s* electron and all of the 3*d* electrons are lost.

8.3 S^{2-}: [Ne] $3s^2 3p^6$
Cl^-: [Ne] $3s^2 3p^6$
The electron configurations are identical.

8.4

$$:\ddot{I}\cdot \quad \cdot Ca \cdot \quad :\ddot{I}\cdot \longrightarrow Ca^{2+} + 2[:\ddot{I}:]^-$$

8.5

$$\cdot\ddot{O}\cdot \quad \cdot Mg \cdot \longrightarrow Mg^{2+} + [:\ddot{O}:]^{2-}$$

8.6 $\mu = q \times r$

$$q = 0.167\ e^- \left(\frac{1.602\times10^{-19}\ C}{1\ e^-}\right) = 2.675 \times 10^{-20}\ C$$

$r = 154.6\ pm = 154.6 \times 10^{-12}\ m$
$q = (2.675 \times 10^{-20}\ C) \times (154.6 \times 10^{-12}\ m) = 4.136 \times 10^{-30}\ C\ m$
in debey units:

$$q = 4.136 \times 10^{-30}\ C\ m\left(\frac{1\ D}{3.34\times10^{-30}\ C\ m}\right) = 1.24\ D$$

8.7 $q = \frac{\mu}{r}$

$$\mu = 9.00\ D\left(\frac{3.34\times10^{-30}\ C\ m}{1\ D}\right) = 3.006 \times 10^{-29}\ C\ m$$

$r = 236\ pm = 236 \times 10^{-12}\ m$

$$q = \left(\frac{3.006\times10^{-29}\ C\ m}{326\times10^{-12}\ m}\right) = 1.27 \times 10^{-19}\ C$$

In electron charges

$$q = 1.27 \times 10^{-19}\ C\left(\frac{1\ e^-}{1.602\times10^{-19}\ C}\right) = 0.795\ e^-$$

On the sodium the charge is $+0.795\ e^-$ and on the chlorine, the charge is $-0.795\ e^-$.
This would be 79.5% positive charge on the Na and 79.5% negative charge on the Cl.

8.8 The bond is polar and the Cl carries the negative charge.

8.9 (a) Br (b) Cl (c) Cl

8.10

O

H O P O H

O

Number of valence electrons:

O 6 each and 4 O atoms total 24 electrons

P 5 electrons

H 1 each and 2 H atoms total 2 electrons

Negative charge 1 electron

Total 32 valence electrons

8.11 SO_2

O S O

NO_3^-

O

O N O

$HBrO_3$

O

O Br O H

H_3AsO_4

H

O

H O As O H

O

8.12 SO_2 has 18 valence electrons

SeO_4^{2-} has 32 valence electrons

NO^+ has 10 valence electrons

8.13

:F—O—F:

[H—N(H)(H)—H]+

O=S—O:

[:O: | O=N—O:]−

8.14 The negative sign should be on the oxygen, so two of the oxygen atoms should have a single bond and three lone pairs and the sulfur should have one double bond, two single bonds, and a lone pair.

8.15 (a) (b)

8.16 In each of these problems, we try to minimize the formal charges in order to determine the preferred Lewis structure. This frequently means violating the octet rule by expanding the octet. Of course, this can only be done for atoms beyond the second period as the atoms in the first and second periods will never expand the octet.

(a) SO_2

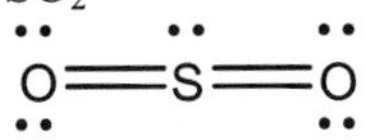

(b) $HClO_3$

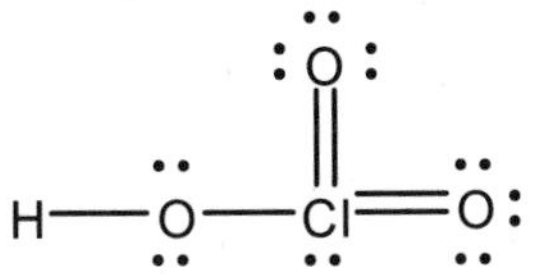

(c) H_3PO_4

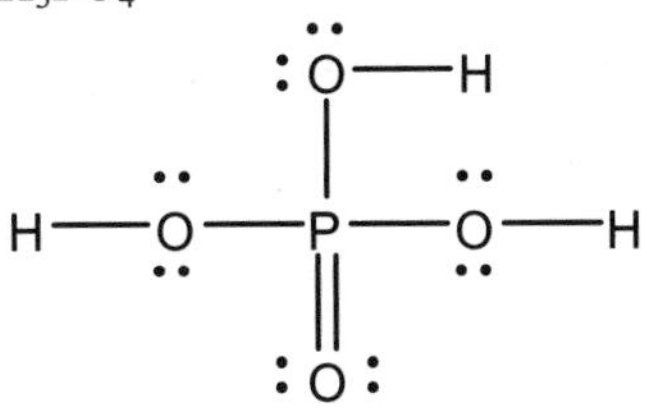

8.17

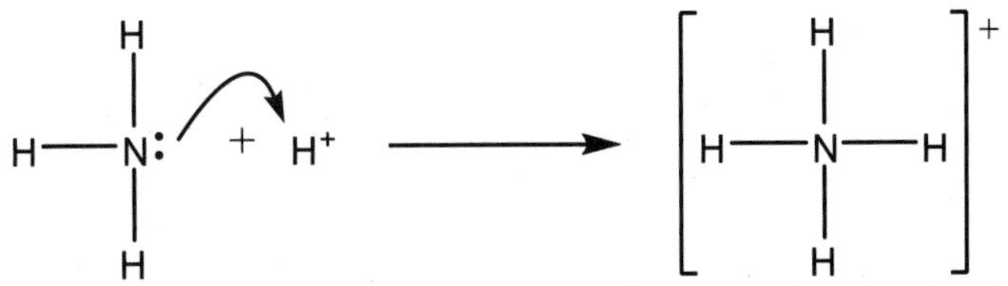

There is no difference between the coordinate covalent bond and the other covalent bonds.

8.18

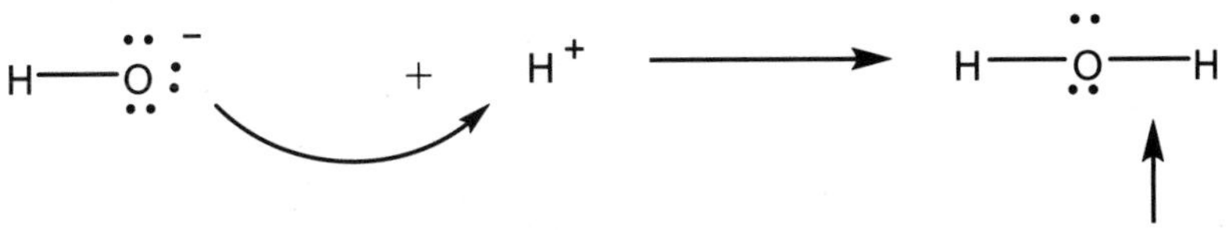

coordinate covalent bond

8.19 There are four resonance structures.

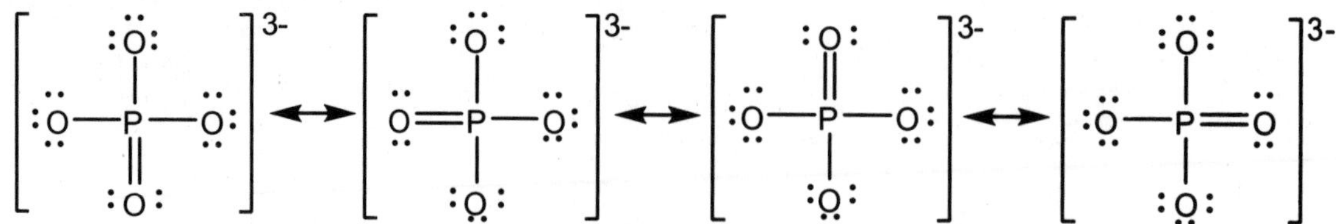

8.20

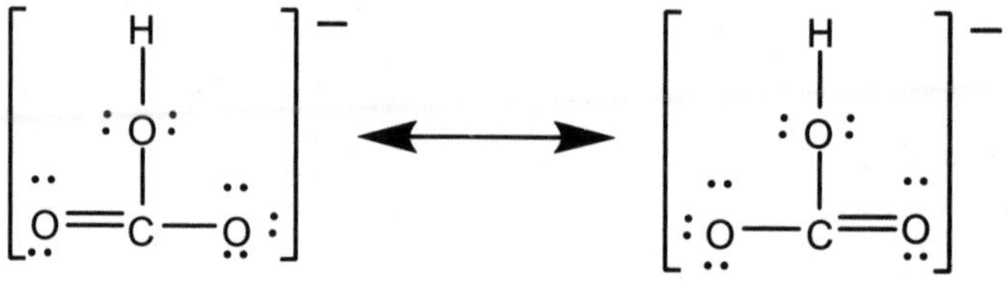

8.21

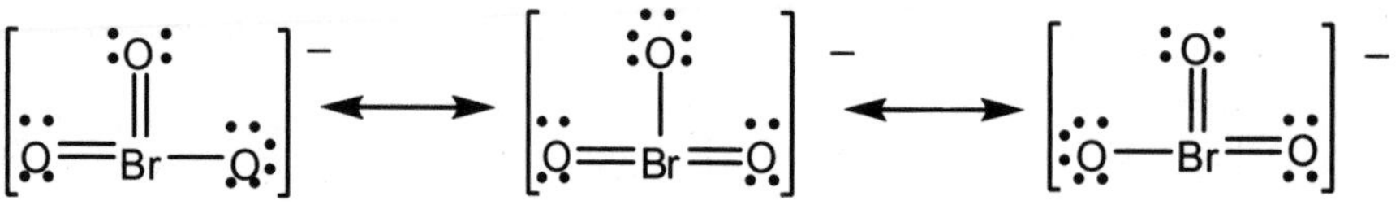

Review Problems

8.55 Magnesium loses two electrons:
$Mg \rightarrow Mg^{2+} + 2e^-$
$[Ne]3s^2 \rightarrow [Ne]$

Bromine gains an electron:
$Br + e^- \rightarrow Br^-$
$[Ar]3d^{10}4s^24p^5 \rightarrow [Kr]$

To keep the overall change of the formula unit neutral, two Br^- ions combine with one Mg^{2+} ion to form $MgBr_2$:
$Mg^{2+} + 2Br^- \rightarrow MgBr_2$

8.57 Pb^{2+}: $[Xe]\,4f^{14}5d^{10}6s^2$
Pb^{4+}: $[Xe]4f^{14}5d^{10}$

8.59 Mn^{3+}: $[Ar]3d^4$ 4 unpaired electrons

8.61 (a) 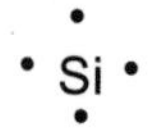 (b) 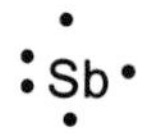 (c)

(d)  (e)

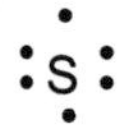

8.63 (a)

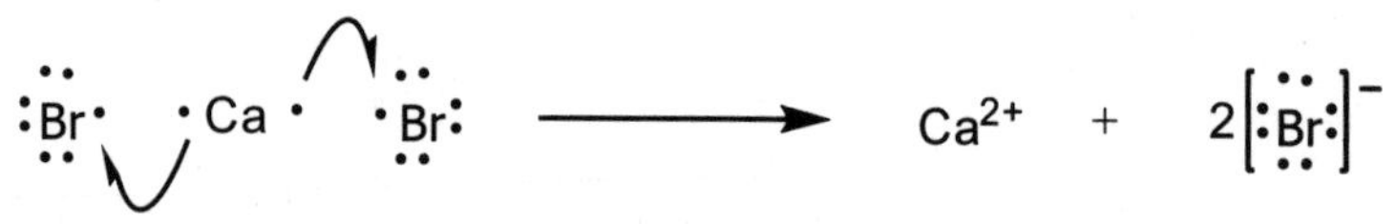

(b)

$$\cdot\ddot{\underset{..}{O}}\cdot \quad \cdot\dot{Al}\cdot \quad \cdot\ddot{\underset{..}{O}}\cdot \quad \cdot\dot{Al}\cdot \quad \cdot\ddot{\underset{..}{O}}\cdot \longrightarrow 2\,Al^{+} + 3[:\ddot{\underset{..}{O}}:]^{2-}$$

(c)

$$K\cdot \quad \cdot\ddot{\underset{..}{S}}\cdot \quad \cdot K \longrightarrow 2\,K^{+} + [:\ddot{\underset{..}{S}}:]^{2-}$$

8.65 $\mu = q \times r$

$$\mu = 0.16\ \text{D}\left(\frac{3.34\times10^{-30}\ \text{C}\cdot\text{m}}{1\ \text{D}}\right) = q \times (115\ \text{pm})\left(\frac{1\ \text{m}}{10^{12}\ \text{pm}}\right)$$

$5.34 \times 10^{-31}\ \text{C·m} = q \times (115 \times 10^{-12}\ \text{m})$

$q = 4.64 \times 10^{-21}\ \text{C}$

$$q = 4.64 \times 10^{-21}\ \text{C}\left(\frac{1\ e^-}{1.60\times10^{-19}\ \text{C}}\right) = 0.029\ e^-$$

The charge on the oxygen is $-0.029\ e^-$ and the charge on the nitrogen is $+0.029\ e^-$. The nitrogen atom is positive.

8.67 $\mu = q \times r$

$$\mu = 1.83\ \text{D}\left(\frac{3.34\times10^{-30}\ \text{C}\cdot\text{m}}{1\ \text{D}}\right) = q \times (91.7\ \text{pm})\left(\frac{1\ \text{m}}{10^{12}\ \text{pm}}\right)$$

$6.11 \times 10^{-30}\ \text{C·m} = q \times (91.7 \times 10^{-12}\ \text{m})$

$q = 6.66 \times 10^{-20}\ \text{C}$

$$q = 6.69 \times 10^{-20}\ \text{C}\left(\frac{1\ e^-}{1.60\times10^{-19}\ \text{C}}\right) = 0.42\ e^-$$

The charge on the fluorine is $-0.42\ e^-$ and the charge on the hydrogen is $+0.42\ e^-$. The fluorine atom is negative.

8.69 Let q be the amount of energy released in the formation of 1 mol of H_2 molecules from H atoms: 435 kJ/mol, the single bond energy for hydrogen.

q = specific heat × mass × ΔT

∴ mass = q ÷ (specific heat × ΔT)

$$\text{g } H_2O = \frac{(435 \times 10^3\ \text{J})}{\left(4.184\ {}^{\text{J}}/_{\text{g}\ {}^\circ\text{C}}\right)\left(100\ {}^\circ\text{C} - 25\ {}^\circ\text{C}\right)} = 1.4 \times 10^3\ \text{g}$$

8.71 $E = h\nu = \frac{hc}{\lambda} \qquad \lambda = \frac{hc}{E}$

$$E = (348 \times 10^3\ \text{J/mol})\left(\frac{1\ \text{mol}}{6.022 \times 10^{23}\ \text{molecules}}\right) = 5.78 \times 10^{-19}\ \text{J/molecule}$$

$$\lambda = \frac{\left(6.626\times10^{-34}\ \text{J sec}\right)\left(3.00\times10^{8}\ \text{m s}^{-1}\right)}{\left(5.78\times10^{-19}\ \text{J}\right)} = 3.44 \times 10^{-7}\ \text{m} = 344\ \text{nm}$$

Ultraviolet region

8.73 (a)

$$:\ddot{Br}\cdot \quad + \quad \cdot\ddot{Br}: \longrightarrow :\ddot{Br}—\ddot{Br}:$$

(b)

$$2\,H\cdot \quad + \quad \cdot\ddot{O}\cdot \longrightarrow H—\ddot{O}—H$$

(c)

$$3\,H\cdot \quad + \quad \cdot\ddot{N}\cdot \longrightarrow H—\ddot{N}—H \text{ (with a third H bonded to N)}$$

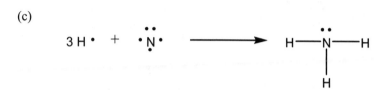

8.75 (a) We predict the formula H_2Se because selenium, being in Group VIA, needs only two additional electrons (one each from two hydrogen atoms) in order to complete its octet.
(b) Arsenic, being in Group VA, needs three electrons from hydrogen atoms in order to complete its octet, and we predict the formula H_3As.
(c) Silicon is in Group IVB, and it needs four electrons (and hence four hydrogen atoms) to complete its octet: SiH_4.

8.77 Here we choose the atom with the smaller electronegativity:
(a) S (b) Si (c) Br (d) C

8.79 Here we choose the linkage that has the greatest difference in electronegativities between the atoms of the bond: N—S.

8.81 (a)

```
     Cl

Cl   Si   Cl

     Cl
```

(b)

```
     F

F    P    F
```

(c)

```
     H

H    P    H
```

(d)

```
Cl   S   Cl
```

8.83 (a) 32 (b) 26 (c) 8 (d) 20

8.85 (a)

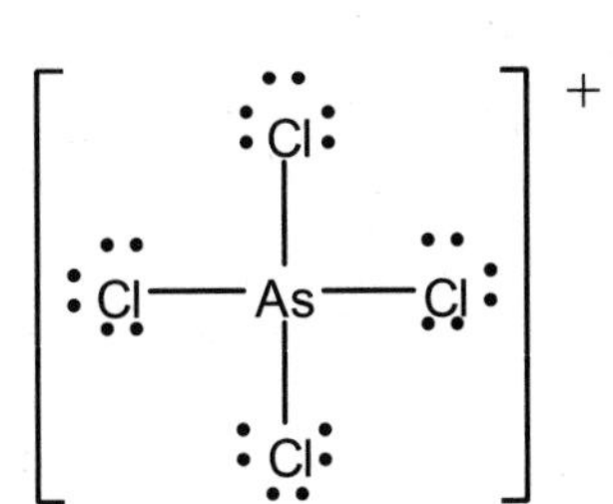

(b)

[O—Cl—O]$^-$

(c)

H—O—N=O

(d)

F—Xe—F

8.87 (a)

Cl—Si(—Cl)(—Cl)—Cl

(b)

F—P(—F)—F

(c)

H—P(—H)—H

(d)

Cl—S—Cl

8.89 (a)

S=C=S

(b)

[C≡N]$^-$

8.91 (a)

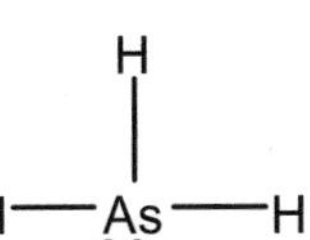

(b)

H—O—Cl—O

(c)

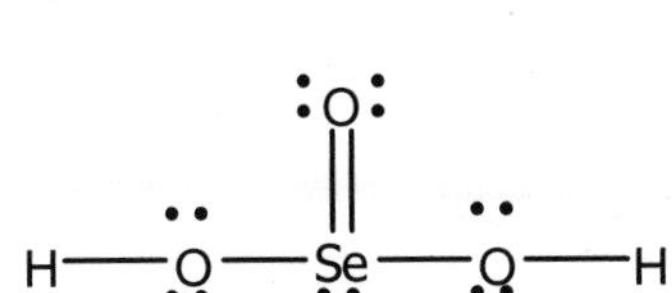

(d)

H—O—As(—O—H)(—O)—O—H

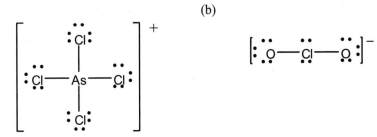

8.93 (a) (b)

8.95 (a) (b)

(c)

8.97

8.99 The formal charges on all of the atoms of the left structure are zero, therefore, the potential energy of this molecule is lower and it is more stable.

8.101 The average bond order is 4/3

8.103 The Lewis structure for NO_3^- is given in the answer to practice exercise 13, and that for NO_2^- is given below.

Resonance causes the average number of bonds in each N—O linkage of NO_3^- to be 1.33. Resonance causes the average number of electron pair bonds in each linkage of NO_2^- to be 1.5. We conclude that the N—O bond in NO_2^- should be shorter than that in NO_3^-.

8.105

These are not preferred structures, because in each Lewis diagram, one oxygen bears a formal charge of +1 whereas the other bears a formal charge of –1. The structure with the formal charges of zero has a lower potential energy and is more stable.

8.107

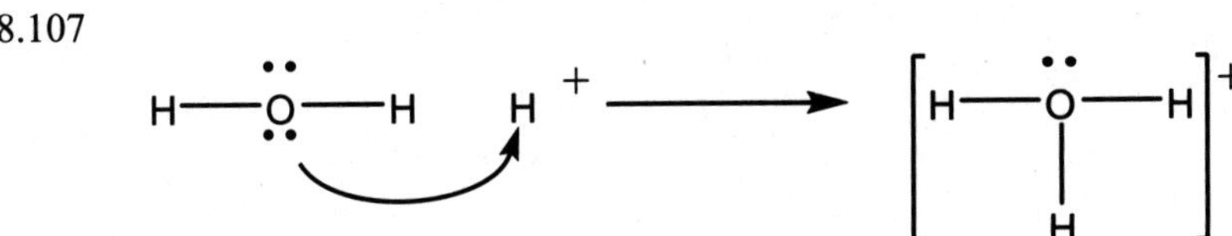

Practice Exercises

9.1 SeF_6 should have an octahedral shape (Figure 9.4) because is has six electron pairs around the central atom.

9.2 $SbCl_5$ should have a trigonal bipyramidal shape (Figure 9.4) because, like PCl_5, it has five electron pairs around the central atom.

9.3 HArF should have a linear shape (Figure 9.7) because although it has five electron pairs around the central Ar atom, only two are being used for bonding.

9.4 I_3^- should have a linear shape (Figure 9.7) because although it has five electron pairs around the central I atom, only two are being used for bonding.

9.5 XeF_4 should have a square planar shape (Figure 9.10) because although it has six electron pairs around the central Xe atom, only four are being used for bonding.

9.6 In SO_3^{2-}, there are three bond pairs and one lone pair of electrons at the sulfur atom, and as shown in Figure 9.5, this ion has a trigonal pyramidal shape.

In CO_3^{2-}, there are three bonding domains (2 single bonds and 1 double bond) at the carbon atom, and as shown in Figure 9.4, this ion has a planar triangular shape.

In XeO_4, there are four bond pairs of electrons around the Xe atom, and as shown in Figure 9.5, this molecule is tetrahedral.

In OF_2, there are two bond pairs and two lone pairs of electrons around the oxygen atom, and as shown in Figure 9.5, this molecule is bent.

9.7 SF_4 is distorted tetrahedral and has one lone pair of electrons on the sulfur, therefore it is polar.

9.8 a) SF_6 is octahedral, and it is not polar.
b) SO_2 is bent, and it is polar.
c) BrCl is polar because there is a difference in electronegativity between Br and Cl.
d) AsH_3, like NH_3, is pyramidal, and it is polar.
e) CF_2Cl_2 is polar, because there is a difference in electronegativity between F and Cl.

9.9 The H–Cl bond is formed by the overlap of the half–filled 1*s* atomic orbital of a H atom with the half–filled 3*p* valence orbital of a Cl atom:

Cl atom in HCl (x = H electron):

$\underline{\uparrow\downarrow}$ $\quad$ $\underline{\uparrow\downarrow}$ $\underline{\uparrow\downarrow}$ $\underline{\uparrow x}$

3*s* $\quad$ 3*p*

The overlap that gives rise to the H–Cl bond is that of a 1*s* orbital of H with a 3*p* orbital of Cl:

9.10 The half–filled 1*s* atomic orbital of each H atom overlaps with a half–filled 3*p* atomic orbital of the P atom, to give three P–H bonds. This should give a bond angle of 90°.

P atom in PH_3 (x = H electron):

$\underline{\uparrow\downarrow}$ $\quad$ $\underline{\uparrow x}$ $\underline{\uparrow x}$ $\underline{\uparrow x}$

3*s* $\quad$ 3*p*

The orbital overlap that forms the P–H bond combines a 1*s* orbital of hydrogen with a 3*p* orbital of phosphorus (note: only half of each p orbital is shown):

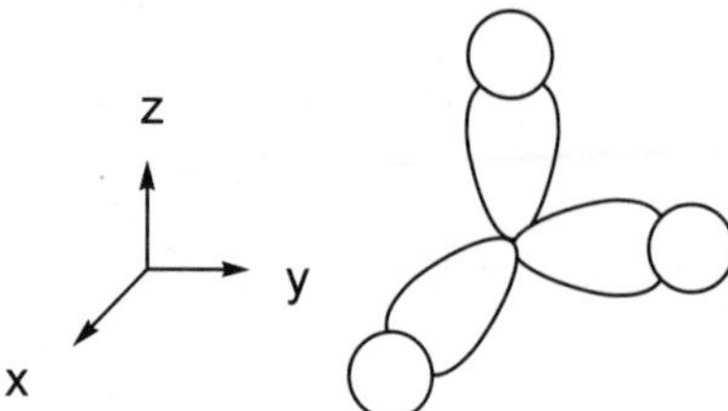

9.11 BCl_3 uses sp^2 hybridized orbitals since it has only three bonding electron pairs and no lone pairs of electrons.

The sp^2 hybrid orbitals on the B, x = Cl electron

↿x ↿x ↿x __

sp^2 2*p*

9.12 BeF_2 uses *sp* hybridized orbitals since it has only two bonding electron pairs and no lone pairs of electrons.

The *sp* hybrid orbitals on the Be; x = F electron

↿x ↿x __ __

sp 2*p*

9.13 Since there are five bonding pairs of electrons on the central phosphorous atom, we choose sp^3d hybridization for the P atom. Each of phosphorous's five sp^3d hybrid orbitals overlaps with a 3*p* atomic orbital of a chlorine atom to form a total of five P–Cl single bonds. Four of the 3*d* atomic orbitals of P remain unhybridized.

9.14 sp^3

↿x ↿x ↿x ↿x

sp^3

9.15 VSEPR theory predicts that $AlCl_5$ will be trigonal bipyramidal. Since there are five bonding pairs of electrons on the central arsenic atom, we choose sp^3d hybridization for the As atom as a trigonal bipyramid. Each of arsenic's five sp^3d hybrid orbitals overlaps with a 3*p* atomic orbital of a chlorine atom to form a total of five As–Cl single bonds. Four of the 4*d* atomic orbitals of As remain unhybridized.

9.16 sp^3d^2

9.17 (a) sp^3 (b) sp^3d

9.18 NH_3 is sp^3 hybridized. Three of the electron pairs are use for bonding with the three hydrogens. The fourth pair of electrons is a lone pair of electrons. This pair of electrons is used for the formation of the bond between the nitrogen of NH_3 and the hydrogen ion, H^+.

9.19 Since there are six bonding pairs of electrons on the central phosphorous atom, we choose sp^3d^2 hybridization for the P atom. Each of phosphorous's six sp^3d^2 hybrid orbitals overlaps with a 3*p* atomic orbital of a chlorine atom to form a total of six P–Cl single bonds. Three of the 3*d* atomic orbitals of P remain unhybridized.

P atom in PCl_6^- (x = Cl electron):

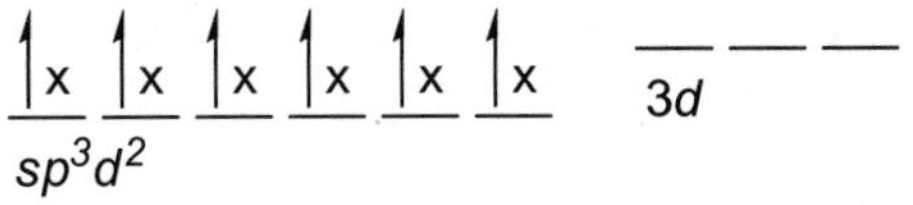

↿x ↿x ↿x ↿x ↿x ↿x   — — —
sp^3d^2 3d

The ion is octahedral because six atoms and no lone pairs surround the central atom.

9.20 atom 1: sp^2 atom 2: sp^3 atom 3: sp^2
There are 10 σ bonds and 2 π bonds in the molecule.

9.21 atom 1: sp atom 2: sp^2 atom 3: sp^3
There are 9 σ bonds and 3 π bonds in the molecule.

9.22 CN^- has 10 valence electons and the MO diagram is similar to that of C and N. The bond order of the ion is 3 and this does agree with the Lewis structure.

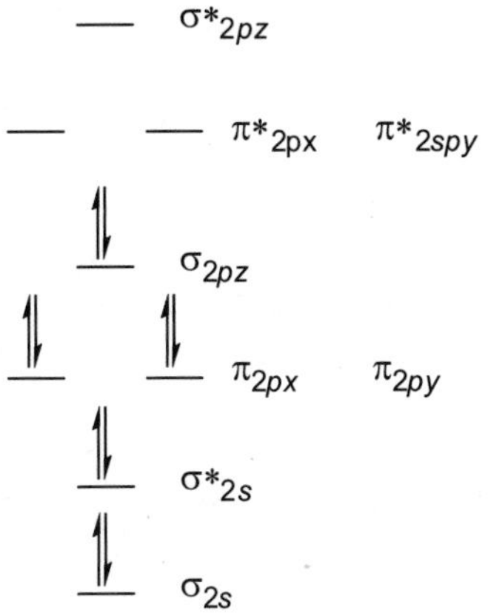

— σ^*_{2pz}

— — π^*_{2px} π^*_{2spy}

⇅ σ_{2pz}

⇅ ⇅ π_{2px} π_{2py}

⇅ σ^*_{2s}

⇅ σ_{2s}

9.23 NO has 11 valence electrons, and the MO diagram is similar to that shown in Table 9.1 for O_2, except that one fewer electron is employed at the highest energy level

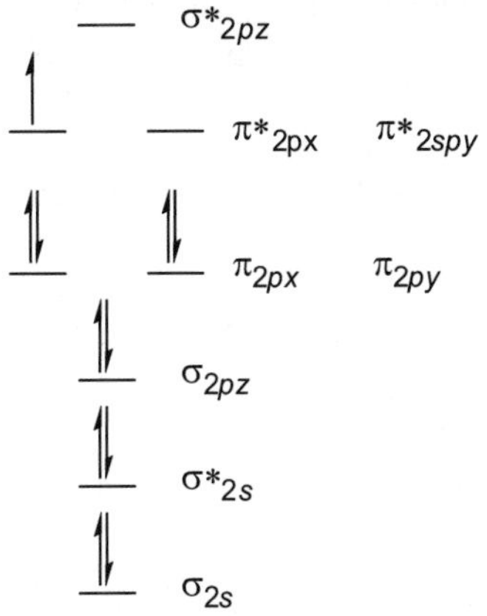

— σ^*_{2pz}

↿ — π^*_{2px} π^*_{2spy}

⇅ ⇅ π_{2px} π_{2py}

⇅ σ_{2pz}

⇅ σ^*_{2s}

⇅ σ_{2s}

$$\text{Bond Order} = \frac{\left(8 \text{ bonding e}^-\right) - \left(3 \text{ antibonding e}^-\right)}{2} = \frac{5}{2}$$

The bond order is calcula ted to be 5/2:

Review Problems

9.50 (a) bent (central N atom has two single bonds and two lone pairs)
(b) planar triangular (central C atom has three bonding domains—a double bond and a single bond)
(c) T-shaped (central I atom has three single bonds and two lone pairs)
(d) Linear (central Br atom has two single bonds and three lone pairs)
(e) planar triangular (central Ga atom has three single bonds and no lone pairs)

9.52 (a) nonlinear (b) trigonal bipyramidal
(c) trigonal pyramidal (d) trigonal pyramidal
(e) nonlinear

9.54 (a) tetrahedral (b) square planar
(c) octahedral (d) tetrahedral
(e) linear

9.56 BrF_4^+

9.58 180°

9.60 (a) 109.5° (b) 109.5°
(c) 120° (d) 180°
(e) 109.5°

9.62 The ones that are polar are (a), (b), and (c). The last two have symmetrical structures, and although individual bonds in these substances are polar bonds, the geometry of the bonds serves to cause the individual dipole moments of the various bonds to cancel one another.

9.64 All are polar. (a), (b), (c) and (e) have asymmetrical structures, and (d) only has one bond, which is polar.

9.66 In SF_6, although the individual bonds in this substance are polar bonds, the geometry of the bonds is symmetrical which serves to cause the individual dipole moments of the various bonds to cancel one another. In SF_5Br, one of the six bonds has a different polarity so the individual dipole moments of the various bonds do not cancel one another.

9.68 The 1*s* atomic orbitals of the hydrogen atoms overlap with the mutually perpendicular *p* atomic orbitals of the selenium atom.

Se atom in H_2Se (x = H electrons):

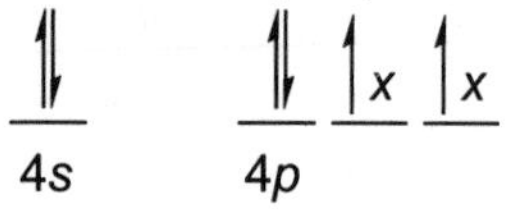

⇅ ⇅ ↑x ↑x

4s 4p

9.70 Atomic Be:

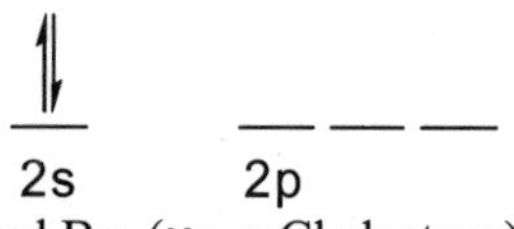

⇅ __ __ __

2s 2p

Hybridized Be: (x = a Cl electron)

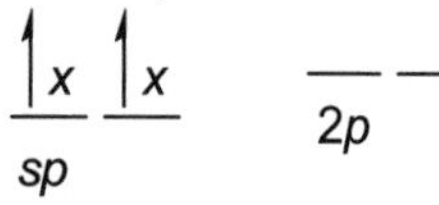

↑x ↑x __ __

sp 2p

9.72 (a) There are three bonds to the central Cl atom, plus one lone pair of electrons. The geometry of the electron pairs is tetrahedral so the Cl atom is to be sp^3 hybridized:

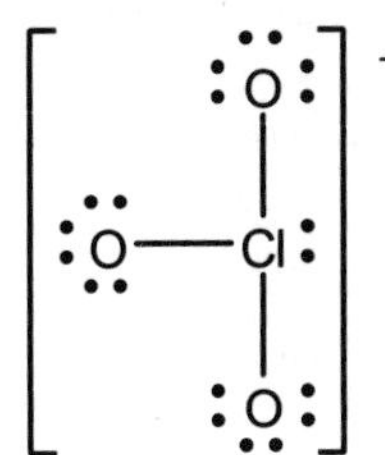

(b) There are three atoms bonded to the central sulfur atom, and no lone pairs on the central sulfur. The geometry of the electron pairs is that of a planar triangle, and the hybridization of the S atom is sp^2:

Two other resonance structures should also be drawn for SO_3.

(c) There are two bonds to the central O atom, as well as two lone pairs. The O atom is to be sp^3 hybridized, and the geometry of the electron pairs is tetrahedral.

9.74 (a) There are three bonds to As and one lone pair at As, requiring As to be sp^3 hybridized.

The Lewis diagram

The hybrid orbital diagram for As: (x = a Cl electron)

(b) There are three atoms bonded to the central Cl atom, and it also has two lone pairs of electrons. The hybridization of Cl is thus sp^3d.

The Lewis diagram

:F̤—C̤l—F̤:
|
:F̤:

The hybrid orbital diagram for Cl: (x = a F electron)

⇅ ⇅ ↑x ↑x — — — — —
sp^3d $3d$

9.76 We can consider that this ion is formed by reaction of SbF_5 with F^-. The antimony atom accepts a pair of electrons from fluoride:

Sb in SbF_6^-: (xx = an electron pair from the donor F^-)

⇅ ⇅ ⇅ ⇅ ⇅ xx
sp^3d^2

9.78 (a) N in the C=N system:

⇅ ↑ ↑ ↑
sp^2 $2p$

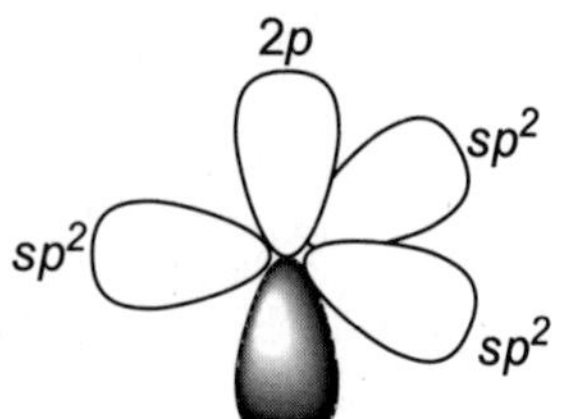

(b) sigma bond

C N

pi bond

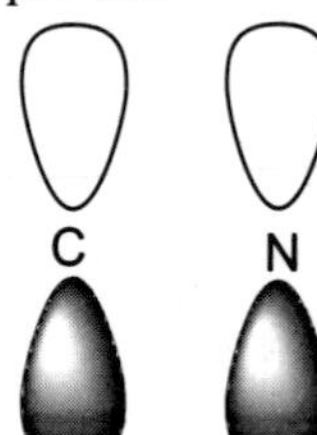

(c)

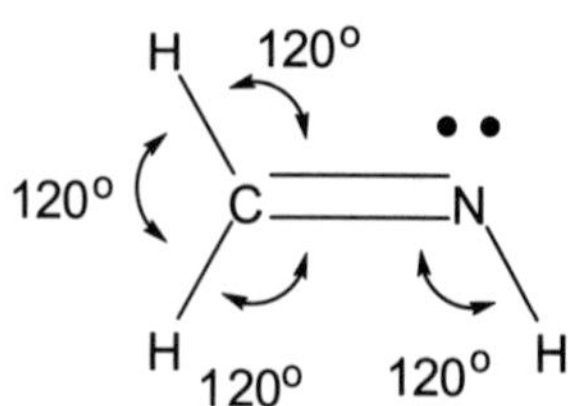

9.80 Each carbon atom is sp^2 hybridized, and each C–Cl bond is formed by the overlap of an sp^2 hybrid of carbon with a p atomic orbital of a chlorine atom. The C=C double bond consists first of a C–C σ bond formed by "head on" overlap of sp^2 hybrids from each C atom. Secondly, the C=C double bond consists of a side–to–side overlap of unhybridized p orbitals of each C atom, to give one π bond. The molecule is planar, and the expected bond angles are all 120°.

9.82 1. sp^3 2. sp 3. sp^2 4. sp^2

9.84 Here we pick the one with the higher bond order.

(a) O_2^+ (b) O_2 (c) N_2

9.86 (a) N_2^+ (b) NO (c) O_2^-

9.88

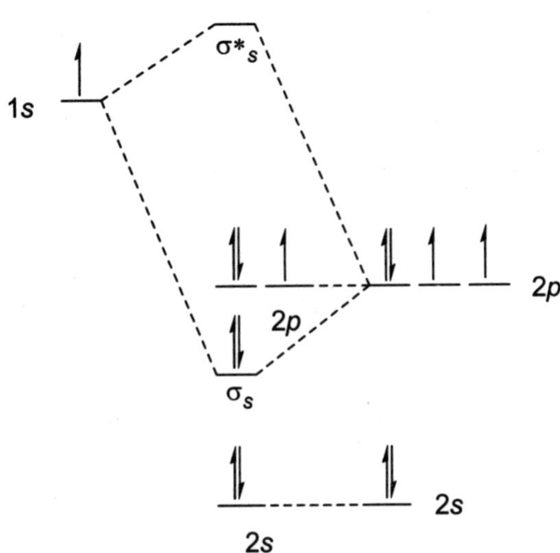

There are two electrons in bonding MOs and three electrons in nonbonding MOs.
The net bond order is 1.

Practice Exercises

10.1 $$\text{psi} = 730 \text{ mm Hg}\left(\frac{14.7 \text{ psi}}{760 \text{ mm Hg}}\right) = 14.1 \text{ psi}$$

$$\text{in. Hg} = 730 \text{ mm Hg}\left(\frac{29.921 \text{ in. Hg}}{760 \text{ mm Hg}}\right) = 28.7 \text{ in. Hg}$$

10.2 $$\text{Pascals} = 888 \text{ mbar}\left(\frac{1 \text{ bar}}{1000 \text{ mbar}}\right)\left(\frac{1 \text{ atm}}{1.013 \text{ bar}}\right)\left(\frac{101{,}325 \text{ Pascal}}{1 \text{ atm}}\right) = 88{,}800 \text{ Pascal}$$

$$\text{torr} = 888 \text{ mbar}\left(\frac{1 \text{ bar}}{1000 \text{ mbar}}\right)\left(\frac{1 \text{ atm}}{1.013 \text{ bar}}\right)\left(\frac{760 \text{ torr}}{1 \text{ atm}}\right) = 666 \text{ torr}$$

10.3 $$\text{mm Hg} = 25 \text{ cm Hg}\left(\frac{10 \text{ mm Hg}}{1 \text{ cm Hg}}\right) = 250 \text{ mm Hg}$$

$$\text{mm Hg} = 770 \text{ torr}\left(\frac{760 \text{ mm Hg}}{760 \text{ torr}}\right) = 770 \text{ mm Hg}$$

The maximum pressure = 770 mm Hg + [(250 mm Hg) × 2] = 1270 mm Hg
The minimum pressure = 770 mm Hg – [(250 mm Hg) × 2] = 270 mmHg

10.4 The pressure of the gas in the manometer is the pressure of the atmosphere less the pressure of the mercury, 10.7 cm Hg.
Using the pressure in the atmosphere from the previous example:

$$\text{mm Hg} = 10.7 \text{ cm Hg}\left(\frac{10 \text{ mm Hg}}{1 \text{ cm Hg}}\right) = 107 \text{ mm Hg}$$

770 mm Hg – 107 mm Hg = 663 mm Hg

10.5 $$\frac{P_1V_1}{T_1} = \frac{P_2V_2}{T_2}$$

$V_2 = 3V_1$ and $T_2 = 2T_1$

$$\frac{P_1V_1}{T_1} = \frac{P_2 3V_1}{2T_1}$$

$P_2 = 2/3\ P_1$
The pressure must change by 2/3.

10.6 Since volume is to decrease, pressure must increase, and we multiply the starting pressure by a volume ratio that is larger than one. Also, since $P_1V_1 = P_2V_2$, we can solve for P_2:

$$P_2 = \frac{P_1V_1}{V_2} = \frac{(740 \text{ torr})(880 \text{ mL})}{(870 \text{ mL})} = 750 \text{ torr}$$

10.7 In general the combined gas law equation is: $\frac{P_1V_1}{T_1} = \frac{P_2V_2}{T_2}$, and in particular, for this problem, we have:

$$P_2 = \frac{P_1V_1T_2}{T_1V_2} = \frac{(745 \text{ torr})(950 \text{ m}^3)(333.2 \text{ K})}{(1150 \text{ m}^3)(298.2 \text{ K})} = 688 \text{ torr}$$

10.8 When gases are held at the same temperature and pressure, and dispensed in this fashion during chemical reactions, then they react in a ratio of volumes that is equal to the ratio of the coefficients (moles) in the balanced chemical equation for the given reaction. We can, therefore, directly use the stoichiometry of the balanced chemical equation to determine the combining ratio of the gas volumes:

$$\text{L } O_2 = (4.50 \text{ L } CH_4)\left(\frac{2 \text{ volume } O_2}{1 \text{ volume } CH_4}\right) = 9.00 \text{ L } O_2$$

10.9 $$\text{L } O_2 = (6.75 \text{ L } CH_4)\left(\frac{2 \text{ volume } O_2}{1 \text{ volume } CH_4}\right) = 13.50 \text{ L } O_2$$

$$21\%\ O_2 = \frac{13.50 \text{ L } O_2}{\text{x L air}} \times 100\%$$

$$\text{L air} = (13.50 \text{ L } O_2)\left(\frac{100\%}{21\%}\right) = 64.3 \text{ L air}$$

10.10 Starting with the combined gas law:

$$\frac{P_1V_1}{T_1} = \frac{P_2V_2}{T_2}$$

Rearrange the equation to find a new volume (V_2)

$$V_2 = \frac{P_1V_1T_2}{T_1P_2}$$

Calculate the new volume for the butane (C_4H_{10}) using the temperature and pressure of the oxygen (O_2)

$$V_2 = \frac{(760 \text{ torr})(75.0 \text{ mL})(308 \text{ K})}{(318 \text{ K})(725 \text{ torr})} = 76.1 \text{ mL } C_4H_{10}$$

Now calculate the number of mL of O_2 require to react with 76.1 mL C_4H_{10}

$$\text{mL } O_2 = 76.1 \text{ mL } C_4H_{10}\left(\frac{13 \text{ volume } O_2}{2 \text{ volume } C_4H_{10}}\right) = 495 \text{ mL } O_2$$

10.11 First determine the number of moles of CO_2 in the tank:

$$n = \frac{PV}{RT}$$

$$P = 2000 \text{ psig}\left(\frac{1 \text{ atm}}{14.696 \text{ psig}}\right) = 140 \text{ atm}$$

$$V = 6.0 \text{ ft}^3\left(\frac{(30.48 \text{ cm})^3}{(1 \text{ ft})^3}\right)\left(\frac{1 \text{ mL}}{1 \text{ cm}^3}\right)\left(\frac{1 \text{ L}}{1000 \text{ mL}}\right) = 170 \text{ L}$$

$$R = 0.0821\ \frac{\text{L}\cdot\text{atm}}{\text{mol}\cdot\text{K}}$$

$$T = 22\ °\text{C} + 273 = 295 \text{ K}$$

$$\text{mol } CO_2 \text{ in the tank} = \frac{(140 \text{ atm})(170 \text{ L})}{\left(0.0821\ \frac{\text{L}\cdot\text{atm}}{\text{mol}\cdot\text{K}}\right)(295 \text{ K})} = 980 \text{ mol } CO_2$$

Then find the total number of grams of CO_2 in the tank, MW CO_2 = 44.01 g/mol

$$\text{g } CO_2 \text{ in the tank} = 980 \text{ mol } CO_2\left(\frac{44.01 \text{ g } CO_2}{1 \text{ mol } CO_2}\right) = 43{,}000 \text{ g } CO_2$$

Amount of solid CO_2 = 43,000 g CO_2 × 0.35 = 15,000 g solid CO_2

10.12 $$n = \frac{PV}{RT} = \frac{(57.8\ \text{atm})(12.0\ \text{L})}{\left(0.0821\ \frac{\text{L atm}}{\text{mol K}}\right)(298\ \text{K})} = 28.3\ \text{moles gas}$$

28.3 mol Ar (39.95 g Ar/mol) = 1,130 g Ar

10.13 Find the number of moles of air

$$n = \frac{PV}{RT} = \frac{(1.0000\ \text{atm})(0.54423\ \text{L})}{\left(0.082057\ \frac{\text{L atm}}{\text{mol K}}\right)(273.15\ \text{K})} = 0.024281\ \text{mol air}$$

$$\text{Mass of the air} = (0.024281\ \text{mol air})\left(\frac{28.8\ \text{g air}}{1\ \text{mol air}}\right) = 0.699\ \text{g air}$$

The mass of the flask = 735.6898 g – 0.699 g air = 734.991 g
Mass of the organic compound = 735.6220 g – 734.991 g = 0.631 g
The number of moles of the organic compound equals the number of moles of air:

$$MW = \frac{0.631\ \text{g organic compound}}{0.024281\ \text{mole organic compound}} = 26.0\ \text{g/mol}$$

10.14 Since PV = nRT, then n = PV/RT

$$n = \frac{PV}{RT} = \frac{(685\ \text{torr})\left(\frac{1\ \text{atm}}{760\ \text{torr}}\right)(0.300\ \text{L})}{\left(0.0821\ \frac{\text{L atm}}{\text{mol K}}\right)(300.2\ \text{K})} = 0.0110\ \text{moles gas}$$

$$\text{molar mass} = \frac{1.45\ \text{g}}{0.0110\ \text{mol}} = 132\ \text{g mol}^{-1}$$

The gas must be xenon.

10.15 $$\text{Density of air} = 1\ \text{mol air}\left(\frac{28.8\ \text{g air}}{1\ \text{mol air}}\right)\left(\frac{1\ \text{mol air}}{22.4\ \text{L}}\right) = 1.29\ \text{g/L}$$

$$\text{Density of radon at STP} = 1\ \text{mol Rn}\left(\frac{222.0\ \text{g Rn}}{1\ \text{mol Rn}}\right)\left(\frac{1\ \text{mol Rn}}{22.4\ \text{L}}\right) = 9.91\ \text{g/L}$$

Since radon is almost eight times denser than air, the sensor should be in the lowest point in the house: the basement.

10.16 d = m/V

Taking 1.00 mol SO_2:

m = 64.1 g

$$V = \frac{nRT}{P} = \frac{(1.00\ \text{mol})\left(0.0821\ \frac{\text{L atm}}{\text{mol K}}\right)(253.15\ \text{K})}{\left(96.5\ \text{kPa}\frac{1\ \text{atm}}{101.325\ \text{kPa}}\right)} = 21.8\ \text{L} = 21{,}800\ \text{mL}$$

$$\text{density} = \frac{64.1\ \text{g}}{21.8\ \text{L}} = 2.94\ \text{g/L}$$

10.17 In general PV = nRT, where n = mass × formula mass. Thus

$$PV = \frac{\text{mass}}{\text{formula mass}} RT$$

We can rearrange this equation to get;

$$\text{formula mass } \frac{(\text{mass/V})RT}{P} = \frac{dRT}{P}$$

$$\text{formula mass} = \frac{\left(5.60 \text{ g L}^{-1}\right)\left(0.0821 \ \frac{\text{L atm}}{\text{mol K}}\right)\left(296.2 \text{ K}\right)}{\left(750 \text{ torr}\right)\left(\dfrac{1 \text{ atm}}{760 \text{ torr}}\right)} = 138 \text{ g mol}^{-1}$$

The empirical mass is 69 g mol^{-1}. The ratio of the molecular mass to the empirical mass is

$$\frac{138 \text{ g mol}^{-1}}{69 \text{ g mol}^{-1}} = 2$$

Therefore, the molecular formula is 2 times the empirical formula, i.e., P_2F_4.

10.18 $$\text{formula mass } \frac{(\text{mass/V})RT}{P} = \frac{dRT}{P}$$

$$\text{formula mass} = \frac{\left(5.55 \text{ g L}^{-1}\right)\left(0.0821 \ \frac{\text{L atm}}{\text{mol K}}\right)\left(313.2 \text{ K}\right)}{\left(1.25 \text{ atm}\right)} = 114 \text{ g mol}^{-1}$$

9 C and 6 H
8 C and 18 H
7 C and 30 H
6 C and 42 H
5 C and 54 H
4 C and 66 H
3 C and 78 H
2 C and 90 H
1 C and 102 H

The most probable compound is C_8H_{18} also known as octane.

10.19 $CS_2(g) + 3O_2(g) \rightarrow 2SO_2(g) + CO_2(g)$

$$\text{mol of } CS_2 = 10.0 \text{ g } CS_2\left(\frac{1 \text{ mol } CS_2}{76.15 \text{ g } CS_2}\right) = 0.131 \text{ mol } CS_2$$

$$\text{mol } CO_2 = 0.131 \text{ mol } CS_2\left(\frac{1 \text{ mol } CO_2}{1 \text{ mol } CS_2}\right) = 0.131 \text{ mol } CO_2$$

$$\text{mol } SO_2 = 0.131 \text{ mol } CS_2\left(\frac{2 \text{ mol } SO_2}{1 \text{ mol } CS_2}\right) = 0.262 \text{ mol } SO_2$$

$$\text{L } CO_2 = \frac{\left(0.131 \text{ mol } CO_2\right)\left(0.0821 \frac{\text{L·atm}}{\text{mol·K}}\right)\left(301 \text{ K}\right)}{\left(880 \text{ torr}\left(\dfrac{1 \text{ atm}}{760 \text{ torr}}\right)\right)} = 2.80 \text{ L } CO_2$$

$$\text{L } SO_2 = \frac{\left(0.262 \text{ mol } SO_2\right)\left(0.0821 \frac{\text{L·atm}}{\text{mol·K}}\right)\left(301 \text{ K}\right)}{\left(880 \text{ torr}\left(\dfrac{1 \text{ atm}}{760 \text{ torr}}\right)\right)} = 5.60 \text{ L } SO_2$$

The total number of liters is = 8.40 L.

10.20 $CaCO_3(s) \rightarrow CaO(s) + CO_2(s)$

$$\text{mol } CO_2 = \frac{PV}{RT} = \frac{(738 \text{ torr})\left(\frac{1 \text{ atm}}{760 \text{ torr}}\right)(0.250 \text{ L})}{\left(0.0821 \frac{\text{L atm}}{\text{mol K}}\right)(296 \text{ K})} = 0.00999 \text{ mol } CO_2$$

$$\text{mol } CaCO_3 = 0.00999 \text{ mol } CO_2\left(\frac{1 \text{ mol } CaCO_3}{1 \text{ mol } CO_2}\right) = 0.00999 \text{ mol } CaCO_3$$

$$\text{g } CaCO_3 = 0.00999 \text{ mol } CaCO_3\left(\frac{100.09 \text{ g } CaCO_3}{1 \text{ mol CaCO3}}\right) = 1.00 \text{ g } CaCO_3$$

10.21 $$\text{mol Ar} = 10.0 \text{ g Ar}\left(\frac{1 \text{ mol Ar}}{39.95 \text{ g Ar}}\right) = 0.250 \text{ mol Ar}$$

$$P_{Ar} = \frac{(0.250 \text{ mol Ar})\left(0.0821 \frac{\text{L·atm}}{\text{mol·K}}\right)(293 \text{ K})}{1.00 \text{ L}} = 6.01 \text{ atm}$$

$$\text{mol } N_2 = 10 \text{ g } N_2\left(\frac{1 \text{ mol } N_2}{28.02 \text{ g } N_2}\right) = 0.357 \text{ mol } N_2$$

$$P_{N_2} = \frac{(0.357 \text{ mol } N_2)\left(0.0821 \frac{\text{L·atm}}{\text{mol·K}}\right)(293 \text{ K})}{1.00 \text{ L}} = 8.59 \text{ atm}$$

$$\text{mol } O_2 = 10 \, O_2\left(\frac{1 \text{ mol } O_2}{32.00 \text{ g } O_2}\right) = 0.313 \text{ mol } O_2$$

$$P_{O_2} = \frac{(0.313 \text{ mol } O_2)\left(0.0821 \frac{\text{L·atm}}{\text{mol·K}}\right)(293 \text{ K})}{1.00 \text{ L}} = 7.53 \text{ atm}$$

$P_{total} = P_{Ar} + P_{N_2} + P_{O_2} = 6.01 \text{ atm} + 8.59 \text{ atm} + 7.53 \text{ atm} = 22.13 \text{ atm}$

10.22 We can determine the pressure due to the oxygen since $P_{total} = P_{N_2} + P_{O_2}$.
$P_{O_2} = P_{total} - P_{N_2} = 237.0 \text{ atm} - 115.0 \text{ atm} = 122.0 \text{ atm}$. We can now use the ideal gas law to determine the number of moles of O_2:

$$n = \frac{PV}{RT} = \frac{(122.0 \text{ atm})(17.00 \text{ L})}{\left(0.0821 \frac{\text{L atm}}{\text{mol K}}\right)(298 \text{ K})} = 84.8 \text{ mol } O_2$$

$$\text{g } O_2 = (84.8 \text{ mol } O_2)\left(\frac{32.0 \text{ g } O_2}{1 \text{ mol } O_2}\right) = 2713 \text{ g } O_2$$

10.23 The total pressure is the pressure of the methane and the pressure of the water. We can determine the pressure of the methane by subtracting the pressure of the water from the total pressure.
The pressure of the water is determined by the temperature of the sample. At 30 °C, the partial pressure of water is 31.82 torr.
$P_{CH_4} = T_{total} - P_{water} = 775 \text{ torr} - 31.82 \text{ torr} = 743.18 \text{ torr}$
The pressure in the flask is 743.18 torr.

$$\text{mol } CH_4 = \frac{PV}{RT} = \frac{(743.18 \text{ torr})\left(\frac{1 \text{ atm}}{760 \text{ torr}}\right)(2.50 \text{ L})}{\left(0.0821 \frac{\text{L atm}}{\text{mol K}}\right)(303 \text{ K})} = 0.0983 \text{ mol } CH_4$$

10.24 First we find the partial pressure of nitrogen, using the vapor pressure of water at 15 °C:
$P_{N_2} = P_{total} - P_{water} = 745 \text{ torr} - 12.79 \text{ torr} = 732 \text{ torr}$.

To calculate the volume of the nitrogen we can use the combined gas law

$$\frac{P_1V_1}{T_1} = \frac{P_2V_2}{T_2}$$

For this problem,

$$V_2 = \frac{P_1V_1T_2}{P_2T_1} = \frac{(732 \text{ torr})(0.310 \text{ L})(273 \text{ K})}{(760 \text{ torr})(288 \text{ K})} = 283 \text{ mL}$$

10.25 Since the stoichiometric ratio of the SO_2 and SO_3 are the same, the pressure in the flask after the reaction when the only substance in the flask is SO_3 will be the same as the pressure in flask when there is just SO_2, 0.750 atm. The pressure will be 1.125 atm when the O_2 is just added.

10.26 Find the number of moles of both the H_2 and NO then find the mol fractions.

$$\text{mol } H_2 = (2.15 \text{ g } H_2)\left(\frac{1 \text{ mol } H_2}{2.016 \text{ g } H_2}\right) = 1.07 \text{ mol } H_2$$

$$\text{mol NO} = (34.0 \text{ g NO})\left(\frac{1 \text{ mol NO}}{30.01 \text{ g NO}}\right) = 1.13 \text{ mol NO}$$

$$\chi_{H_2} = \frac{1.07 \text{ mol } H_2}{1.13 \text{ mol NO} + 1.07 \text{ mol } H_2} = 0.486$$

$$\chi_{NO} = \frac{1.13 \text{ mol NO}}{1.13 \text{ mol NO} + 1.07 \text{ mol } H_2} = 0.514$$

$$P_{H_2} = (P_{total})(\chi_{H_2}) = (2.05 \text{ atm})(0.486) = 0.996 \text{ atm}$$

$$P_{NO} = (P_{total})(\chi_{NO}) = (2.05 \text{ atm})(0.514) = 1.05 \text{ atm}$$

10.27 The mole fraction is defined in Equation 10.5:

$$X_{O_2} = \frac{P_{O_2}}{P_{total}} = \frac{116 \text{ torr}}{760 \text{ torr}} = 0.153 \text{ or } 15.3\%$$

10.28 $$\frac{\text{effusion rate (Br} - 81)}{\text{effusion rate (Br} - 79)} = \sqrt{\frac{M_{Br-79}}{M_{Br-81}}} = \sqrt{\frac{78.9}{80.9}} = 0.988$$

10.29 Use Equation 10.7;

$$\frac{\text{effusion rate (HX)}}{\text{effusion rate (HCl)}} = \sqrt{\frac{M_{HCl}}{M_{HX}}}$$

$$M_{HX} = M_{HCl} \times \left(\frac{\text{effusion rate (HX)}}{\text{effusion rate (HCl)}}\right)^2 = 36.46 \text{ g mol}^{-1} \times (1.88)^2 = 128.9 \text{ g mol}^{-1}$$

The unknown gas must be HI.

Review Problems

10.25 (a) $$\text{torr} = (1.26 \text{ atm})\left(\frac{760 \text{ torr}}{1 \text{ atm}}\right) = 958 \text{ torr}$$

(b) $$\text{atm} = (740 \text{ torr})\left(\frac{1 \text{ atm}}{760 \text{ torr}}\right) = 0.974 \text{ atm}$$

(c) $$\text{mm Hg} = 738 \text{ torr}\left(\frac{760 \text{ torr}}{760 \text{ mm Hg}}\right) = 738 \text{ mm Hg}$$

(d) $$\text{torr} = (1.45 \times 10^3 \text{ Pa})\left(\frac{760 \text{ torr}}{1.01325 \times 10^5 \text{ Pa}}\right) = 10.9 \text{ torr}$$

10.27 (a) $$\text{torr} = (0.329 \text{ atm})\left(\frac{760 \text{ torr}}{1 \text{ atm}}\right) = 250 \text{ torr}$$

(b) $$\text{torr} = (0.460 \text{ atm})\left(\frac{760 \text{ torr}}{1 \text{ atm}}\right) = 350 \text{ torr}$$

10.29 $765 \text{ torr} - 720 \text{ torr} = 45 \text{ torr}$ $$45 \text{ torr}\left(\frac{760 \text{ mm Hg}}{760 \text{ torr}}\right) = 45 \text{ mm Hg}$$

$$\text{cm Hg} = (45 \text{ mm Hg})\left(\frac{1 \text{ cm}}{10 \text{ mm}}\right) = 4.5 \text{ cm Hg}$$

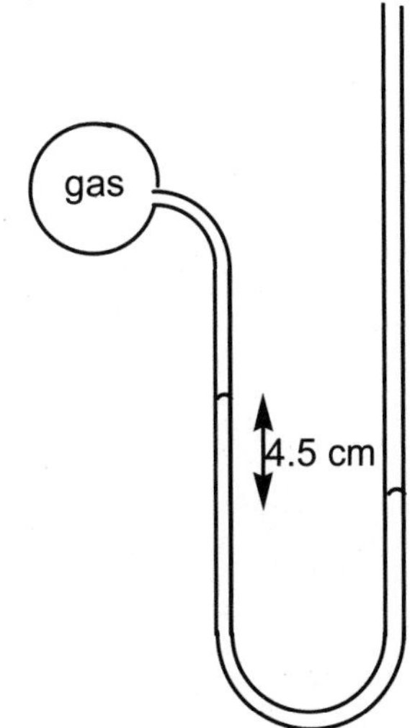

10.31 $$65 \text{ mm Hg}\left(\frac{760 \text{ torr}}{760 \text{ mm Hg}}\right) = 65 \text{ torr}$$ $748 \text{ torr} + 65 \text{ torr} = 813 \text{ torr}$

10.33 In a closed-end manometer the difference in height of the mercury levels in the two arms corresponds to the pressure of the gas. Therefore, the pressure of the gas is 125 mm Hg.

$$125 \text{ mm Hg}\left(\frac{760 \text{ torr}}{760 \text{ mm Hg}}\right) = 125 \text{ torr}$$

10.35 Use Boyle's Law to solve for the second volume:

$$V_2 = \frac{P_1V_1}{P_2} = \frac{(255\text{ mL})(725\text{ torr})}{365\text{ torr}} = 507\text{ mL}$$

10.37 Use Charles's Law to solve the second volume:

$$V_2 = \frac{V_1T_2}{T_1} = \frac{3.86\text{ L }(353\text{ K})}{318\text{ K}} = 4.28\text{ L}$$

10.39 Compare pressure change to temperature to solve for temperature change:

$$T_2 = \frac{P_2T_1}{P_1} = \frac{(1700\text{ torr})(558\text{ K})}{850\text{ torr}} = 1116\text{ K} = 1120\text{ K} \qquad 1116\text{ K} - 273\text{ K} = 843\ °\text{C}$$

10.41 In general the combined gas law equation is: $\frac{P_1V_1}{T_1} = \frac{P_2V_2}{T_2}$, and in particular, for this problem, we have:

$$P_2 = \frac{P_1V_1T_2}{T_1V_2} = \frac{(740\text{ torr})(2.58\text{ L})(348.2\text{ K})}{(297.2\text{ K})(2.81\text{ L})} = 796\text{ torr}$$

10.43 In general the combined gas law equation is $\frac{P_1V_1}{T_1} = \frac{P_2V_2}{T_2}$, and in particular, for this problem, we have:

$$V_2 = \frac{P_1V_1T_2}{T_1P_2} = \frac{(745\text{ torr})(2.68\text{ L})(648.2\text{ K})}{(297.2\text{ K})(760\text{ torr})} = 5.73\text{ L}$$

10.45 In general the combined gas law equation is: $\frac{P_1V_1}{T_1} = \frac{P_2V_2}{T_2}$, and in particular, for this problem, we have:

$$T_2 = \frac{P_2V_2T_1}{P_1V_1} = \frac{(373\text{ torr})(9.45\text{ L})(293.2\text{ K})}{(761\text{ torr})(6.18\text{ L})} = 219.8\text{ K} = -53.4\ °\text{C}$$

10.47 If PV = nRT, then R = PV/nT.

Let P = 1 atm = 101,325 Pa, T = 273 K, and n = 1.
Next, express the volume of the standard mole using the units m^3, instead of L, remembering that 22.4 L = 22,400 cm^3:

$$\text{mL} = 22{,}400\text{ cm}^3 \times \left(\frac{1\text{ mL}}{1\text{ cm}}\right) = 22{,}400\text{ mL}$$

$$R = \left(\frac{(760\text{ torr})(22{,}400\text{ mL})}{(1\text{ mole})(273\text{ K})}\right) = 6.24 \times 10^4\text{ mL torr mol}^{-1}\text{ K}^{-1}$$

10.49 $$V = \frac{nRT}{P} = \frac{\left(0.136\text{ g}\left(\frac{1\text{ mol}}{32.0\text{ g}}\right)\right)\left(0.0821\ \frac{\text{L atm}}{\text{mol K}}\right)(293.2\text{ K})}{\left(748\text{ torr}\left(\frac{1\text{ atm}}{760\text{ torr}}\right)\right)} = 0.104\text{ L}$$

10.51 $$P = \frac{nRT}{V} = \frac{\left(10.0\text{ g}\left(\frac{1\text{ mol}}{32.0\text{ g}}\right)\right)\left(0.0821\frac{\text{L atm}}{\text{mol K}}\right)(300\text{ K})}{(2.50\text{ L})} = 3.08\text{ atm}\left(\frac{760\text{ torr}}{1\text{ atm}}\right) = 2340\text{ torr}$$

10.53 $$n = \frac{PV}{RT} = \frac{(624\text{ torr})\left(\frac{\text{atm}}{760\text{ torr}}\right)(0.0265\text{ L})}{\left(0.0821\frac{\text{L atm}}{\text{mol K}}\right)(293\text{ K})} = \left(9.04\times10^{-4}\text{ mol}\right)\left(\frac{44.0\text{ g}}{1\text{ mol}}\right) = 0.0398\text{ g}$$

10.55 (a) $$\text{density } C_2H_6 = \left(\frac{30.1\text{ g }C_2H_6}{1\text{ mol }C_2H_6}\right)\left(\frac{1\text{ mol}}{22.4\text{ L}}\right) = 1.34\text{ g L}^{-1}$$

(b) $$\text{density } N_2 = \left(\frac{28.0\text{ g }N_2}{1\text{ mol }N_2}\right)\left(\frac{1\text{ mol}}{22.4\text{ L}}\right) = 1.25\text{ g L}^{-1}$$

(c) $$\text{density } Cl_2 = \left(\frac{70.9\text{ g }Cl_2}{1\text{ mol }Cl_2}\right)\left(\frac{1\text{ mol}}{22.4\text{ L}}\right) = 3.17\text{ g L}^{-1}$$

(d) $$\text{density Ar} = \left(\frac{39.9\text{ g Ar}}{1\text{ mol Ar}}\right)\left(\frac{1\text{ mol}}{22.4\text{ L}}\right) = 1.78\text{ g L}^{-1}$$

10.57 In general PV = nRT, where n = mass ÷ formula mass. Thus

$$PV = \frac{\text{mass}}{(\text{formula mass})}RT$$

and we arrive at the formula for the density (mass divided by volume) of a gas:

$$d = \frac{P\times(\text{formula mass})}{RT}$$

$$d = \frac{(742\text{ torr})\left(\frac{1\text{ atm}}{760\text{ torr}}\right)(32.0\text{ g/mol})}{\left(0.0821\ \frac{\text{L atm}}{\text{mol K}}\right)(297.2\text{ K})}$$

$d = 1.28$ g/L for O_2

10.59 First determine the number of moles from the ideal gas law:

$$n = \frac{PV}{RT} = \frac{(10.0\text{ torr})\left(\frac{1\text{ atm}}{760\text{ torr}}\right)(255\text{ mL})\left(\frac{1\text{ L}}{1000\text{ mL}}\right)}{\left(0.0821\ \frac{\text{L atm}}{\text{mol K}}\right)(298.2\text{ K})} = 1.37\times10^{-4}\text{ mol}$$

Now calculate the molecular mass:

$$\text{molecular mass} = \frac{\text{mass}}{\#\text{ of moles}} = \frac{(12.1\text{ mg})\left(\frac{1\text{ g}}{1000\text{ mg}}\right)}{1.37\times10^{-4}\text{ mol}} = 88.2\text{ g/mol}$$

10.61 $$\text{molecular mass} = \frac{dRT}{P} = \frac{\text{mass RT}}{PV} = \frac{(1.13\text{ g/L})\left(0.0821\ \frac{\text{L atm}}{\text{mol K}}\right)(295\text{ K})}{(755\text{ torr})\left(\frac{1\text{ atm}}{760\text{ torr}}\right)} = 27.6\text{ g/mol}$$

10.63 The balanced equation is

$2C_4H_{10} + 13O_2 \rightarrow 8CO_2 + 10H_2O$

$$\text{mL }O_2 = (175\text{ mL }C_4H_{10})\left(\frac{13\text{ mL }O_2}{2\text{ mL }C_4H_{10}}\right) = 1.14\times10^3\text{ mL }O_2$$

10.65 mol C_3H_6 = (18.0 g C_3H_6)$\left(\frac{1 \text{ mol } C_3H_6}{42.08 \text{ g } C_3H_6}\right)$ = 0.428 mol C_3H_6

mol H_2 = (0.428 mol C_3H_6)$\left(\frac{1 \text{ mol } H_2}{1 \text{ mol } C_3H_6}\right)$ = 0.428 mol H_2

$$V = \frac{nRT}{P} = \frac{(0.428 \text{ mol } H_2)\left(0.0821 \frac{\text{L atm}}{\text{mol K}}\right)(297.2 \text{ K})}{(740 \text{ torr})\left(\frac{1 \text{ atm}}{760 \text{ torr}}\right)} = 10.7 \text{ L } H_2$$

10.67 $CH_4 + 2O_2 \rightarrow CO_2 + 2H_2O$

$$n_{CH_4} = \frac{PV}{RT} = \frac{(725 \text{ torr})\left(\frac{1 \text{ atm}}{760 \text{ torr}}\right)(16.8 \times 10^{-3} \text{ L})}{\left(0.0821 \frac{\text{L atm}}{\text{mol K}}\right)(308 \text{ K})} = 6.34 \times 10^{-4} \text{ mol } CH_4$$

mol O_2 = (6.34 × 10^{-4} mol CH_4)$\left(\frac{2 \text{ mol } O_2}{1 \text{ mol } CH_4}\right)$ = 1.27 × 10^{-3} mol O_2

$$V_{O_2} = \frac{nRT}{P} = \frac{(1.27 \times 10^{-3} \text{ moles})\left(0.0821 \frac{\text{L atm}}{\text{mol K}}\right)(300 \text{ K})}{(654 \text{ torr})\left(\frac{1 \text{ atm}}{760 \text{ torr}}\right)} = 3.63 \times 10^{-2} \text{ L} = 36.3 \text{ mL } O_2$$

10.69 $2CO + O_2 \rightarrow 2CO_2$

$$\text{moles CO} = \frac{(683 \text{ torr})\left(\frac{1 \text{ atm}}{760 \text{ torr}}\right)(0.300 \text{ L})}{\left(0.0821 \frac{\text{L atm}}{\text{mol K}}\right)(298 \text{ K})} = 1.10 \times 10^{-2} \text{ moles}$$

$$\text{moles } O_2 = \frac{(715 \text{ torr})\left(\frac{1 \text{ atm}}{760 \text{ torr}}\right)(0.150 \text{ L})}{\left(0.0821 \frac{\text{L atm}}{\text{mol K}}\right)(398 \text{ K})} \quad 4.32 \times 10^{-3} \text{ moles}$$

∴ O_2 is the limiting reactant

moles CO_2 = (4.32 × 10^{-3} moles O_2)$\left(\frac{2 \text{ mol } CO_2}{1 \text{ mol } O_2}\right)$ = 8.64 × 10^{-3} moles CO_2

$$V = \frac{(8.64 \times 10^{-3} \text{ mol})\left(0.0821 \frac{\text{L atm}}{\text{mol K}}\right)(300 \text{ K})}{(745 \text{ torr})\left(\frac{1 \text{ atm}}{760 \text{ torr}}\right)} \quad 2.17 \times 10^{-1} \text{ L} = 217 \text{ mL}$$

10.71 $P_{Tot} = P_{N_2} + P_{O_2} + P_{He}$

P_{Tot} = 200 torr + 150 torr + 300 torr = 650 torr

10.73 Assume all gases behave ideally and recall that 1 mole of an ideal gas at 0 °C and 1 atm occupies a volume of 22.4 L. So,

$$P_{N_2} = 0.30\text{ atm}\left(\frac{760\text{ torr}}{1\text{ atm}}\right) = 228\text{ torr}$$

$$P_{O_2} = 0.20\text{ atm}\left(\frac{760\text{ torr}}{1\text{ atm}}\right) = 152\text{ torr}$$

$$P_{He} = 0.40\text{ atm}\left(\frac{760\text{ torr}}{1\text{ atm}}\right) = 304\text{ torr}$$

$$P_{CO_2} = 0.10\text{ atm}\left(\frac{760\text{ torr}}{1\text{ atm}}\right) = 76\text{ torr}$$

10.75 $P_{total} = (P_{CO} + P_{H_2O})$
$P_{H_2O} = 17.54$ torr at 20 °C, from Table 10.2.
$P_{CO} = 754 - 17.54\text{ torr} = 736\text{ torr}$
The temperature stays constant so, $P_1V_1 = P_2V_2$, and

$$V_2 = \frac{P_1V_1}{P_2} = \frac{(736\text{ torr})(268\text{ mL})}{(760\text{ torr})} = 260\text{ mL}$$

10.77 From Table 10.2, the vapor pressure of water at 20 °C is 17.54 torr. Thus only (742 – 17.54) = 724 torr is due to "dry" methane. In other words, the fraction of the wet methane sample that is pure methane is 724/742 = 0.976. The question can now be phrased: What volume of wet methane, when multiplied by 0.976, equals 244 mL?

Volume "wet" methane × 0.976 = 244 mL

Volume "wet" methane = 244 mL/0.976 = 250 mL

In other words, one must collect 250 total mL of "wet methane" gas in order to have collected the equivalent of 244 mL of pure methane.

10.79 Effusion rates for gases are inversely proportional to the square root of the gas density, and the gas with the lower density ought to effuse more rapidly. Nitrogen in this problem has the higher effusion rate because it has the lower density:

$$\frac{\text{rate}(N_2)}{\text{rate}(CO_2)} = \sqrt{\frac{1.96\text{ g L}^{-1}}{1.25\text{ g L}^{-1}}} = 1.25$$

10.81 The relative rates are inversely proportional to the square roots of their molecular masses:

$$\frac{\text{rate}(^{235}UF_6)}{\text{rate}(^{238}UF_6)} = \sqrt{\frac{\text{molar mass }(^{238}UF_6)}{\text{molar mass }(^{235}UF_6)}} = \sqrt{\frac{352\text{ g mol}^{-1}}{349\text{ g mol}^{-1}}} = 1.0043$$

Meaning that the rate of effusion of the $^{235}UF_6$ is only 1.0043 times faster than the $^{238}UF_6$ isotope.

Practice Exercises

11.1 (a) $CH_3CH_2CH_2CH_2CH_3 < CH_3CH_2OH < Ca(OH)_2$
(b) $CH_3–O–CH_3 < CH_3CH_2NH_2 < HOCH_2CH_2CH_2CH_2OH$

11.2 Propylamine would have a substantially higher boiling point because of its ability to form hydrogen bonds (there are N–H bonds in propylamine, but not in trimethylamine.)

11.3 The piston should be pushed in. This will decrease the volume and increase the pressure, and when equilibrium is re–established, there will be fewer molecules in the gas phase.

11.4 The number of molecules in the vapor will decrease, and the number of molecules in the liquid will increase, but the sum of the molecules in the vapor and the liquid remains the same.

11.5 The boiling point is most likely (a) less than 10 °C above 100 °C.

11.6 We use the curve for water, and find that at 330 torr, the boiling point is approximately 75 °C.

11.7 Adding heat will shift the equilibrium to the right, producing more vapor. This increase in the amount of vapor causes a corresponding increase in the pressure, such that the vapor pressure generally increases with increasing temperature.

11.8

Boiling	Endothermic
Melting	Endothermic
Condensing	Exothermic
Subliming	Endothermic
Freezing	Exothermic

No, each physical change is always exothermic, or always endothermic as shown.

11.9 For calcium:

8 corners × 1/8 Ca^{2+} per corner = 1 Ca^{2+}

6 faces × 1/2 Ca^{2+} per face = 3 Ca^{2+}

For fluoride:

8 inside the unit cell × 1 F^- = 8 F^-

Thus, the ratio is 4 Ca^{2+} to 8 F^-.

11.10 For cesium:

8 corners × 1/8 Cs^+ per corner = 1 Cs^+

For chloride:

1 Cl^- in center, Total: 1 Cl^-

Thus, the ratio is 1 to 1.

11.11 The compound is an organic molecule and the solid is held together by dipole–dipole attractions and London forces. It is also a soft solid with a low melting point, so it is a molecular crystal.

11.12 Because this is a high melting, hard material, it must be a covalent or network solid. Covalent bonds link the various atoms of the crystal.

11.13 Since the melt does not conduct electricity, it is not an ionic substance. The softness and the low melting point suggest that this is a molecular solid, and indeed the formula is most properly written S_8.

11.14 The line from the triple point to the critical point is the vapor pressure curve, see Figure 11.21.

11.15 Refer to the phase diagram for water, Figure 11.46. We "move" along a horizontal line marked for a pressure of 2.15 torr. At –20 °C, the sample is a solid. If we bring the temperature from –20 °C to 50 °C, keeping the pressure constant at 2.15 torr, the sample becomes a gas. The process is thus solid → gas, i.e. sublimation.

11.16 As diagramed in Figure 11.46, this falls in the liquid region.

Review Problems

11.78 Diethyl ether has the faster rate of vaporization, since it does not have hydrogen bonds, as does butanol.

11.80 London forces are possible in them all. Where another intermolecular force can operate, it is generally stronger than London forces, and this other type of interaction overshadows the importance of the London force. The substances in the list that can have dipole–dipole attractions are those with permanent dipole moments: (a), (b), and (d). SF_6, (c), is a non–polar molecular substance. HF, (a), has hydrogen bonding.

11.82 Chloroform would be expected to display larger dipole-dipole attractions because it has a larger dipole moment than bromoform. (Chlorine has a higher electronegativity which results in each C–Cl bond having a larger dipole than each C–Br bond.) On the other hand, bromoform would be expected to show stronger London forces due to having larger electron clouds which are more polarizable than those of chlorine.

Since bromoform in fact has a higher boiling point that chloroform, we must conclude that it experiences stronger intermolecular attractions than chloroform, which can only be due to London forces. Therefore, London forces are more important in determining the boiling points of these two compounds.

11.84 Ethanol, because it has H-bonding.

11.86 ether < acetone < benzene < water < acetic acid

11.88 $$\text{kJ} = (125 \text{ g } H_2O)\left(\frac{1 \text{ mol } H_2O}{18.015 \text{ g } H_2O}\right)\left(\frac{43.9 \text{ kJ}}{1 \text{ mol } H_2O}\right) = 305 \text{ kJ}$$

11.90 We can approach this problem by first asking either of two equivalent questions about the system: how much heat energy (q) is needed in order to melt the entire sample of solid water (105 g), or how much energy is lost when the liquid water (45.0 g) is cooled to the freezing point? Regardless, there is only one final temperature for the combined (150.0 g) sample, and we need to know if this temperature is at the melting point (0 °C, at which temperature some solid water remains in equilibrium with a certain amount of liquid water) or above the melting point (at which temperature all of the solid water will have melted).

Heat flow supposing that all of the solid water is melted:

$$q = 6.01 \text{ kJ/mole} \times 105 \text{ g} \times 1 \text{ mol}/18.0 \text{ g} = 35.1 \text{ kJ}$$

Heat flow on cooling the liquid water to the freezing point:

$$q = 45.0 \text{ g} \times 4.18 \text{ J/g °C} \times 85 \text{ °C} = 1.60 \times 10^4 \text{ J} = 16.0 \text{ kJ}$$

The lesser of these two values is the correct one, and we conclude that 16.0 kJ of heat energy will be transferred from the liquid to the solid, and that the final temperature of the mixture will be 0 °C. The system will be an equilibrium mixture weighing 150 g and having some solid and some liquid in equilibrium with one another. The amount of solid that must melt in order to decrease the temperature of 45.0 g of water from 85 °C to 0 °C is: 16.0 kJ ÷ 6.01 kJ/mol = 2.66 mol of solid water. 2.66 mol × 18.0 g/mol = 47.9 g of water must melt.

(a) The final temperature will be 0 °C.
(b) 47.9 g of water must melt.

11.92 For zinc:

4 surrounding center $= 4\ Zn^{2+}$

For sulfide:

8 corners × 1/8 S^{2-} per corner = 1 S^{2-}

6 faces × 1/2 S^{2-} per face = 3 S^{2-}

Total $= 4\ S^{2-}$

11.94 From figure 11.31, we can see that the length of the diagonal of the cell = 4r, where r = radius of the atom. According to the Pythagorean theorem,

$$a^2 + b^2 = c^2$$

for a right triangle. Since a = b here, we may re–write this as

$$2l^2 = c^2,$$

where l = length of the edge of the unit cell. As mentioned above, the diagonal of the unit cell = 4r, so we may say that

$$2l^2 = (4r)^2$$
$$l^2 = (4r)^2/2$$
$$l^2 = 16r^2/2$$
$$l^2 = 8r^2$$
$$l = \sqrt{8r^2}$$

Finally, substituting the value provided for r in the problem, $l = \sqrt{8(1.24\ Å)^2} = 3.51\ Å$. Using the conversion factor 1pm = 100 Å, this is 351 pm.

11.96 Each edge is composed of 2 × radius of the cation plus 2 × radius of the anion. The edge is therefore

2 × 133 + 2 × 195 = 656 pm.

11.98 Using the Bragg equation (eqn. 11.2), $n\lambda = 2d \sin\theta$

(a) $n(229\text{pm}) = 2(1,000)\sin\theta$

$0.1145n = \sin\theta$

$\theta = 6.57°$

(b) $n(229\text{pm}) = 2(100)\sin\theta$

$0.458n = \sin\theta$

$\theta = 27.3°$

11.100 According to the Pythagorean theorem,

$$a^2 + b^2 = c^2$$

for a right triangle. First, we need to find the length of a diagonal on a face of the unit cell. Since a = b here, we may re-write this as

$$2l^2 = c^2,$$

where l = length of the edge of the unit cell and c = the diagonal length. Using the given 412.3 pm as the length of the edge, c = 583.1 pm. The diagonal length inside the cell from corner to opposite corner may now be found by the same theorem:

$$a^2 + b^2 = c^2$$
$$(412.3)^2 + (583.1)^2 = c^2$$
$$c = 714.1\text{ pm}$$

This diagonal length inside the cell from corner to opposite corner is due to 1 Cs^+ ion and 1 Cl^- ion (see Figure 11.34). Therefore:

$2r_{Cs+} + 2r_{Cl-} = 714.1\text{pm}$

$2r_{Cs+} + 2(181\text{pm}) = 714.1\text{pm}$ $\qquad 2r_{Cs+} = 352$ pm

$r_{Cs+} = 176$ pm

11.102 This must be a molecular solid, because if it were ionic it would be high–melting, and the melt would conduct.

11.104 This is a metallic solid.

11.106 (a) molecular (d) metallic (g) ionic
(b) ionic (e) covalent
(c) ionic (f) molecular

11.108

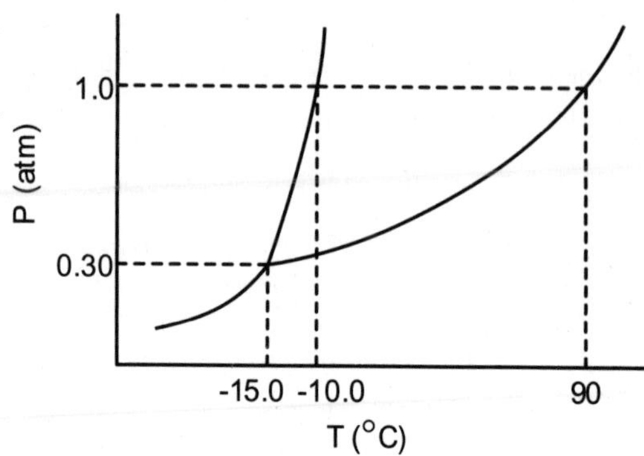

11.110 (a) solid (b) gas (c) liquid (d) solid, liquid, and gas

Practice Exercises

12.1 $C_{H_2S} = k_H P_{H_2S}$

$P_{H_2S} = 1.0$ atm $\quad C_{H_2S} = \left(\frac{0.1 \text{ mol } H_2S}{L}\right)\left(\frac{34.08 \text{ g } H_2S}{1 \text{ mol } H_2S}\right) = 3.4 \text{ g L}^{-1}$

3.4 g L^{-1} H_2S = k_H (1.0 atm H_sS)
$k_H = 3.4 \text{ g L}^{-1}$
Hydrogen sulfide is more soluble in water than nitrogen and oxygen. Hydrogen sulfide reacts with the water to form hydronium ions and HS^-.

12.2 With one atmosphere of air, the concentrations of the gases in the water depends on the partial pressures of the gases.

Oxygen: $C_{O_2} = \left(\frac{0.00430 \text{ g } O_2}{100 \text{ mL } H_2O}\right)\left(\frac{159 \text{ mm Hg}}{760 \text{ mm Hg}}\right) = 0.899 \text{ mg } O_2/\ 100 \text{ mL}$

Nitrogen $C_{N_2} = \left(\frac{0.00190 \text{ g } N_2}{100 \text{ mL } H_2O}\right)\left(\frac{593 \text{ mm Hg}}{760 \text{ mm Hg}}\right) = 1.48 \text{ mg } N_2/\ 100 \text{ mL}$

12.3 A 10% w/w solution of sucrose will need 10 grams of sucrose for each 100 g of solution. For a solution with 45.0 g of sucrose:

$$10\% \text{ solution} = \frac{45.0 \text{ g sucrose}}{x \text{ g solution}}$$

x = 450 g solution
g water = 450 g solution – 45.0 g sucrose
g water = 405 g water

$$\text{mL water} = (405 \text{ g water})\left(\frac{1 \text{ cm}^3}{0.9982 \text{ g}}\right) = 405.7 \text{ mL water}$$

12.4 The total mass of the solution is to be 250 g. If the solution is to be 1.00 % (w/w) NaBr, then the mass of NaBr will be: 250 g × 1.00 g NaBr/100 g solution = 2.50 g NaBr. We therefore need 2.50 g of NaBr and (250 – 2.50) = 248 g H_2O. The volume of water that is needed is: 248 g × 0.988 g/mL = 251 mL H_2O.

12.5 An HCl solution that is 37 % (w/w) has 37 grams of HCl for every 1.0×10^2 grams of solution.

$$\text{g solution} = (7.5 \text{ g HCl})\left(\frac{1.0 \times 10^2 \text{ g solution}}{37 \text{ g HCl}}\right) 2.0 \times 10^1 \text{ g solution}$$

12.6 We need to know the number of moles of Na_2SO_4 and the number of kg of water.
44.00 g Na_2SO_4 ÷ 142.0 g/mol = 0.3099 mol NaOH
250 g H_2O × 1 kg/1000 g = 0.250 kg H_2O

The molality is thus given by:
m = 0.3099 mol/0.25 kg = 1.239 mol NaOH/kg H_2O = 1.239 m

12.7 $\text{g } CH_3OH = (2000 \text{ g } H_2O)\left(\frac{0.250 \text{ mol } CH_3OH}{1000 \text{ g } H_2O}\right)\left(\frac{32.0 \text{ g } CH_3OH}{1 \text{ mol } CH_3OH}\right)$

= 16.0 g CH_3OH

12.8 First we need to find the number of grams of $Fe(NO_3)_3$ for each kg of solvent.

$$0.853 \text{ g } Fe(NO_3)_3 = 0.853 \text{ mol } Fe(NO_3)_3 \left(\frac{241.86 \text{ g } Fe(NO_3)_3}{1 \text{ mol } Fe(NO_3)_3}\right) = 206.3 \text{ g } Fe(NO_3)_3$$

Then we need to find how the ratio of the moles of $Fe(NO_3)_3$ to the mass of the solution:

$$\text{ratio} = \frac{0.853 \text{ mol } Fe(NO_3)_3}{1000 \text{ g } H_2O + 206.3 \text{ g } Fe(NO_3)_3} = 7.072 \times 10^{-4} \text{ mol } Fe(NO_3)_3 \text{ / g solution}$$

(a) $$\text{g solution} = (0.0200 \text{ mol } Fe(NO_3)_3)\left(\frac{1 \text{ g solvent}}{7.07\times10^{-4} \text{ mol } Fe(NO_3)_3}\right) = 28.3 \text{ g solution}$$

(b) $$\text{g sol'n} = (0.0500 \text{ mol } Fe^{3+})\left(\frac{1 \text{ mol } Fe(NO_3)_3}{1 \text{ mol } Fe^{3+}}\right)\left(\frac{1 \text{ g solvent}}{7.07\times10^{-4} \text{ mol } Fe(NO_3)_3}\right) = 70.7 \text{ g sol'n}$$

(c) $$\text{g sol'n} = (0.00300 \text{ mol } NO_3^-)\left(\frac{1 \text{ mol } Fe(NO_3)_3}{3 \text{ mol } NO_3^-}\right)\left(\frac{1 \text{ g solvent}}{7.07\times10^{-4} \text{ mol } Fe(NO_3)_3}\right) = 1.41 \text{ g sol'n}$$

12.9 If a solution is 50% NaOH, then it has 50 g of NaOH for each 100 g of solution. The mass of water is 50 g of water for 50 g of NaOH. To calculate the molality of the solution, we need to find the moles of NaOH for each kilogram of water.

$$m = \left(\frac{50 \text{ g NaOH}}{50 \text{ g } H_2O}\right)\left(\frac{1 \text{ mol NaOH}}{40.0 \text{ g NaOH}}\right)\left(\frac{1000 \text{ g } H_2O}{1 \text{ kg } H_2O}\right) = 25 \text{ mol NaOH/kg } H_2O = 25\ m \text{ NaOH}$$

12.10 If a solution is 37.0% (w/w) HCl, then 37.0% of the mass of any sample of such a solution is HCl and (100.0 – 37.0) = 63.0% of the mass is water. In order to determine the molality of the solution, we can conveniently choose 100.0 g of the solution as a starting point. Then 37.0 g of this solution are HCl and 63.0 g are H_2O. For molality, we need to know the number of moles of HCl and the mass in kg of the solvent:

37.0 g HCl ÷ 36.46 g/mol = 1.01 mol HCl
63.0 g H_2O × 1 kg/1000 g = 0.0630 kg H_2O
molality = mol HCl/kg H_2O = 1.01 mol/0.0630 kg = 16.1 m

12.11 We first determine the mass of one L (1000 mL) of this solution, using the density:

1000 mL × 1.38 g/mL = 1.38×10^3 g

Next, we use the fact that 40.0 % of this total mass is due to HBr, and calculate the mass of HBr in the 1000 mL of solution:

0.400 × (1.38×10^3) = 552 g HBr

This is converted to the number of moles of HBr in 552 g:

552 g HBr ÷ 80.91 g/mol = 6.82 mol HBr

Last, the molarity is the number of moles of HBr per liter of solution

6.82 mol/1 L = 6.82 M

12.12 First determine the number of moles of $Al(NO_3)_3$ dissolved in the liter of water.

$$\text{mol } Al(NO_3)_3 = (1.00 \text{ g } Al(NO_3)_3)\left(\frac{1 \text{ mol } Al(NO_3)_3}{212.996 \text{ g } Al(NO_3)_3}\right) = 0.00469 \text{ mol } Al(NO_3)_3$$

Next find the mass of the water:

$$\text{g } H_2O = (1.00 \text{ L } H_2O)\left(\frac{1000 \text{ mL } H_2O}{1 \text{ L } H_2O}\right)\left(\frac{0.9982 \text{ g } H_2O}{1 \text{ mL } H_2O}\right) = 998.2 \text{ g } H_2O$$

To find the molarity of the solution, first we need to find the mass of the solution, and then the volume of the solution:

g solution = 998.2 g H_2O + 1.00 g $Al(NO_3)_3$ = 999.2 g solution

$$\text{L solution} = (999.2\ \text{solution})\left(\frac{1\ \text{mL solution}}{0.9989\ \text{g solution}}\right)\left(\frac{1\ \text{L solution}}{1000\ \text{mL solution}}\right) = 1.0003\ \text{L}$$

$$M\ \text{of solution} = \frac{0.00469\ \text{mol Al}(\text{NO}_3)_3}{1.0003\ \text{L solution}} = 0.00469\ M\ \text{Al(NO}_3)_3$$

The molality of the solution can also be determined

$$m\ \text{of solution} = \frac{0.00469\ \text{mol Al}(\text{NO}_3)_3}{0.9982\ \text{kg H}_2\text{O}} = 0.00470\ m\ \text{Al(NO}_3)_3$$

12.13 First determine the number of moles of each component of the solution:

For $C_{16}H_{22}O_4$, 20.0 g/278 g/mol = 0.0719 mol

For C_8H_{18}, 50.0 g/114 g/mol = 0.439 mol

The mole fraction of solvent is:

0.439 mol/(0.439 mol + 0.0719 mol) = 0.859

Using Raoult's Law, we next find the vapor pressure to expect for the solution, which arises only from the solvent (since the solute is known to be nonvolatile):

$P_{solvent} = \chi_{solvent} \times P°_{solvent} = 0.859 \times 10.5\ \text{torr} = 9.02\ \text{torr}$

12.14 $P_{acetone} = \chi_{acetone} \times P°_{acetone}$

First, solve for the mole fraction of acetone:

150 torr = $\chi_{acetone} \times$ 162 torr

$\chi_{acetone} = 0.926$

Now solve for the moles of steric acid:

$$\chi_{acetone} = \frac{\text{mol acetone}}{\text{mol acetone} + \text{mol stearic acid}}$$

$$\text{mol acetone} = (156\ \text{g acetone})\left(\frac{1\ \text{mol acetone}}{58.0\ \text{mol acetone}}\right) = 2.69\ \text{mol acetone}$$

$$\chi_{acetone} = 0.926 = \frac{2.69\ \text{mol acetone}}{2.69\ \text{mol acetone} + x\ \text{mol stearic acid}}$$

(0.926)(2.69 mol acetone + x mol stearic acid) = 2.69 mol acetone

2.49 mol acetone + 0.926x mol stearic acid = 2.69 mol acetone

0.926x mol stearic acid = 0.199

x mol stearic acid = 0.215 mol stearic acid

Finally solve to find the number of grams of stearic acid

$$\text{g stearic acid} = (0.215\ \text{mol stearic acid})\left(\frac{284.5\ \text{g stearic acid}}{1\ \text{mol stearic acid}}\right) = 61.2\ \text{g stearic acid}$$

12.15 $P_{cyclohexane} = \chi_{cyclohexane} \times P°_{cyclohexane} = 0.750 \times 66.9\ \text{torr} = 50.2\ \text{torr}$

$P_{toluene} = \chi_{toluene} \times P°_{toluene} = 0.250 \times 21.1\ \text{torr} = 5.28\ \text{torr}$

$P_{total} = P_{cyclohexane} + P_{toluene} = 33.5\ \text{torr} + 10.6\ \text{torr} = 55.4\ \text{torr}$

12.16 First we need to find the moles of the cyclohexane and the moles of toluene.

$$\text{mol cylcohexane: } = (100\ \text{g cyclohexane})\left(\frac{1\ \text{mol cyclohexane}}{84.15\ \text{g cyclohexane}}\right) = 1.188\ \text{mol cyclohexane}$$

$$\text{mol toluene} = (100\ \text{g toluene})\left(\frac{1\ \text{mol toluene}}{92.14\ \text{g toluene}}\right) = 1.085\ \text{mol toluene}$$

Now, find the $\chi_{cyclohexane}$ and the $\chi_{toluene}$

$$\chi_{cyclohexane} = \frac{1.188 \text{ mol cyclohexane}}{1.188 \text{ mol cyclohexane} + 1.085 \text{ mol toluene}} = 0.523$$

$\chi_{toluene} = 1 - \chi_{cyclohexane} = 1 - 0.523 = 0.477$
$P_{cyclohexane} = \chi_{cyclohexane} \times P°_{cyclohexane} = 0.523 \times 66.9 \text{ torr} = 35.0 \text{ torr}$
$P_{toluene} = \chi_{toluene} \times P°_{toluene} = 0.477 \times 21.1 \text{ torr} = 10.1 \text{ torr}$
$P_{total} = P_{cyclohexane} + P_{toluene} = 35.0 \text{ torr} + 10.1 \text{ torr} = 45.1 \text{ torr}$

12.17 A 10 % solution contains 10 g sugar and 90 g water.
$10 \text{ g } C_{12}H_{22}O_{11} \div 342 \text{ g/mol} = 0.029 \text{ mol } C_{12}H_{22}O_{11}$
$90 \text{ g } H_2O \times 1 \text{ kg}/1000 \text{ g} = 0.090 \text{ kg } H_2O$
$\text{m} = 0.029 \text{ mol}/0.090 \text{ kg} = 0.32 \text{ mol/kg}$
$\Delta T_b = K_b \times m = 0.51 \text{ °C } m^{-1} \times 0.32\ m = 0.16 \text{ °C}$
$T_b = 100.16 \text{ °C}$

12.18 $\Delta T_b = K_b \times m = 0.51 \text{ °C } m^{-1} \times x\ m = 2.36 \text{ °C}$
$x\ m = 4.63\ m$
To find the number of grams of glucose, first we need to find the number of moles of glucose.
mol glucose = (m solution)(kg solvent)
mol glucose = (4.63 m)(0.250 kg H_2O) = 1.16 mol glucose

$$\text{g glucose} = (1.16 \text{ mol glucose})\left(\frac{180.9 \text{ g glucose}}{1 \text{ mol glucose}}\right) = 209 \text{ g glucose}$$

12.19 It is first necessary to obtain the values of the freezing point of pure benzene and the value of K_f for benzene from Table 12.4 of the text. We proceed to determine the number of moles of solute that are present and that have caused this depression in the freezing point: $\Delta T = K_f m$
$\therefore m = \Delta T/K_f = (5.45 \text{ °C} - 4.13 \text{ °C})/(5.07 \text{ °C kg mol}^{-1}) = 0.260\ m$
Next, use this molality to determine the number of moles of solute that must be present:
0.260 mol solute/kg solvent × 0.0850 kg solvent = 0.0221 mol solute
Last, determine the formula mass of the solute:
3.46 g/0.0221 mol = 157 g/mol

12.20 To find the molar mass of the substance, first, we need to find the molality of the solution from the freezing point depression, and then using the 5.0% (wt/wt) amount, determine the moles of the solute.

$$m = \frac{\Delta T}{K_f} = \frac{80.2\,°C - 77.3\,°C}{6.9\ °C\ m^{-1}} = 0.420\ m$$

Assume there is 100 g of solution:

$$5.0\% \text{ (wt/wt)} = \frac{5 \text{ g unknown substance}}{5 \text{ g unknown substance} + 95 \text{ g naphthalene}}$$

We have 95 g of naphthalene, or 0.095 kg naphthalene and 5 g of the unknown.
Using the equation for molality, we can determine the number of moles of the unknown
mol unknown = (m solution)(kg solvent) = (0.420 m)(0.095 kg naphthalene) = 0.0399 mol unknown

$$\text{molar mass} = \frac{5.0 \text{ g unknown}}{0.0399 \text{ mol unknown}} = 125 \text{ g mol}^{-1}$$

12.21 Use the equation $\Pi = MRT$:

$$M = \frac{(5 \text{ g protein})\left(\frac{1 \text{ mol protein}}{230{,}000 \text{ g protein}}\right)}{0.1000 \text{ L solution}} = 2.17 \times 10^{-4}\ M \text{ solution}$$

R = 0.0821 L atm/K mol
T = 4.0 + 273.2 = 277.2 K
$\Pi = (2.17 \times 10^{-4}\ M \text{ solution})(0.0821 \text{ L atm/K mol})(277.2 \text{ K}) = 4.95 \times 10^{-3} \text{ atm}$

$$\text{mm Hg} = 4.94 \times 10^{-3}\ \text{atm}\left(\frac{760\ \text{mm Hg}}{1\ \text{atm}}\right) = 3.76\ \text{mm Hg}$$

$$\text{mm } H_2O = 3.76\ \text{mm Hg}\left(\frac{13.6\ \text{mm } H_2O}{1\ \text{mm Hg}}\right) = 51.1\ \text{mm } H_2O$$

12.22 We can use the equation $\Pi = M\text{RT}$:

$$42.5\ \text{torr}\left(\frac{1\ \text{atm}}{760\ \text{torr}}\right) = (M)(0.0821\ \text{L atm/K mol})(298\ \text{K})$$

$$M = 2.29 \times 10^{-3}\ M$$

$$\text{mass} = (125\ \text{mL})\left(\frac{2.29\times10^{-3}\ \text{mol}}{1000\ \text{mL}}\right)\left(\frac{180.2\ \text{g } C_6H_{12}O_6}{1\ \text{mol } C_6H_{12}O_6}\right) = 0.0516\ \text{g } C_6H_{12}O_6$$

12.23 $\Pi = M\text{RT}$

$$\Pi = 6.45\ \text{cm water}\left(\frac{10\ \text{mm } H_2O}{1\ \text{cm } H_2O}\right)\left(\frac{1.00\ \text{g mL}^{-1}}{13.6\ \text{g mL}^{-1}}\right)\left(\frac{1\ \text{atm}}{760\ \text{mm Hg}}\right) = 6.24 \times 10^{-3}\ \text{atm}$$

R = 0.0821 L atm / mol K

T = 277 K

$$M = \frac{\Pi}{\text{RT}} = \frac{(6.24\times10^{-3}\ \text{atm})}{(0.0821\ \text{L atm/mol K})(277\ \text{K})} = 2.74 \times 10^{-3}\ \text{mol L}^{-1}$$

$$\text{mol protein} = (2.74 \times 10^{-3}\ \text{mol L}^{-1})(0.1000\ \text{L}) = 2.74 \times 10^{-4}\ \text{mol}$$

$$\text{molar mass} = \frac{0.137\ \text{g protein}}{2.74\times10^{-4}\ \text{mol protein}} = 4.99 \times 10^{4}\ \text{g mol}^{-1}$$

12.24 We can use the equation $\Pi = M\text{RT}$, remembering to convert pressure to atm:

$$\text{atm} = (25.0\ \text{torr})\left(\frac{1\ \text{atm}}{760\ \text{torr}}\right) = 0.0329\ \text{atm}$$

$$\Pi = 0.0329\ \text{atm} = M \times (0.0821\ \text{L atm/K mol})(298\ \text{K})$$

$$M = 1.34 \times 10^{-3}\ \text{mol L}^{-1}$$

$$\text{mol} = 1.34 \times 10^{-3}\ \text{mol L}^{-1} \times 0.100\ \text{L} = 1.34 \times 10^{-4}\ \text{mol}$$

$$\text{formula mass} = \frac{72.4 \times 10^{-3}\ \text{g}}{1.34 \times 10^{-4}\ \text{mol}} = 5.38 \times 10^{2}\ \text{g mol}^{-1}$$

12.25 For the solution as if the solute were 100% dissociated:

$\Delta T = (1.86\ °\text{C m}^{-1})(2 \times 0.237\ \text{m}) = 0.882\ °\text{C}$ and the freezing point should be $-0.882\ °\text{C}$.

For the solution as if the solute were 0% dissociated:

$\Delta T = (1.86\ °\text{C m}^{-1})(1 \times 0.237\ \text{m}) = 0.441\ °\text{C}$ and the freezing point should be $-0.441\ °\text{C}$.

12.26 Use the freezing point depression equation:

$\Delta T = K_f m$

Remember that there are two moles of ions for each mole of $MgSO_4$.

K_f water = 1.86 °C m^{-1}

(a) For 0.1 m $MgSO_4$ $\quad m = 0.2\ m$

$\Delta T = (1.86\ °\text{C}\ m^{-1})(0.2\ m) = 0.372\ °\text{C}$

(b) For 0.01 m $MgSO_4$ $\quad m = 0.02\ m$

$\Delta T = (1.86\ °\text{C}\ m^{-1})(0.02\ m) = 0.0372\ °\text{C}$

(c) For 0.001 *m* $MgSO_4$ $m = 0.002\ m$

$\Delta T = (1.86\ °C\ m^{-1})(0.002\ m) = 0.00372\ °C$

The first freezing point depression could be measured using a laboratory thermometer that can measure 0.1 °C increments.

Review Problems

12.38 This is to be very much like that shown in Figure 12.5:

(a) $KCl(s) \rightarrow K^+(g) + Cl^-(g)$, $\Delta H° = +690\ kJ\ mol^{-1}$

(b) $K^+(g) + Cl^-(g) \rightarrow K^+(aq) + Cl^-(aq)$, $\Delta H° = -686\ kJ\ mol^{-1}$

$KCl(s) \rightarrow K^+(aq) + Cl^-(aq)$, $\Delta H° = +4\ kJ\ mol^{-1}$

12.40 $\dfrac{C_1}{P_1} = \dfrac{C_2}{P_2}$

$$C_2 = \frac{C_1 \times P_2}{P_1} = \frac{(0.025\ g\ L^{-1}) \times (1.5\ atm)}{(1.0\ atm)} = 0.038\ g/L.$$

12.42 $\dfrac{C_1}{P_1} = \dfrac{C_2}{P_2}$

$$C_2 = \frac{C_1 \times P_2}{P_1} = \frac{(0.010\ g\ L^{-1}) \times (2.0\ atm)}{(1.0\ atm)} = 0.020\ g\ L^{-1}$$

12.44 One liter of solution has a mass of:

$$g\ solution = 1\ L\ solution\left(\frac{1000\ mL\ solution}{1\ L\ solution}\right)\left(\frac{1.07\ g\ solution}{1\ mL\ solution}\right) = 1{,}070\ g$$

According to the given molarity, it contains 3.000 mol NaCl. This has a mass of:

$$g\ NaCl = 3.000\ mol\ NaCl\left(\frac{58.45\ g\ NaCl}{1\ mol\ NaCl}\right) = 175.4\ g\ NaCl$$

Thus, the mass of water in 1 L solution must be:
1,070 g – 175.4 g = 895 g water

$$m = \left(\frac{3.000\ mol\ NaCl}{0.895\ kg\ solvent}\right) = 3.35\ m$$

12.46 24.0 g glucose ÷ 180 g/mol = 0.133 mol glucose
molality = 0.133 mol glucose/1.00 kg solvent = 0.133 molal
mole fraction = moles glucose/total moles
moles glucose = 0.133

$$moles\ H_2O = (1.00 \times 10^3\ g\ H_2O)\left(\frac{1\ mole\ H_2O}{18\ g\ H_2O}\right) = 55.6\ mol\ H_2O$$

$$\chi_{glucose} = \frac{0.133}{55.6 + 0.133} = 2.39 \times 10^{-3}$$

$$mass\ \% = \frac{24.0\ g\ glucose}{1000\ g\ H_2O + 24\ g\ glucose} \times 100\% = 2.34\%$$

12.48 We need to know the mole amounts of both components of the mixture. It is convenient to work from an amount of solution that contains 1.25 mol of ethyl alcohol and, therefore, 1.00 kg of solvent. Convert the number of moles into mass amounts as follows:

For CH_3CH_2OH:

$$\frac{\text{g ethanol}}{\text{g solution}} = \left(\frac{1.25 \text{ mol ethanol}}{1 \text{ kg water}}\right)\left(\frac{46.08 \text{ g ethanol}}{1 \text{ mol ethanol}}\right)\left(\frac{1 \text{ kg water}}{1000 \text{ g water}}\right) = 57.6 \text{ g ethanol/1000 g water}$$

Mass % ethanol = (mass ethanol/(total solution mass) × 100%

Mass % ethanol = (57.6 g ethanol/(1,000 g water + 57.5 g ethanol) × 100%

= 5.45%

12.50 If we assume 100 g of solution we have 7.50 g NH_3 and 92.50 g H_2O.

$$\text{mol } NH_3 = (7.50 \text{ g } NH_3)\left(\frac{1 \text{ mole } NH_3}{17.03 \text{ g } NH_3}\right) = 0.440 \text{ mol } NH_3$$

$$\text{kg } H_2O = (92.5 \text{ g } H_2O)\left(\frac{1 \text{ kg}}{1000 \text{ g}}\right) = 0.0925 \text{ kg } H_2O$$

$$m = \frac{0.440 \text{ moles } NH_3}{0.0925 \text{ kg } H_2O} = 4.76\ m$$

$$\text{mol } H_2O = (92.5 \text{ g } H_2O)\left(\frac{1 \text{ mole } H_2O}{18.0 \text{ g } H_2O}\right) = 5.14 \text{ mol } H_2O$$

$$\text{mole percent} = \frac{0.440 \text{ mol } NH_3}{(0.440 \text{ mol } NH_3 + 5.14 \text{ mol } H_2O)} \times 100\% = 7.89\%$$

12.52 If we choose, for convenience, an amount of solution that contains 1 kg of solvent, then it also contains 0.363 moles of $NaNO_3$. The number of moles of solvent is:

$$\text{mol } H_2O = (1000 \text{ g})\left(\frac{1 \text{ mole } H_2O}{18.02 \text{ g } H_2O}\right) = 55.6 \text{ mol } H_2O$$

Now, convert the number of moles to a number of grams: for $NaNO_3$, 0.363 mol × 85.0 g/mol = 30.9 g; for H_2O, 1000 g was assumed and the percent (w/w) values are:

% $NaNO_3$ = 30.9 g/1030.9 g × 100% = 3.00%

% H_2O = 1000 g/1030.9 g × 100 = 97.0%

To determine the molar concentration of $NaNO_3$ assume 1 kg of solvent which would then contain 0.363 mole of $NaNO_3$ or 30.9 g $NaNO_3$. The total mass of the solution would be 1000 g + 30.9 g = 1031 g of solution. Now, the ratio of moles of solute to grams of solution is 0.363 mol $NaNO_3$/1031 g solution. From this calculate the molarity of the solution

$$M \text{ of solution} = \left(\frac{0.363 \text{ mol } NaNO_3}{1031 \text{ g solution}}\right)\left(\frac{1.0185 \text{ g soln}}{1 \text{ mL soln}}\right)\left(\frac{1000 \text{ mL soln}}{1 \text{ L soln}}\right) = 0.359\ M\ NaNO_3$$

$$\chi_{NaNO_3} = \frac{0.363 \text{ mol } NaNO_3}{55.6 \text{ mol } H_2O + 0.363 \text{ mol } NaNO_3} = 6.49 \times 10^{-3}$$

12.54 $P_{solution} = P^{\circ}_{solvent} \times \chi_{solvent}$

We need to determine $\chi_{solvent}$:

$$\text{mol glucose} = (65.0 \text{ g})\left(\frac{1 \text{ mol}}{180.2 \text{ g}}\right) = 0.361 \text{ mol glucose}$$

$$\text{mol } H_2O = (150 \text{ g } H_2O)\left(\frac{1 \text{ mol } H_2O}{18.02 \text{ g } H_2O}\right) = 8.32 \text{ mol } H_2O$$

The total number of moles is thus: 8.32 mol + 0.361 mol = 8.69 mol and the mole fraction of the solvent is:

$\chi_{solvent} = \left(\dfrac{8.32 \text{ mol solvent}}{8.69 \text{ mol solution}}\right) = 0.957$. Therefore,

$P_{solution} = 23.8 \text{ torr} \times 0.957 = 22.8 \text{ torr}$.

12.56 $P_{benzene} = \chi_{benzene} \times P°_{benzene}$

$P_{toluene} = \chi_{toluene} \times P°_{toluene}$

$P_{Tot} = P_{benzene} + P_{Toluene}$

$$\text{mol benzene} = (35.0 \text{ g})\left(\frac{1 \text{ mol}}{78.11 \text{ g}}\right) = 0.448 \text{ mol benzene}$$

$$\text{mol toluene} = (65.0 \text{ g})\left(\frac{1 \text{ mol}}{92.14 \text{ g}}\right) = 0.705 \text{ mol toluene}$$

$$\chi_{benzene} = \frac{0.448}{0.448 + 0.705} = 0.389$$

$$\chi_{toluene} = \frac{0.705}{0.448 + 0.705} = 0.611$$

$P_{benzene} = (0.389)(93.4 \text{ torr}) = 36.3 \text{ torr}$

$P_{toluene} = (0.611)(26.9 \text{ torr}) = 16.4 \text{ torr}$

$P_{Total} = 36.3 \text{ torr} + 16.4 \text{ torr} = 52.7 \text{ torr}$

12.58 The following relationships are to be established: $P_{Total} = 96 \text{ torr} = (P°_{benzene} \times \chi_{benzene}) + (P°_{toluene} \times \chi_{toluene})$. The relationship between the two mole fractions is: $\chi_{benzene} = 1 - \chi_{toluene}$, since the sum of the two mole fractions is one. Substituting this expression for $\chi_{benzene}$ into the first equation gives:

$96 \text{ torr} = [P°_{benzene} \times (1 - \chi_{Toluene})] + [P°_{Toluene} \times \chi_{Toluene}]$,

$96 \text{ torr} = [180 \text{ torr} \times (1 - \chi_{Toluene})] + [60 \text{ torr} \times \chi_{Toluene}]$.

Solving for $\chi_{Toluene}$ we get: $120 \times \chi_{Toluene} = 84$,

$\chi_{Toluene} = 0.70$ and $\chi_{benzene} = 0.30$. The mole % values are to be 70 mol% toluene and 30 mol% benzene.

12.60 (a) $\chi_{solvent} = \dfrac{P}{P°} = \dfrac{511 \text{ torr}}{526 \text{ torr}} = 0.971$

$\chi_{solute} = 1 - \chi_{solvent} = 0.029$

(b) We know $0.971 = \dfrac{1 \text{ mol}}{1 \text{ mol} + x \text{ mol}}$ $\qquad x = 2.99 \times 10^{-2}$ moles

(b) $\text{molar mass} = \dfrac{8.3 \text{ g}}{2.99 \times 10^{-2} \text{ mol}} = 278 \text{ g/mol}$

12.62 $\Delta T_f = K_f m$

$m = \Delta T_f / K_f = 3.00 \text{ °C} / 1.86 \text{ °C kg/mol} = 1.61 \text{ mol/kg}$

$$\text{kg} = (100 \text{ g})\left(\frac{1 \text{ kg}}{1000 \text{ g}}\right) = 0.1 \text{ kg}$$

$$\text{mol} = \left(\frac{1.61 \text{ mol}}{1 \text{ kg}}\right)(0.1 \text{ kg}) = 0.161 \text{ mol}$$

$$\text{g} = (0.161 \text{ mol})\left(\frac{342.3 \text{ g}}{1 \text{ mol}}\right) = 55.1 \text{ g}$$

12.64 $\Delta T = (5.45 - 3.45) = 2.00\ °C = K_f \times m = 5.07\ °C\ kg\ mol^{-1} \times m$
$m = 0.394$ mol solute/kg solvent
0.394 mol/kg benzene × 0.200 kg benzene = 0.0788 mol solute and the molecular mass is: 12.00 g/0.0788 mol = 152 g/mol

12.66 $\Delta T_f = K_f m$
$m = \Delta T_f/K_f = 0.307\ °C/5.07\ °C\ kg/mol = 0.0606\ mol/kg$

$$\text{mol} = \left(\frac{0.0606\ \text{mol}}{1\ \text{kg}}\right)(0.5\ \text{kg}) = 0.0303\ \text{mol}$$

$$\text{molar mass} = \frac{3.84\ \text{g}}{0.0303\ \text{mol}} = 127\ \text{g/mol}$$

The empirical formula has a mass of 64.1 g/mol. So the molecular formula is $C_8H_4N_2$.

12.68 (a) If the equation is correct, the units on both sides of the equation should be g/mol. The units on the right side of this equation are:

$$\frac{(\text{g}) \times (\text{L atm mol}^{-1}\ \text{K}^{-1}) \times (\text{K})}{\text{L} \times \text{atm}} = \text{g/mol}$$

which is correct.

(b) $\Pi = MRT = (n/V)RT, \quad n = \Pi V/RT$
This means that we can calculate the number of moles of solute in one L of solution, as follows:

$$n = \frac{(0.021\ \text{torr})\left(^{1\ \text{atm}}/_{760\ \text{torr}}\right)(1.0\ \text{L})}{\left(0.0821\ \text{L atm mol}^{-1}\ \text{K}^{-1}\right)(298\ \text{K})} = 1.1 \times 10^{-6}\ \text{mol}$$

The molecular mass is the mass in 1 L divided by the number of moles in 1 L:
$2.0\ g/1.1 \times 10^{-6}\ mol = 1.8 \times 10^6\ g/mol$

12.70 The equation for the vapor pressure is:

$$P_{solution} = P°_{H_2O} \times \chi_{H_2O}$$

Where $P°_{H_2O}$ is 17.5 torr. To calculate the vapor pressure we need to find the mole fraction of water first.

$$\chi_{H_2O} = \text{moles } H_2O/(\text{moles } H_2O + \text{moles NaCl})$$

Calculate the moles of NaCl in 10.0 g

$$\text{mol NaCl} = (23.0\ \text{g NaCl})\left(\frac{1\ \text{mol NaCl}}{58.44\ \text{g NaCl}}\right) = 0.394\ \text{moles NaCl}$$

When NaCl dissolves in water, Na^+ and Cl^- are formed. So, for every mole of NaCl that dissolves, two moles of ions are formed. For this solution, the number of moles of ions is 0.788.
The number of moles of solvent (water) is:

$$\text{mol } H_2O = (100\ \text{g } H_2O)\left(\frac{1\ \text{mol } H_2O}{18.02\ \text{g } H_2O}\right) = 5.55\ \text{moles } H_2O$$

Calculate the mole fraction as

$$\chi_{H_2O} = \frac{(\text{moles } H_2O)}{(\text{moles } H_2O + \text{moles NaCl})} = \frac{5.55\ \text{mol}}{(5.55\ \text{mol} + 0.788\ \text{mol})} = 0.876$$

The vapor pressure is then $P_{solution} = P°_{H_2O} \times \chi_{H_2O} = 17.5\ torr \times 0.876 = 15.3\ torr$

12.72 Assume 100 mL of solution, that is, 2.0 g NaCl and 0.100 L of solution:
$\Pi = M\text{RT}$

$$\text{M} = \frac{(2.0\text{ g NaCl})\left(\dfrac{1\text{ mol NaCl}}{58.45\text{ g NaCl}}\right)}{0.100\text{ L}} = 0.34\text{ M}$$

For every NaCl there are two ions produces so $M = 0.68\ M$

$$\Pi = (0.68\text{ M})(0.0821\text{ L atm/mol K})(298\text{ K})\left(\frac{760\text{ torr}}{1\text{ atm}}\right) = 1.3 \times 10^4\text{ torr}$$

12.74 $CaCl_2 \rightarrow Ca^{2+} + 2Cl^-$; van't Hoff factor, i = 3
$\Delta T_f = i \times K_f \times m = (3)(1.86\ °C\ m^{-1})(0.20\ m) = 1.1\ °C$
The freezing point is –1.1 °C.

12.76 Any electrolyte such as $NiSO_4$, that dissociated to give 2 ions, if fully dissociated should have a van't Hoff factor of 2.

12.78 $\Delta T_f = i \times K_f \times m$

$i = \Delta T_f/K_f \times m = 0.415°C/(1.86\ °C\ m^{-1})(0.118\ m) = 1.89$

Practice Exercises

13.1 Using the coefficients of the reaction, we find that the ratio of iodide production to sulfite disappearance in 1:3, and the ratio of sulfate production to sulfite disappearance is 3:3.

$$\text{Rate of production of } I^- = (2.4 \times 10^{-4} \text{ mol L}^{-1} \text{ s}^{-1})\left(\frac{1 \text{ mol } I^-}{3 \text{ mol } SO_3^{2-}}\right) = 8.0 \times 10^{-5} \text{ mol L}^{-1} \text{ s}^{-1}$$

$$\text{Rate of production of } SO_4^{2-} = (2.4 \times 10^{-4} \text{ mol L}^{-1} \text{ s}^{-1})\left(\frac{3 \text{ mol } SO_4^{2-}}{3 \text{ mol } SO_3^{2-}}\right) = 2.4 \times 10^{-4} \text{ mol L}^{-1} \text{ s}^{-1}$$

13.2 From the coefficients in the balanced equation we see that, for every two moles of SO_2 that is produced, 2 moles of H_2S are consumed, three moles of O_2 are consumed, and two moles of H_2O are produced.

$$\text{Rate of disappearance of } O_2 = \left(\frac{3 \text{ mol } O_2}{2 \text{ mol } SO_2}\right)\left(\frac{0.30 \text{ mol}}{\text{L s}}\right) = 0.45 \text{ mol L}^{-1} \text{ s}^{-1}$$

$$\text{Rate of disappearance of } H_2S = \left(\frac{2 \text{ mol } H_2S}{2 \text{ mol } SO_2}\right)\left(\frac{0.30 \text{ mol}}{\text{L s}}\right) = 0.30 \text{ mol L}^{-1} \text{ s}^{-1}$$

13.3 The rate of the reaction at 2.00 minutes (120 s) is equal to the slope of the tangent to the curve at 120 s. After drawing the tangent, the slope can be estimated as follows:

$$\text{rate}_{\text{with respect to HI}} = \left\{\frac{[\text{HI}]_{\text{final}} - [\text{HI}]_{\text{initial}}}{t_{\text{final}} - t_{\text{initail}}}\right\} = \left\{\frac{0 \text{ mol L}^{-1} - 0.075 \text{ mol L}^{-1}}{360 \text{ s} - 0 \text{ s}}\right\} = -2.1 \times 10^{-4} \text{ mol L}^{-1} \text{ s}^{-1}$$

13.4 The rate of the reaction after 250 seconds have elapsed is equal to the slope of the tangent to the curve at 250 seconds. First draw the tangent, and then estimate its slope as follows, where A is taken to represent one point on the tangent, and B is taken to represent another point on the tangent:

$$\text{rate} = \left(\frac{\text{A (mol/L)} - \text{B (mol/L)}}{\text{A (s)} - \text{B (s)}}\right) = \frac{\text{change in concentration}}{\text{change in time}}$$

A value near 1×10^{-4} mol L^{-1} s^{-1} is correct.

13.5 (a) Using the rate law:
Rate = $k[NO]^2[H_2]$
Substitute in the concentration and the rate and solve for the rate constant:
$7.86 \times 10^{-3} \text{ mol L}^{-1} \text{ s}^{-1} = k(2 \times 10^{-6} \text{ mol L}^{-1})^2(2 \times 10^{-6} \text{ mol L}^{-1})$
$k = 9.8 \times 10^{14} \text{ L}^2 \text{ mol}^{-2} \text{ s}^{-1}$

(b) The units can be derived from the equation:
$\text{mol L}^{-1} \text{ s}^{-1} = k(\text{mol L}^{-1})^2(\text{mol L}^{-1})$
$\text{mol L}^{-1} \text{ s}^{-1} = k (\text{mol}^3 \text{ L}^{-3})$

$$k = \frac{\text{mol L}^{-1} \text{ s}^{-1}}{\text{mol}^3 \text{ L}^{-3}} = \text{L}^2 \text{ mol}^{-2} \text{ s}^{-1}$$

13.6 (a) First use the given data in the rate law:
Rate = $k[\text{HI}]^2$
$2.5 \times 10^{-4} \text{ mol L}^{-1} \text{ s}^{-1} = k[5.58 \times 10^{-2} \text{ mol/L}]^2$
$k = 8.0 \times 10^{-2} \text{ L mol}^{-1} \text{ s}^{-1}$

(b) L mol^{-1} s^{-1}

13.7 The order of the reaction with respect to a given substance is the exponent to which that substance is raised in the rate law:
order of the reaction with respect to $[BrO_3^-] = 1$
order of the reaction with respect to $[SO_3^{2-}] = 1$
overall order of the reaction = 1 + 1 = 2

13.8 The rate law is second order with respect to Cl_2 and first order with respect to NO. Therefore the exponent for the Cl_2 is two and the exponent for NO is one:
Rate = $k[Cl_2]^2[NO]$

13.9 In each case, $k = \text{rate}/[A][B]^2$, and the units of k are $L^2\ mol^{-2}\ s^{-1}$.

Each calculation is performed as follows, using the second data set as the example:

$$k = \frac{0.40\ mol\ L^{-1}\ s^{-1}}{\left(0.20\ mol\ L^{-1}\right)\left(0.10\ mol\ L^{-1}\right)^2} = 2.0 \times 10^2\ L^2\ mol^{-2}\ s^{-1}$$

Each of the other data sets also gives the same value:
$k = 2.0 \times 10^2\ L^2\ mol^{-2}\ s^{-1}$.

13.10 The rate law is: rate = $k[A][B]^2$

(a) If the concentration of B is tripled, then the rate will increase nine–fold,
rate = $k[A][3B]^2$
rate = $9k[A][B]^2$

(b) If the concentration of A is tripled, then the rate will increase three–fold,
rate = $k[3A][B]^2$
rate = $3k[A][B]^2$

(c) If the concentration of A is tripled, and the concentration of B is halved, then the rate will decrease by three fourths

$$\text{rate} = k[3A][\tfrac{1}{2}B]^2$$

$$\text{rate} = \frac{3}{4}k[A][B]^2$$

13.11 $$\text{rate} = k[NO]^n[H_2]^m$$

(a) To find the rate law, take two reactions in which the concentration of one of the reactants is held constant, compare the two reactions and solve for the exponent:
For NO, use the first two reactions:

$$\frac{k[NO]_1^n[H_2]_1^m}{k[NO]_2^n[H_2]_2^m} = \frac{\text{rate}_1}{\text{rate}_2}$$

$$\frac{k\left[0.40\times10^{-4}\ mol\ L^{-1}\right]^n\left[0.30\times10^{-4}\ mol\ L^{-1}\right]^m}{k\left[0.80\times10^{-4}\ mol\ L^{-1}\right]^n\left[0.30\times10^{-4}\ mol\ L^{-1}\right]^m} = \frac{1.0\times10^{-8}\ mol\ L^{-1}\ s^{-1}}{4.0\times10^{-8}\ mol\ L^{-1}\ s^{-1}}$$

$$\frac{\left[0.40\times10^{-4}\ mol\ L^{-1}\right]^n}{\left[0.80\times10^{-4}\ mol\ L^{-1}\right]^n} = \frac{1.0\times10^{-8}\ mol\ L^{-1}\ s^{-1}}{4.0\times10^{-8}\ mol\ L^{-1}\ s^{-1}}$$

$$\left(\frac{1}{2}\right)^n = \frac{1}{4} \qquad n = 2$$

For H_2, use the second two reactions:

$$\frac{k[NO]_2^{\,n}[H_2]_2^{\,m}}{k[NO]_3^{\,n}[H_2]_3^{\,m}} = \frac{rate_2}{rate_3}$$

$$\frac{k\left[0.80\times10^{-4}\ \text{mol L}^{-1}\right]^n\left[0.30\times10^{-4}\ \text{mol L}^{-1}\right]^m}{k\left[0.80\times10^{-4}\ \text{mol L}^{-1}\right]^n\left[0.60\times10^{-4}\ \text{mol L}^{-1}\right]^m} = \frac{4.0\times10^{-8}\ \text{mol L}^{-1}\ \text{s}^{-1}}{8.0\times10^{-8}\ \text{mol L}^{-1}\ \text{s}^{-1}}$$

$$\frac{\left[0.30\times10^{-4}\ \text{mol L}^{-1}\right]^m}{\left[0.60\times10^{-4}\ \text{mol L}^{-1}\right]^m} = \frac{4.0\times10^{-8}\ \text{mol L}^{-1}\ \text{s}^{-1}}{8.0\times10^{-8}\ \text{mol L}^{-1}\ \text{s}^{-1}}$$

$$\left(\frac{1}{2}\right)^m = \frac{1}{2} \qquad m = 1$$

$$\text{rate} = k[NO]^2[H_2]^1$$

(b) To find the value for the rate constant, choose one of the reactions and use the values for the concentrations and rate and solve for the rate constant:

$$\text{rate} = k[NO]^2[H_2]^1$$

$$1.0\times10^{-8}\ \text{mol L}^{-1}\ \text{s}^{-1} = k\left[0.40\times10^{-4}\ \text{mol L}^{-1}\right]^2\left[0.30\times10^{-4}\ \text{mol L}^{-1}\right]^1$$

$$k = 2.1\times10^5\ \text{L}^2\ \text{mol}^{-2}\ \text{s}^{-1}$$

(c) The units for the rate constant can be determined from the rate law and cancelling the units:

$$\text{mol L}^{-1}\ \text{s}^{-1} = k\left[\text{mol L}^{-1}\right]^2\left[\text{mol L}^{-1}\right]^1$$

$$k = \frac{\left[\text{mol L}^{-1}\ \text{s}^{-1}\right]}{\left[\text{mol L}^{-1}\right]^2\left[\text{mol L}^{-1}\right]^1} = \text{L}^2\ \text{mol}^{-2}\ \text{s}^{-1}$$

13.12 (a) When the concentration of sucrose is doubled, the rate doubles.
When the concentration of sucrose is raised five times, the rate goes up five times.
The concentration and rate are directly proportional; therefore the reaction is first order with respect to sucrose.

(b) rate = k[sucrose]
$(6.17\times10^{-4}\ \text{mol L}^{-1}\ \text{s}^{-1}) = k(0.10\ \text{mol L}^{-1})$
$k = 6.17\times10^{-4}\ \text{s}^{-1}$
The rate constant is $6.17\times10^{-4}\ \text{s}^{-1}$

13.13 (a) The rate law will likely take the form rate = $k[A]^n[B]^{n'}$, where n and n' are the order of the reaction with respect to A and B, respectively. On comparing the first two lines of data, in which the concentration of B is held constant, we note that increasing the concentration of A by a factor of 2 (from 0.40 to 0.80) causes an increase in the rate by a factor of 4 (from 1.0×10^{-4} to 4.0×10^{-4}). Thus, we have a rate increase by 2^2, caused by a concentration increase by a factor of 2. This corresponds to the case in Table 13.3 for which n = 2, and we conclude that the reaction is second–order with respect to A.

On comparing the second and third lines of data, neither the concentration of A nor the concentration of B are held constant, but we know that the reaction is second–order with respect to A and we can solve the two equations for the order with respect to B:

$$\frac{rate_2}{rate_3} = \frac{k[A]_2^{\,2}[B]_2^{\,m}}{k[A]_3^{\,2}[B]_3^{\,m}}$$

$$\frac{2.25\times10^{-4}\ \text{mol L}^{-1}\ \text{s}^{-1}}{1.60\times10^{-3}\ \text{mol L}^{-1}\ \text{s}^{-1}}=\frac{k(0.60\ \text{M})^2(0.30\ \text{M})^m}{k(0.80\ \text{M})^2(0.60\ \text{M})^m}$$

$$\frac{2.25\times10^{-4}}{1.60\times10^{-3}}=\frac{(0.36)(0.30\ \text{M})^m}{(0.64)(0.60\ \text{M})^m}$$

$$0.141=0.563\left(\frac{1}{2}\right)^m$$

$$0.25=\left(\frac{1}{2}\right)^m$$

$$\frac{1}{4}=\left(\frac{1}{2}\right)^m$$

$m = 2$
Rate = $k[A]^2[B]^2$

(b) For the rate constant, use one of the experiments and insert the values and solve for k
Rate = $k[A]^2[B]^2$
1.00×10^{-4} mol L^{-1} s^{-1} = $k[0.40 \text{ mol L}^{-1}]^2[0.30 \text{ mol L}^{-1}]^2$
$k = 6.9 \times 10^{-3}$ L^3 mol^{-3} s^{-1}

(c) The units for the rate constant are

$$\frac{\text{mol L}^{-1}\ \text{s}^{-1}}{(\text{mol L}^{-1})^2(\text{mol L}^{-1})^2}=\text{L}^3\ \text{mol}^{-3}\ \text{s}^{-1}$$

(d) The overall order for this reaction is 2 + 2 = 4

13.14 Since this is a first order reaction, then we can use the integrated rate law for a first order reaction:

$$\ln\frac{[A]_0}{[A]_t}=kt$$

If only 5% of the active ingredient can decompose in two years, then 95% must remain, therefore, $[A]_0$ = 100, $[A]_t$ = 95, and t = 2 yr

$$\ln\frac{[A]_0}{[A]_t}=kt$$

$$\ln\frac{100}{95}=k(2\ \text{yr})$$

$k = 2.56 \times 10^{-2}$ yr^{-1}

13.15 (a) In order to find the concentration at specific time for a first order reaction, we substitute into equation 13.3, first converting the time to seconds:
t = 2 hr × 3600 s/hr = 7200 s

$$\ln\frac{[A]_0}{[A]_t}=kt$$

$$\text{antiln}\left[\ln\frac{[A]_0}{[A]_t}\right]=\text{antiln}\,[kt]$$

$$\frac{[A]_0}{[A]_t}=\text{antiln}\,[kt]$$

$$\frac{[A]_0}{[A]_t} = \text{antiln}\left[\left(6.2 \times 10^{-5}\ s^{-1}\right)\left(7200\ s\right)\right]$$

$$\frac{0.40\ M}{[A]_t} = 1.56$$

$$[A]_t = \frac{0.40\ M}{1.56}$$

$$= 0.26\ M$$

(b) Again, we use equation 13.3, this time solving for time:

$$\ln\frac{[A]_0}{[A]_t} = kt$$

$$t = \frac{1}{k} \times \ln\frac{[A]_0}{[A]_t} = \frac{1}{6.2 \times 10^{-5}\ s^{-1}} \times \ln\frac{0.40\ M}{0.30\ M} = 4600\ s$$

4.6×10^3 s × 1 min/60 s = 77 min

13.16 For a first–order reaction:

$$t_{1/2} = \frac{0.693}{k} = \frac{0.693}{6.17 \times 10^{-4}\ s^{-1}} = 1.12 \times 10^3\ s$$

$$t_{1/2} = 1.12 \times 10^3\ s \times \frac{1\ min}{60\ s}$$

$$= 18.7\ min$$

If we refer to the chart given in the text in example 13.7, we see that two half lives will have passed if there is to be only one quarter of the original amount of material remaining. This corresponds to:

18.7 min per half–life × 2 half lives = 37.4 min

13.17 From problem 13.14, the rate constant is: $k = 2.56 \times 10^{-2}\ yr^{-1}$, and use the half–life of a first order reaction equation:

$$t_{½} = \frac{\ln 2}{k} = \frac{\ln 2}{2.56 \times 10^{-2}\ yr^{-1}} = 27.1\ yr$$

13.18 Recall that for a first order process

$$k = \frac{0.693}{t_{1/2}}$$

So $k = \dfrac{0.693}{5730\ yr} = 1.21 \times 10^{-4}$/yr. Also,

$$\ln\frac{[A]_0}{[A]_t} = kt$$

$$t = \frac{1}{k}\ln\frac{[A]_0}{[A]_t} = \frac{1}{1.21 \times 10^{-4}\ /yr}\ln\frac{8}{1} = 1.72 \times 10^4\ yrs$$

13.19 Use the value of the rate constant for C–14 from the previous problem: $1.21 \times 10^{-4}\ y^{-1}$. To find the upper and lower limits of dates before present, set the concentration of the C–14 to 5% and 95%, respectively, and then solve for the time:

For samples with less than 5% of the C–14 remaining:

$$\ln\frac{[A]_0}{[A]_t} = kt$$

$$t = \frac{1}{k}\ln\frac{[A]_0}{[A]_t} = \frac{1}{1.21 \times 10^{-4}\ yr^{-1}}\ln\frac{100}{5} = 2.48 \times 10^4\ yr$$

Therefore the upper limit of dates is 24,800 years before present.

For samples with more than 95% of the C–14 remaining:

$$\ln\frac{[A]_0}{[A]_t} = kt$$

$$t = \frac{1}{k}\ln\frac{[A]_0}{[A]_t} = \frac{1}{1.21 \times 10^{-4}\ yr^{-1}}\ln\frac{100}{95} = 4.24 \times 10^2\ yr$$

Therefore the lower limit of dates is 424 years before present.

13.20 This is a second–order reaction, and we use equation 13.8:

$$\frac{1}{[NOCl]_t} - \frac{1}{[NOCl]_0} = kt$$

$$\frac{1}{[0.010\ M]} - \frac{1}{[0.040\ M]} = (0.020\ L\ mol^{-1}\ s^{-1}) \times t$$

$t = 3.8 \times 10^3\ s$

$t = 3.8 \times 10^3\ s \times 1\ min/60\ s = 63\ min$

13.21 This is the same reaction as in the previous problem, so it is a second–order reaction, and we use equation 13.8 and $k = 0.020\ L\ mol^{-1}\ s^{-1}$:

$$\frac{1}{[NOCl]_t} - \frac{1}{[NOCl]_0} = kt$$

$$\frac{1}{[0.00035\ M]} - \frac{1}{[x\ M]} = (0.020\ L\ mol^{-1}\ s^{-1}) \times t$$

We need to find the time in seconds:

From 10:35 to 3:15, 4 hours and 40 minutes has elapsed, or 280 minutes

$$s = (280\ min)\left(\frac{60\ s}{1\ min}\right) = 1.68 \times 10^4\ s$$

$$\frac{1}{[0.00035\ M]} - \frac{1}{[x\ M]} = (0.020\ L\ mol^{-1}\ s^{-1}) \times (1.68\times10^4\ s)$$

$$-\frac{1}{[x\ M]} = (3.36\times10^2\ L\ mol^{-1}) - (2.86\times10^3\ L\ mol^{-1})$$

$4.0 \times 10^{-4}\ M$

13.22 rate = $k[NO_2]^2$
To find the rate constant solve the rate law with the given data:
4.42×10^{-7} mol $L^{-1}s^{-1} = k(6.54 \times 10^{-4}$ mol $L^{-1})^2$

$$k = \frac{4.42\times10^{-7}\ \text{mol L}^{-1}\ \text{s}^{-1}}{\left(6.54\times10^{-4}\ \text{mol L}^{-1}\right)^2} = 1.03\ \text{L mol}^{-1}\ \text{s}^{-1}$$

The half life of the system is found using the half–life equation for a second–order reaction

$$t_{1/2} = \frac{1}{k\times(\text{initial concentration of reactant})}$$

$$t_{1/2} = \frac{1}{\left(1.03\ \text{L mol}^{-1}\ \text{s}^{-1}\right)\left(6.54\times10^{-4}\ \text{mol L}^{-1}\right)} = 1.48 \times 10^3\ \text{s}$$

13.23 The reaction is first–order. A second–order reaction should have a half–life that depends on the initial concentration according to equation 13.9.

13.24 Use the equation 13.12:

$$\ln\frac{k_2}{k_1} = \frac{-E_a}{R}\left[\frac{1}{T_2} - \frac{1}{T_1}\right]$$

For the 5% decomposition over 2 years, or 104.4 weeks

$$\frac{1}{[B]_t} - \frac{1}{[B]_0} = k_1 t$$

$$\frac{1}{[95\ M]} - \frac{1}{[100\ M]} = k_1(104.4\ \text{weeks})$$

$k_1 = 5.04 \times 10^{-6}\ M^{-1}$ weeks^{-1}
For the 5% decomposition over 1 week

$$\frac{1}{[B]_t} - \frac{1}{[B]_0} = k_2 t$$

$$\frac{1}{[95\ M]} - \frac{1}{[100\ M]} = k_2(1\ \text{week})$$

$k_2 = 5.26 \times 10^{-4}\ M^{-1}$ weeks^{-1}
Now use the values for k_1, k_2, $T_1 = 298$ K, and $E_a = 154 \times 10^3$ J mol^{-1}

$$\ln\frac{k_2}{k_1} = \frac{-E_a}{R}\left[\frac{1}{T_2} - \frac{1}{T_1}\right]$$

$$\ln\left[\frac{5.26\times10^{-4}\ \text{M}^{-1}\ \text{weeks}^{-1}}{5.04\times10^{-6}\ \text{M}^{-1}\ \text{weeks}^{-1}}\right] = \frac{-154\times10^3\ \text{J mol}^{-1}}{8.314\ \text{J mol}^{-1}\ \text{K}^{-1}}\left[\frac{1}{T_2} - \frac{1}{298\ \text{K}}\right]$$

$$\ln(1.04 \times 10^2) = (-1.85 \times 10^4\ \text{K})\left[\frac{1}{T_2} - \left(3.36\times10^{-3}\ \text{K}^{-1}\right)\right]$$

$$-2.51 \times 10^{-4}\ \text{K}^{-1} = \left[\frac{1}{T_2} - \left(3.36\times10^{-3}\ \text{K}^{-1}\right)\right]$$

$$\frac{1}{T_2} = 3.11 \times 10^{-3}\ \text{K}^{-1}$$

$T_2 = 322$ K or 49 °C

13.25 (a) Use equation 13.12:

$$\ln \frac{k_2}{k_1} = \frac{-E_a}{R}\left[\frac{1}{T_2} - \frac{1}{T_1}\right]$$

$$\ln\left[\frac{23 \text{ L mol}^{-1} \text{ s}^{-1}}{3.2 \text{ L mol}^{-1} \text{ s}^{-1}}\right] = \frac{-E_a}{8.314 \text{ J mol}^{-1} \text{ K}^{-1}}\left[\frac{1}{673 \text{ K}} - \frac{1}{623 \text{ K}}\right]$$

Solving for E_a gives 1.4×10^5 J/mol = 1.4×10^2 kJ/mol

(b) We again use equation 13.12, substituting the values:

$k_1 = 3.2 \text{ L mol}^{-1} \text{ s}^{-1}$ at $T_1 = 623$ K
$k_2 = ?$ at $T_2 = 573$ K

$$\ln \frac{k_2}{k_1} = \frac{-E_a}{R}\left[\frac{1}{T_2} - \frac{1}{T_1}\right]$$

$$\ln\left[\frac{k_2}{3.2 \text{ L mol}^{-1} \text{ s}^{-1}}\right] = \frac{-1.4 \times 10^5 \text{ J mol}^{-1}}{8.314 \text{ J mol}^{-1} \text{ K}^{-1}}\left[\frac{1}{573 \text{ K}} - \frac{1}{623 \text{ K}}\right]$$

Solving for k_2 gives $0.30 \text{ L mol}^{-1} \text{ s}^{-1}$.

13.26 (a), (b), and (e) may be elementary processes.
Equations (c), (d), and (f) are not elementary processes because they have more than two molecules colliding at one time, and this is very unlikely.

13.27 If the reaction occurs in a single step, one molecule of each reactant must be involved, according to the balanced equation. Therefore, the rate law is expected to be: Rate = $k[NO][O_3]$.

13.28 The slow step (second step) of the mechanism determines the rate law:

$$\text{Rate} = k[NO_2Cl]^1[Cl]^1$$

However, Cl is an intermediate and cannot be part of the rate law expression. We need to solve for the concentration of Cl by using the first step of the mechanism. Assuming that the first step is an equilibrium step, the rates of the forward and reverse reactions are equal:

$$\text{Rate} = k_{forward}[NO_2Cl] = k_{reverse}[Cl][NO_2]$$

Solving for [Cl] we get

$$[Cl] = \frac{k_f}{k_r}\frac{[NO_2Cl]}{[NO_2]}$$

Substituting into the rate law expression for the second step yields:

$$\text{Rate} = \frac{k[NO_2Cl]^2}{[NO_2]}$$, where all the constants have been combined into one new constant.

Review Problems

13.47 Since they are in a 1-to-1 mol ratio, the rate of formation of SO_2 is *equal* and *opposite* to the rate of consumption of SO_2Cl_2. This is equal to the slope of the curve at any point on the graph (see below). At 200 min, we obtain a value of about 1×10^{-4} *M*/s. At 600 minutes, this has decreased to about 7×10^{-5} *M*/s.

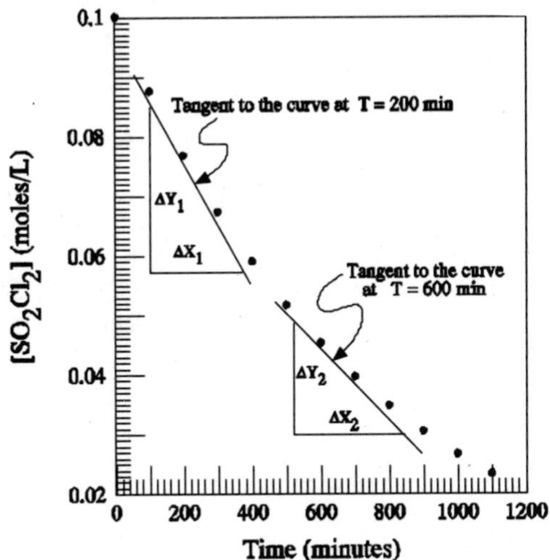

13.49 This is determined by the coefficients of the balanced chemical equation. For every mole of N_2 that reacts, 3 mol of H_2 will react. Thus the rate of disappearance of hydrogen is three times the rate of disappearance of nitrogen. Similarly, the rate of disappearance of N_2 is half the rate of appearance of NH_3, or NH_3 appears twice as fast as N_2 disappears.

13.51 (a) rate for $O_2 = -1.20$ mol L^{-1} $s^{-1} \times 19/2 = -11.4$ mol L^{-1} s^{-1}
By convention, this is reported as a positive number: 11.4 mol L^{-1} s^{-1}
(b) rate for $CO_2 = +1.20$ mol L^{-1} $s^{-1} \times 12/2 = 7.20$ mol L^{-1} s^{-1}
(c) rate for $H_2O = +1.20$ mol L^{-1} $s^{-1} \times 14/2 = 8.40$ mol L^{-1} s^{-1}

13.53 The rate can be found by simply inserting the given concentration values:

rate = $(5.0 \times 10^5\ L^5\ mol^{-5}\ s^{-1})[H_2SeO_3][I^-]^3[H^+]^2$
rate = $(5.0 \times 10^5\ L^5\ mol^{-5}\ s^{-1})(2.0 \times 10^{-2}\ mol/L)(2.0 \times 10^{-3}\ mol/L)^3(1.0 \times 10^{-3}\ mol/L)^2$
rate = 8.0×10^{-11} mol L^{-1} s^{-1}

13.55 rate = $(7.1 \times 10^9\ L^2\ mol^{-2}\ s^{-1})(1.0 \times 10^{-3}\ mol/L)^2(3.4 \times 10^{-2}\ mol/L)$
rate = 2.4×10^2 mol L^{-1} s^{-1}

13.57 On comparing the data of the first and second experiments, we find that, the concentration of N is unchanged, and the concentration of M has been doubled, causing a doubling of the rate. This corresponds to the fourth case in Table 13.3, and we conclude that the order of the reaction with respect to M is 1. In the second and third experiments, we have a different result. When the concentration of M is held constant, the concentration of N is tripled, causing an increase in the rate by a factor of nine. This constitutes the eighth case in Table 13.3, and we conclude that the order of the reaction with respect to N is 2. This means that the overall rate expression is: rate = $k[M][N]^2$ and we can solve for the value of k by substituting the appropriate data:
5.0×10^{-3} mol L^{-1} $s^{-1} = k \times [0.020\ mol/L][0.010\ mol/L]^2$
$k = 2.5 \times 10^3\ L^2\ mol^{-2}\ s^{-1}$

13.59 The reaction is first–order in OCl^-, because an increase in concentration by a factor of two, while holding the concentration of I^- constant (compare the first and second experiments of the table), has caused an increase in rate by a factor of $2^1 = 2$. The order of reaction with respect to I^- is also 1, as is demonstrated by a comparison of the first and third experiments.

rate = k[OCl^-][I^-]
Using the last data set:
3.5×10^4 mol L^{-1} s^{-1} = k[1.7×10^{-3} mol/L][3.4×10^{-3} mol/L]
k = 6.1×10^9 L mol^{-1} s^{-1}

13.61 Compare the first and second experiments. On doubling the ICl concentration, the rate is found to increase by a factor of $2 = 2^1$, and the order of the reaction with respect to ICl is 1 (case number four in Table 13.3). In the first and third experiments, the concentration of ICl is constant, whereas the concentration of H_2 in the first experiment is twice that in the third. This causes a change in the rate by a factor of 2 also, and the rate law is found to be: rate = k[ICl][H_2]. Using the data of the first experiment:
1.5×10^{-3} mol L^{-1} s^{-1} = k[0.10 mol L^{-1}][0.10 mol L^{-1}]
k = 1.5×10^{-1} L mol^{-1} s^{-1}

13.63 A graph of ln $[SO_2Cl_2]_t$ versus t will yield a straight line if the data obeys a first–order rate law.

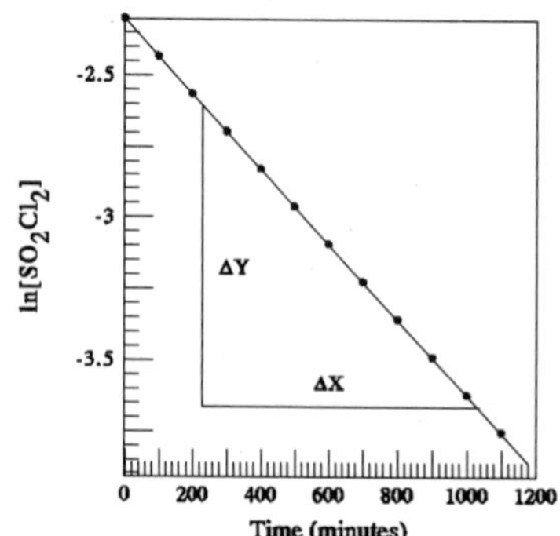

These data do yield a straight line when ln $[SO_2Cl_2]_t$ is plotted against the time, t. The slope of this line equals –k. Plotting the data provided and using linear regression to fit the data to a straight line yields a value of 1.32×10^{-3} min^{-1} for k.

13.65 (a) The time involved must be converted to a value in seconds:
1 hr × 3600 s/hr = 3.6×10^3 s, and then we make use of equation 13.3, where x is taken to represent the desired SO_2Cl_2 concentration:

$$\ln \frac{0.0040\ \text{M}}{\text{x}} = (2.2 \times 10^{-5}\ \text{s}^{-1})(3.6 \times 10^3\ \text{s})$$

x = 3.7×10^{-3} *M*

(b) The time is converted to a value having the units seconds
24 hr × 3600 s/hr = 8.64×10^4 s, and then we use equation 13.3, where x is taken to represent the desired SO_2Cl_2 concentration:

$$\ln \frac{0.0040\ \text{M}}{\text{x}} = (2.2 \times 10^{-5}\ \text{s}^{-1})(8.64 \times 10^4\ \text{s})$$

x = 6.0×10^{-4} *M*

13.67 Any consistent set of units for expressing concentration may be used in equation 13.3, where we let A represent the drug that is involved:

$$\ln \frac{[A]_0}{[A]_t} = kt$$

$$\ln \frac{25.0\ \text{mg/kg}}{15.0\ \text{mg/kg}} = k(120\ \text{min})$$

Solving for k we get $4.26 \times 10^{-3}\ \text{min}^{-1}$

13.69 We use the equation:

$$\frac{1}{[HI]_t} - \frac{1}{[HI]_0} = kt$$

$$\frac{1}{[8.0 \times 10^{-4}\ M]} - \frac{1}{[3.4 \times 10^{-2}\ M]} = (1.6 \times 10^{-3}\ \text{L mol}^{-1}\ \text{s}^{-1}) \times t$$

Solving for t gives:
$t = 7.6 \times 10^5$ s or $t = (7.6 \times 10^5\ \text{s}) \times 1\ \text{min}/60\ \text{s} = 1.3 \times 10^4$ min

13.71 $$\text{half lives} = (2.0\ \text{hrs})\left(\frac{60\ \text{min}}{1\ \text{hr}}\right)\left(\frac{1\ \text{half life}}{15\ \text{min}}\right) = 8.0\ \text{half lives}$$

Eight half lives correspond to the following fraction of original material remaining:

Number of half lives	Fraction remaining
1	1/2
2	1/4
3	1/8
4	1/16
5	1/32
6	1/64
7	1/128
8	1/256

13.73 It requires approximately 500 min (as determined from the graph) for the concentration of SO_2Cl_2 to decrease from 0.100 *M* to 0.050 *M*, i.e., to decrease to half its initial concentration. Likewise, in another 500 minutes, the concentration decreases by half again, i.e. from 0.050 *M* to 0.025 *M*. This means that the half–life of the reaction is independent of the initial concentration, and we conclude that the reaction is first–order in SO_2Cl_2.

13.75 In order to solve this problem, it must be assumed that all of the argon–40 that is found in the rock must have come from the potassium–40, i.e., that the rock contains no other source of argon–40. If the above assumption is valid, then any argon–40 that is found in the rock represents an equivalent amount of potassium–40, since the stoichiometry is 1:1. Since equal amounts of potassium–40 and argon–40 have been found, this indicates that the amount of potassium–40 that remains is exactly half the amount that was present originally. In other words, the potassium–40 has undergone one half–life of decay by the time of the analysis. The rock is thus seen to be 1.3×10^9 years old.

13.77 Using equation 13.7 we may determine how long it has been since the tree died.

$$\ln\left(\frac{^{14}C}{^{12}C}\right) = \left(1.2 \times 10^{-4}\right)t$$

Taking the natural log we determine:

$$\ln\left(\frac{1.2 \times 10^{-12}}{4.8 \times 10^{-14}}\right) = \left(1.2 \times 10^{-4}\right)t$$

$$t = \left(\frac{1}{1.2 \times 10^{-4}}\right)\ln\left(\frac{1.2 \times 10^{-12}}{4.8 \times 10^{-14}}\right) = 2.7 \times 10^{4}\ \text{yr}$$

The tree died 2.7×10^4 years ago. This is when the volcanic eruption occurred.

13.79 The graph is prepared exactly as in example 13.11 of the text. The slope is found using linear regression, to be: -9.5×10^3 K. Thus $-9.5 \times 10^3\ \text{K} = -E_a/R$

$$E_a = -(-9.5 \times 10^3\ \text{K})(8.314\ \text{J K}^{-1}\ \text{mol}^{-1}) = 7.9 \times 10^4\ \text{J/mol} = 79\ \text{kJ/mol}$$

Using the equation, we proceed as follows:

$$\ln\frac{k_2}{k_1} = \frac{-E_a}{R}\left[\frac{1}{T_2} - \frac{1}{T_1}\right]$$

$$\ln\left[\frac{1.94 \times 10^{-3}\ \text{L mol}^{-1}\ \text{s}^{-1}}{2.88 \times 10^{-4}\ \text{L mol}^{-1}\ \text{s}^{-1}}\right] = \frac{-E_a}{8.314\ \text{J mol}^{-1}\ \text{K}^{-1}}\left[\frac{1}{673\ \text{K}} - \frac{1}{593\ \text{K}}\right]$$

$$1.907 = \frac{2.00 \times 10^{-4}\ \text{K}^{-1}}{8.314\ \text{J mol}^{-1}\ \text{K}^{-1}} \times E_a$$

$E_a = 7.93 \times 10^4$ J/mol = 79.3 kJ/mol

13.81 Using the equation we have:

$$\ln\frac{k_2}{k_1} = \frac{-E_a}{R}\left[\frac{1}{T_2} - \frac{1}{T_1}\right]$$

$$\ln\left[\frac{1.0 \times 10^{-3}\ \text{L mol}^{-1}\ \text{s}^{-1}}{9.3 \times 10^{-5}\ \text{L mol}^{-1}\ \text{s}^{-1}}\right] = \frac{-E_a}{8.314\ \text{J mol}^{-1}\ \text{K}^{-1}}\left[\frac{1}{403\ \text{K}} - \frac{1}{373\ \text{K}}\right]$$

$$2.37 = \frac{2.00 \times 10^{-4}\ \text{K}^{-1}}{8.314\ \text{J mol}^{-1}\ \text{K}^{-1}} \times E_a$$

$E_a = 9.89 \times 10^4$ J/mol = 99 kJ/mol

Equation states $k = A\exp\left(\frac{-E_a}{RT}\right)$

$$A = \frac{k}{\exp\left(\frac{-E_a}{RT}\right)}$$

$$= \frac{9.3 \times 10^{-5}\ \text{L mol}^{-1}\ \text{s}^{-1}}{\exp\left(\frac{-9.89 \times 10^{4}\ \text{J/mol}}{(8.314\ \text{J/mol K})(373\ \text{K})}\right)}$$

$$= 6.6 \times 10^{9}\ \text{L mol}^{-1}\ \text{s}^{-1}$$

13.83 Use equation 13.10:

(a)

$$k = A \exp\left(\frac{-E_a}{RT}\right)$$

$$= \left(4.3 \times 10^{13}\ s^{-1}\right)\exp\left(\frac{-103 \times 10^3\ J\ mol^{-1}}{\left(8.314\ {}^{J}\!/\!_{mol\ K}\right)\left(293\ K\right)}\right)$$

$$= 1.9 \times 10^{-5}\ s^{-1}$$

(b)

$$k = A \exp\left(\frac{-E_a}{RT}\right)$$

$$= \left(4.3 \times 10^{13}\ s^{-1}\right)\exp\left(\frac{-103 \times 10^3\ J\ mol^{-1}}{\left(8.314\ {}^{J}\!/\!_{mol\ K}\right)\left(373\ K\right)}\right)$$

$$= 1.6 \times 10^{-1}\ s^{-1}$$

Practice Exercises

14.1 $2N_2O_3 + O_2 \rightleftarrows 4NO_2$

14.2 (a) $\frac{[H_2O]^2}{[H_2]^2[O_2]} = K_c$ (b) $\frac{[CO_2][H_2O]^2}{[CH_4][O_2]^2} = K_c$

14.3 Since the starting equation has been reversed and divided by two, we must invert the equilibrium constant, and then take the square root: $K_c = 1.2 \times 10^{-13}$

14.4 If we divide both equations by 2 and reverse the second we get:

$CO(g) + 1/2O_2(g) \rightarrow CO_2(g)$ $K_c = 5.7 \times 10^{45}$
$H_2O(g) \rightarrow H_2(g) + 1/2O_2(g)$ $K_c = 3.3 \times 10^{-41}$

Note that when we divide the equation by two, we need to take the square root of the rate constant. When we reverse the reaction, we need to take the inverse.

Adding these equations we get the desired equation so we need simply multiply the values for K_c in order to obtain the new value: $K_c = 1.9 \times 10^5$

14.5 $K_P = \frac{\left(P_{N_2O}\right)^2}{\left(P_{N_2}\right)^2\left(P_{O_2}\right)}$

14.6 $K_P = \frac{\left(P_{HI}\right)^2}{\left(P_{H_2}\right)\left(P_{I_2}\right)}$

14.7 Use the equation $K_p = K_c(RT)^{\Delta n_g}$. In this reaction, $\Delta n_g = 3 - 2 = 1$, so

$$K_p = K_c(RT)^{\Delta n_g} = \left(7.3 \times 10^{34}\right)\left(\left(0.0821\ \tfrac{\text{L atm}}{\text{mol K}}\right)(298\ \text{K})\right)^1 = 1.8 \times 10^{36}$$

14.8 We would expect K_P to be smaller than K_c since Δn_g is negative.
Use the equation:

$$K_p = K_c(RT)^{\Delta n_g}$$
$$K_c = \frac{K_p}{(RT)^{\Delta n_g}}$$

In this case, $\Delta n_g = (1 - 3) = -2$, and we have:

$$K_c = \frac{K_p}{(RT)^{\Delta n_g}} = \frac{3.8 \times 10^{-2}}{\left(\left(0.0821\ \tfrac{\text{L atm}}{\text{mol K}}\right)(473\ \text{K})\right)^{-2}} = 57$$

14.9 $K_c = \frac{1}{[NH_3(g)][HCl(g)]}$

14.10 (a) $K_c = \dfrac{1}{[Cl_2\ (g)]}$

(b) $K_c = [Na^+(aq)][OH^-(aq)][H_2(g)]$

(c) $K_c = [Ag^+]^2[CrO_4^{2-}]$

(d) $K_c = \dfrac{[Ca^{+2}(aq)][HCO_3^-(aq)]^2}{[CO_2(aq)]}$

14.11 To solve this problem, the first step is to find Q, the mass action expression, and then compare it to the value for K_c. If Q is larger than K_c, then there are more products than the equilibrium concentration and the reaction will move to reactants. If Q is smaller than K_c, then there are more reactants than the equilibrium concentrations and the reaction will move to products.
For this reaction, the concentrations of H_2, Br_2 and HBr are all equal. If we set them to x, then we see that they cancel to give Q = 1:

$$Q = \frac{[HBr]^2}{[H_2][Br_2]} = \frac{[x]^2}{[x][x]} = 1$$

Q is larger than K_c, therefore the reaction will move to reactants.

14.12 Reaction (b) will proceed farthest to completion since it has the largest value for K_c.

14.13 Only the gases will be affected by the volume of the container. But the stoichiometric ratio of the reactants to products is 5:5, so as the volume is changes, the number of moles of gas will not change, therefore the reaction will not change, and there will be no change in the amount of H_3PO_4.

14.14 (a) The equilibrium will shift to the right, decreasing the concentration of Cl_2 at equilibrium, and consuming some of the added PCl_3. The value of K_p will be unchanged.

(b) The equilibrium will shift to the left, consuming some of the added PCl_5 and increasing the amount of Cl_2 at equilibrium. The value of K_p will be unchanged.

(c) For any exothermic equilibrium, an increase in temperature causes the equilibrium to shift to the left, in order to remove energy in response to the stress. This equilibrium is shifted to the left, making more Cl_2 and more PCl_3 at the new equilibrium. The value of K_p is given by the following:

$$K_p = \frac{P_{PCl_5}}{P_{PCl_3} \times P_{Cl_2}}$$

In this system, an increase in temperature (which causes an increase in the equilibrium concentrations of both PCl_3 and Cl_2 and a decrease in the equilibrium concentration of PCl_5) causes an increase in the denominator of the above expression as well as a decrease in the numerator of the above expression. Both of these changes serve to decrease the value of K_p.

(d) Decreasing the container volume for a gaseous system will produce an increase in partial pressures for all gaseous reactants and products. In order to lower the increase in partial pressures, the equilibrium will shift so as to favor the product side having the smaller number of gaseous molecules, in this case to the right. This shift will decrease the amount of Cl_2 and PCl_3 at equilibrium, and it will increase the amount of PCl_5 at equilibrium. While the change in volume will change the position of the equilibrium, it does not change the value for K_p.

14.15 $2CO(g) + O_2(g) \rightleftarrows 2CO_2(g)$

Using the stoichiometry of the reaction we can see that for every mol of O_2 that is used, twice as much CO will react and twice as much CO_2 will be produced. Consequently, if the $[O_2]$ decreases by 0.030 mol/L, the [CO] decreases by 0.060 mol/L and $[CO_2]$ increases by 0.060 mol/L.

14.16 $K_c = \frac{[CO_2][H_2]}{[CO][H_2O]} = \frac{(0.150)(0.200)}{(0.180)(0.0411)} = 4.06$

14.17 (a) The initial concentrations were:
$[PCl_3]$ = 0.200 mol/1.00 L = 0.200 *M*
$[Cl_2]$ = 0.100 mol/1.00 L = 0.100 *M*
$[PCl_5]$ = 0.00 mol/1.00 L = 0.000 *M*

(b) The change in concentration of PCl_3 was (0.200 – 0.120) *M* = 0.080 mol/L. The other materials must have undergone changes in concentration that are dictated by the coefficients of the balanced chemical equation, namely: $PCl_3 + Cl_2 \rightarrow PCl_5$ or both PCl_3 and Cl_2 have decreased by 0.080 *M* and PCl_5 has increased by 0.080 *M*.

(c) As stated in the problem, the equilibrium concentration of PCl_3 is 0.120 *M*. The equilibrium concentration of PCl_5 is 0.080 *M* since initially there was no PCl_5. The equilibrium concentration of Cl_2 equals the initial concentration minus the amount that reacted, 0.100 *M* – 0.080 *M* = 0.020 *M*.

(d) $K_c = \frac{[PCl_5]}{[PCl_3][Cl_2]} = \frac{(0.080)}{(0.120)(0.020)} = 33$

14.18 $K_c = \frac{[NO_2]^2}{[N_2O_4]}$

$$4.61 \times 10^{-3} = \frac{[NO_2]^2}{\left(\frac{0.0466}{2}\right)}$$

$[NO_2] = 1.04 \times 10^{-2}$ *M*

14.19 $K_c = \frac{[CH_3CO_2C_2H_5][H_2O]}{[CH_3CO_2H][C_2H_5OH]} = \frac{(0.910)(0.00850)}{(0.210)[C_2H_5OH]} = 4.10$

$[C_2H_5OH] = 8.98 \times 10^{-3}$ *M*

14.20 Initially we have $[H_2] = [I_2]$ = 0.200 *M*.

	$[H_2]$	$[I_2]$	[HI]
I	0.200	0.200	–
C	–x	–x	+2x
E	0.200–x	0.200–x	+2x

Substituting the above values for equilibrium concentrations into the mass action expression gives:

$$K_c = \frac{[HI]^2}{[H_2][I_2]} = \frac{(2x)^2}{(0.200-x)(0.200-x)} = 49.5$$

Take the square root of both sides of this equation to get; $\frac{2x}{(0.200-x)} = 7.04$. This equation is easily solved giving x = 0.156. The substances then have the following concentrations at equilibrium: $[H_2] = [I_2]$ = 0.200 – 0.156 = 0.044 *M*, [HI] = 2(0.156) = 0.312 *M*.

14.21 Initially we have $[H_2] = 0.200\ M$, $[I_2] = 0.100\ M$.

	$[H_2]$	$[I_2]$	$[HI]$
I	0.200	0.100	–
C	–x	–x	+2x
E	0.200–x	0.100–x	+2x

Substituting the above values for equilibrium concentrations into the mass action expression gives:

$$K_c = \frac{[HI]^2}{[H_2][I_2]} = \frac{(2x)^2}{(0.200-x)(0.100-x)} = 49.5$$

$4x^2 = 49.5(0.0200 - 0.300x + x^2)$

$45.5x^2 - 14.9x + 0.990 = 0$

Solve the quadratic equation:

$$x = \frac{-b \pm \sqrt{b^2-4ac}}{2a} = \frac{-(-14.9) \pm \sqrt{(14.9)^2 - 4(45.5)(0.990)}}{2(45.5)} = 0.0.0934$$

The substances then have the following concentrations at equilibrium:

$[H_2] = 0.200 - 0.0934 = 0.107\ M$

$[I_2] = 0.100 - 0.0934 = 0.0066\ M$

$[HI] = 2(0.0.0934) = 0.1868\ M$.

14.22 $2NH_3(g) \rightleftarrows N_2(g) + 3H_2(g)$

	$[NH_3]$	$[N_2]$	$[H_2]$
I	0.041	–	–
C	–2x	+x	+3x
E	0.041–x	x	+3x

Substitute the values for the equilibrium concentrations into the mass action expression:

$$K_c = \frac{[N_2][H_2]^3}{[NH_3]^2} = \frac{(x)(3x)^3}{(0.041-x)^2} = 2.3 \times 10^{-9}$$

We will assume that x is small compared to the concentration of NH_3, so the equation will simplify to:

$$K_c = \frac{27x^4}{(0.041)^2} = 1.4 \times 10^{-13}$$

$x^4 = 1.4 \times 10^{-13}$

$x = 6.2 \times 10^{-4}$

$[N_2] = 6.2 \times 10^{-4}$

$[H_2] = 1.9 \times 10^{-3}$

14.23 $N_2(g) + O_2(g) \rightleftarrows 2NO(g)$

	$[N_2]$	$[O_2]$	$[NO]$
I	0.033	0.00810	–
C	–x	–x	+2x
E	0.033–x	0.00810–x	+2x

Substituting the above values for equilibrium concentrations into the mass action expression gives:

$$K_c = \frac{[NO]^2}{[N_2][O_2]} = \frac{(2x)^2}{(0.033-x)(0.00810-x)} = 4.8 \times 10^{-31}$$

If we assume that x << 0.033 and x << 0.00810, we can simplify this equation. (Because the value of K_c is so low, this assumption should be valid.) The equation simplifies as:

$$K_c = \frac{(2x)^2}{(0.033)(0.00810)} = 4.8 \times 10^{-31}$$

This equation is easily solved to give $x = 5.7 \times 10^{-18}$ *M*. The equilibrium concentration of NO is 2x according to the ICE table so, $[NO] = 1.1 \times 10^{-17}$ *M*.

Review Problems

14.19 (a) $K_c = \frac{[POCl_3]^2}{[PCl_3]^2[O_2]}$

(b) $K_c = \frac{[SO_2]^2[O_2]}{[SO_3]^2}$

(c) $K_c = \frac{[NO]^2[H_2O]^2}{[N_2H_4][O_2]^2}$

(d) $K_c = \frac{[NO_2]^2[H_2O]^8}{[N_2H_4][H_2O_2]^6}$

(e) $K_c = \frac{[SO_2][HCl]^2}{[SOCl_2][H_2O]}$

14.21 (a) $K_p = \frac{(P_{POCl_3})^2}{(P_{PCl_3})^2(P_{O_2})}$

(b) $K_p = \frac{(P_{SO_2})^2(P_{O_2})}{(P_{SO_3})^2}$

(c) $K_p = \frac{(P_{NO})^2(P_{H_2O})^2}{(P_{N_2H_4})(P_{O_2})^2}$

(d) $K_p = \frac{(P_{NO_2})^2(P_{H_2O})^8}{(P_{N_2H_4})(P_{H_2O_2})^6}$

(e) $K_p = \frac{(P_{SO_2})(P_{HCl})^2}{(P_{SOCl_2})(P_{H_2O})}$

14.23 (a) $K_c = \frac{[Ag(NH_3)_2{}^+]}{[Ag^+][NH_3]^2}$ (b) $K_c = \frac{[Cd(SCN)_4{}^{2-}]}{[Cd^{2+}][SCN^-]^4}$

14.25 The first equation has been reversed in making the second equation. We therefore take the inverse of the value of the first equilibrium constant in order to determine a value for the second equilibrium constant:
$K = 1 \times 10^{85}$

14.27 (a) $K_c = \frac{[HCl]^2}{[H_2][Cl_2]}$ (b) $K_c = \frac{[HCl]}{[H_2]^{1/2}[Cl_2]^{1/2}}$

K_c for reaction (b) is the square root of K_c for reaction (a).

14.29 $M = P/RT$

$$M = \frac{(745\ \text{torr})\left(\dfrac{1\ \text{atm}}{760\ \text{torr}}\right)}{\left(0.0821\ \frac{\text{L atm}}{\text{mol K}}\right)(318\ \text{K})} = 0.0375\ M$$

14.31 b, since $\Delta n_g = 0$

14.33 $K_p = K_c \times (RT)^{\Delta ng}$
$6.3 \times 10^{-3} = K_c[(0.0821\ \text{L atm K}^{-1}\ \text{mol}^{-1})(498\ \text{K})]^{-2} = (5.98 \times 10^{-4}) \times K_c$
$K_c = 11$

14.35 $K_p = K_c \times (RT)^{\Delta ng}$
$K_p = 4.2 \times 10^{-4}[(0.0821\ \text{L atm K}^{-1}\ \text{mol}^{-1})(773\ \text{K})]^{1} = 2.7 \times 10^{-2}$

14.37 $K_p = K_c \times (RT)^{\Delta ng}$
$K_p = (0.40)[(0.0821\ \text{L atm K}^{-1}\ \text{mol}^{-1})(1046\ \text{K})]^{-2} = 5.4 \times 10^{-5}$

14.39 In each case we get approximately 55.5 *M*:

(a)

$$\text{mol H}_2\text{O} = (18.0\ \text{mL H}_2\text{O})\left(\frac{1\ \text{g}}{1\ \text{mL}}\right)\left(\frac{1\ \text{mol H}_2\text{O}}{18.02\ \text{g H}_2\text{O}}\right) = 0.999\ \text{mol H}_2\text{O}$$

$$M = \left(\frac{0.999\ \text{mol H}_2\text{O}}{18.0\ \text{mL H}_2\text{O}}\right)\left(\frac{1000\ \text{mL}}{1\ \text{L}}\right) = 55.5\ M$$

(b)

$$\text{mol H}_2\text{O} = (100.0\ \text{mL H}_2\text{O})\left(\frac{1\ \text{g}}{1\ \text{mL}}\right)\left(\frac{1\ \text{mol H}_2\text{O}}{18.02\ \text{g H}_2\text{O}}\right) = 5.549\ \text{mol H}_2\text{O}$$

$$M = \left(\frac{5.549\ \text{mol H}_2\text{O}}{100.0\ \text{mL H}_2\text{O}}\right)\left(\frac{1000\ \text{mL}}{1\ \text{L}}\right) = 55.49\ M$$

(c)

$$\text{mol H}_2\text{O} = (1.00\ \text{L H}_2\text{O})\left(\frac{1000\ \text{mL}}{1\ \text{L}}\right)\left(\frac{1\ \text{g}}{1\ \text{mL}}\right)\left(\frac{1\ \text{mol H}_2\text{O}}{18.02\ \text{g H}_2\text{O}}\right) = 55.5\ \text{mol H}_2\text{O}$$

$$M = \left(\frac{55.5\ \text{mol H}_2\text{O}}{1.00\ \text{L H}_2\text{O}}\right) = 55.5\ M$$

14.41 (a) $K_c = \dfrac{[CO]^2}{[O_2]}$

(b) $K_c = [H_2O][SO_2]$

(c) $K_c = \dfrac{[CH_4][CO_2]}{[H_2O]^2}$

(d) $K_c = \dfrac{[H_2O][CO_2]}{[HF]^2}$

(e) $K_c = [H_2O]^5$

14.43

	[HCl]	[HI]	[Cl_2]
I	0.100	–	–
C	–2x	+2x	+x
E	0.100–2x	+2x	+x

Note: Since the $I_2(s)$ has a constant concentration, it may be neglected.

$$K_c = \frac{[HI]^2[Cl_2]}{[HCl]^2} = 1.6 \times 10^{-34}$$

$$K_c = \frac{(2x)^2(x)}{(0.100-2x)^2} = 1.6 \times 10^{-34}$$

Because the value of K_c is so small, we make the simplifying assumption that $(0.100 - 2x) \approx 0.100$, and the above equation becomes:

$$K_c = \frac{[HI]^2[Cl_2]}{[HCl]^2} = 1.6 \times 10^{-34}$$

$$K_c = \frac{(2x)^2(x)}{(0.100)^2} = 1.6 \times 10^{-34}$$

$4x^3 = 1.6 \times 10^{-36}$; $\therefore$ $x = 7.37 \times 10^{-13}$, and the above assumption is seen to have been valid.

$[HI] = 2x = 1.47 \times 10^{-12}$ *M*
$[Cl_2] = x = 7.37 \times 10^{-13}$ *M*
$[HCl] = (0.100 - 2x) \approx 0.100$ *M*

14.45 (a) The system shifts to the right to consume some of the added methane.
(b) The system shifts to the left to consume some of the added hydrogen.
(c) The system shifts to the right to make some more carbon disulfide.
(d) The system shifts to the left to decrease the amount of gaseous moles.
(e) The system shifts to the right to absorb some of the added heat.

14.47 (a) right (b) left (c) left
(d) right (e) no effect (f) left

14.49 The mass action expression for this equilibrium is:

$$K_c = \frac{[PCl_5]}{[PCl_3][Cl_2]} = 0.18$$

and the value for the reaction quotient for this system is:

$$Q = \frac{(0.00500)}{(0.0420)(0.0240)} = 4.96$$

(a) No. This is not the value of the equilibrium constant, and we conclude that the system is not at equilibrium.

(b) Since the value of the reaction quotient for this system is larger than that of the equilibrium constant, the system must shift to the left to reach equilibrium.

14.51 $K_c = \frac{[CH_3OH]}{[CO][H_2]^2} = \frac{[CH_3OH]}{(0.180)(0.220)^2} = 0.500$

$[CH_3OH] = 4.36 \times 10^{-3}$ *M*.

14.53 $K_c = \frac{[CH_3OH]}{[CO][H_2]^2} = \frac{(0.00261)}{(0.105)(0.250)^2} = 0.398$

14.55

	[HBr]	[H$_2$]	[Br$_2$]
I	0.500	–	–
C	–2x	+x	+x
E	0.500–2x	+x	+x

The problem tell us that $[Br_2] = 0.0955$ *M* = x at equilibrium. Using the ICE table as a guide we see that the equilibrium concentrations are; $[H_2] = [Br_2] = 0.0955$ *M* and $[HBr] = 0.500–2(0.0955) = 0.309$ *M*.

$$K_c = \frac{[H_2][Br_2]}{[HBr]^2} = \frac{(0.0955)(0.0955)}{(0.309)^2} = 0.0955$$

14.57 According to the problem, the concentration of NO_2 increases in the course of this reaction. This means our ICE table will look like the following:

	[NO$_2$]	[NO]	[N$_2$O]	[O$_2$]
I	0.0560	0.294	0.184	0.377
C	+x	+x	–x	–x
E	0.0560 + x	0.294 + x	0.184 – x	0.377 – x

The problem tell us that $[NO_2] = 0.118$ *M* = 0.0560 + x at equilibrium. Solving we get; x = 0.062 *M*. Using the ICE table as a guide we see that the equilibrium concentrations are; [NO] = 0.356 *M*, $[N_2O]$ = 0.122 *M* and $[O_2] = 0.315$ *M*.

$$K_c = \frac{[N_2O][O_2]}{[NO_2][NO]} = \frac{(0.122)(0.315)}{(0.118)(0.356)} = 0.915$$

14.59 $2BrCl \rightleftarrows Br_2 + Cl_2$

	[BrCl]	[Br$_2$]	[Cl$_2$]
I	0.050	–	–
C	–2x	+x	+x
E	0.050–2x	+x	+x

Substituting the above values for equilibrium concentrations into the mass action expression gives:

$$K_c = \frac{[Br_2][Cl_2]}{[BrCl]^2} = \frac{(x)(x)}{(0.050-2x)^2} = 0.145$$

Take the square root of both sides to get

$$K_c = \frac{x}{0.050-2x} = 0.381$$

Solving for x gives: x = 0.011 *M* = $[Br_2] = [Cl_2]$

14.61 The initial concentrations are each 0.240 mol/2.00 L = 0.120 *M*.

	$[SO_3]$	[NO]	$[NO_2]$	$[SO_2]$
I	0.120	0.120	–	–
C	–x	–x	+x	+x
E	0.120–x	0.120–x	+x	+x

Substituting the above values for equilibrium concentrations into the mass action expression gives:

$$K_c = \frac{[NO_2][SO_2]}{[SO_3][NO]} = \frac{(x)(x)}{(0.120-x)(0.120-x)} = 0.500$$

Taking the square root of both sides of this equation gives: 0.707 = x/(0.120 – x)
Solving for x we have: 1.707(x) = 0.0848,
x = 0.0497 mol/L = $[NO_2]$ = $[SO_2]$,
[NO] = $[SO_3]$ = 0.120 – x = 0.0703 mol/L

14.63 The initial concentrations are all 1.00 mol/100 L = 0.0100 *M*. Since the initial concentrations are all the same, the reaction quotient is equal to 1.0, and we conclude that the system must shift to the left to reach equilibrium since $Q > K_c$.

	[CO]	$[H_2O]$	$[CO_2]$	$[H_2]$
I	0.0100	0.0100	0.0100	0.0100
C	+x	+x	–x	–x
E	0.0100+x	0.0100+x	0.0100–x	0.0100–x

Substituting the above values for equilibrium concentrations into the mass action expression gives:

$$K_c = \frac{[CO_2][H_2]}{[CO][H_2O]} = \frac{(0.0100-x)(0.0100-x)}{(0.0100+x)(0.0100+x)} = 0.400$$

We take the square root of both sides of the above equation:

$$\frac{(0.0100-x)}{(0.0100+x)} = 0.632$$

and (0.632)(0.0100 + x) = 0.0100 – x
(1.632)x = 3.68×10^{-3}, or
x = 2.25×10^{-3} mol/L
The equilibrium concentrations are then:
$[H_2]$ = $[CO_2]$ = $(0.0100 - 2.25 \times 10^{-3}) = 7.7 \times 10^{-3}$ *M*,
[CO] = $[H_2O]$ = $(0.0100 + 2.25 \times 10^{-3})$ = 0.0123 *M*.

14.65

	[HCl]	$[H_2]$	$[Cl_2]$
I	0.0500	–	–
C	–2x	+x	+x
E	0.0500–2x	+x	+x

$$K_c = \frac{[H_2][Cl_2]}{[HCl]^2} = \frac{(x)(x)}{(0.0500-2x)^2} = 3.2 \times 10^{-34}$$

Because K_c is so exceedingly small, we can make the simplifying assumption that x is also small enough to make $(0.0500 - 2x) \approx 0.0500$. Thus we have: $3.2 \times 10^{-34} = (x)^2/(0.0500)^2$

Taking the square root of both sides, and solving for the value of x gives:
$x = 8.9 \times 10^{-19}\ M = [H_2] = [Cl_2]$
$[HCl] = (0.0500 - x) \approx 0.0500$ mol/L

14.67 $$K_c = \frac{[CO]^2[O_2]}{[CO_2]^2} = 6.4 \times 10^{-7}$$

	$[CO_2]$	$[CO]$	$[O_2]$
I	1.0×10^{-2}	–	–
C	$-2x$	$+2x$	$+x$
E	$1.0 \times 10^{-2} - 2x$	$+2x$	$+x$

$$K_c = \frac{[2x]^2[x]}{[1.0 \times 10^{-2} - 2x]^2} = 6.4 \times 10^{-7}$$

Assume $x << 1.0 \times 10^{-2}$.

$$\frac{4x^3}{(1.0 \times 10^{-2})^2} = 6.4 \times 10^{-7} \qquad x = 2.5 \times 10^{-4}$$

$$[CO] = 2x = 5.0 \times 10^{-4}\ M$$

14.69 We first approach the problem in the normal fashion with an initial concentration of $PCl_5 = 0.013\ M$.

	$[PCl_3]$	$[Cl_2]$	$[PCl_5]$
I	–	–	0.013
C	$+x$	$+x$	$-x$
E	$+x$	$+x$	$0.013-x$

Substituting the above values for equilibrium concentrations into the mass action expression gives:

$$K_c = \frac{[PCl_5]}{[PCl_3][Cl_2]} = \frac{(0.013-x)}{(x)(x)} = 0.18$$

rearranging; $0.18x^2 + x - 0.013 = 0$

We next attempt to use the quadratic equation to solve for the value of x, setting a = 0.18; b = 1; c = –0.013.

However, we find that unless we carry one more significant figure than is allowed, the quadratic formula for this problem gives us a concentration of zero for PCl_5. A better solution is obtained by "allowing" the initial equilibrium to shift **completely** to the left, giving us a new initial situation from which to work:

	$[PCl_3]$	$[Cl_2]$	$[PCl_5]$
I	0.013	0.013	–
C	$-x$	$-x$	$+x$
E	$0.013-x$	$0.013-x$	$+x$

Substituting the above values for equilibrium concentrations into the mass action expression gives:

$$K_c = \frac{[PCl_5]}{[PCl_3][Cl_2]} = \frac{(+x)}{(0.013-x)(0.013-x)} = 0.18$$

Now, we may assume that x << 0.013. The equation is simplified and we solve for x
$x = [PCl_5] = 3.0 \times 10^{-5}$ *M*.

14.71

	$[SO_3]$	[NO]	$[NO_2]$	$[SO_2]$
I	0.0500	0.100	–	–
C	–x	–x	+x	+x
E	0.0500–x	0.100–x	+x	+x

Substituting the above values for equilibrium concentrations into the mass action expression gives:

$$K_c = \frac{[NO_2][SO_2]}{[SO_3][NO]} = \frac{(x)(x)}{(0.0500-x)(0.100-x)} = 0.500$$

Since the equilibrium constant is not much larger than either of the values 0.0500 or 0.100, we cannot neglect the size of x in the above expression. A simplifying assumption is not therefore possible, and we must solve for the value of x using the quadratic equation. Multiplying out the above denominator, collecting like terms, and putting the result into the standard quadratic form gives:

$$0.500x^2 + (7.50 \times 10^{-2})x - (2.50 \times 10^{-3}) = 0$$

$$x = \frac{-7.50 \times 10^{-2} \pm \sqrt{(7.50 \times 10^{-2})^2 - 4(0.500)(-2.50 \times 10^{-3})}}{2(0.500)} = 0.0281\ M$$

using the (+) root. So, $[NO_2] = [SO_2] = 0.0281$ *M*

14.73 $$K_c = \frac{[CO][H_2O]}{[HCHO_2]^2} = 4.3 \times 10^5$$

Since K_c is large, start by assuming all of the $HCHO_2$ decomposes to give CO and H_2O

	$[HCHO_2]$	[CO]	$[O_2]$
I	–	0.200	0.200
C	+x	–x	–x
E	+x	0.200 –x	0.200 –x

$$Kc = \frac{[0.200][0.200]}{[x]} = 4.3 \times 10^5$$

$x = 9.3 \times 10^{-8}$
so, at equilibrium
$[CO] = [H_2O] = 0.200 - x = 0.200$

Practice Exercises

15.1 Conjugate acid base pairs (a), (c), and (f)

(b) The conjugate base of HI is I^-

(d) The conjugate base of HNO_2 is NO_2^- and the conjugate base of NH_4^+ is NH_3

(e) The conjugate acid of CO_3^{2-} is HCO_3^- and the conjugate acid of CN^- is HCN

15.2 In each case the conjugate base is obtained by removing a proton from the acid:

(a) OH^- (b) I^- (c) NO_2^- (d) $H_2PO_4^-$

(e) HPO_4^{2-} (f) PO_4^{3-} (g) HS^- (h) NH_3

15.3 In each case the conjugate acid is obtained by adding a proton to the base:

(a) H_2O_2 (b) HSO_4^- (c) HCO_3^- (d) HCN

(e) NH_3 (f) NH_4^+ (g) H_3PO_4 (h) $H_2PO_4^-$

15.4 HCN and CN^- HCl and Cl^-

15.5 The Brønsted acids are $H_2PO_4^-(aq)$ and $H_2CO_3(aq)$

The Brønsted bases are $HCO_3^-(aq)$ and $HPO_4^{2-}(aq)$

15.6

conjugate pair (PO_4^{3-} and HPO_4^{2-})

$$PO_4^{3-}(aq) + HC_2H_3O_2(aq) \rightleftharpoons HPO_4^{2-}(aq) + C_2H_3O_2^-(aq)$$

base acid acid base

conjugate pair ($HC_2H_3O_2$ and $C_2H_3O_2^-$)

15.7 (a) $H_2PO_4^-$ amphoteric since it can both accept and donate a proton

(b) HPO_4^{2-} amphoteric since it can both accept and donate a proton

(c) H_2S amphoteric since it can both accept and donate a proton

(d) H_3PO_4 not amphoteric: it can only donate protons

(e) NH_4^+ not amphoteric: it can only donate protons

(f) H_2O amphoteric since it can both accept and donate a proton

(g) HI not amphoteric: it can only donate protons

(h) HNO_2 not amphoteric: it can only donate protons

15.8 $HPO_4^{2-}(aq) + OH^-(aq) \rightarrow PO_4^{3-}(aq) + H_2O$; HPO_4^{2-} acting as an acid

$HPO_4^{2-}(aq) + H_3O^+(aq) \rightarrow H_2PO_4^- + H_2O$; HPO_4^{2-} acting as a base

15.9 $HSO_4^-(aq) + HPO_4^{2-}(aq) \rightarrow SO_4^{-2}(aq) + H_2PO_4^-(aq)$

15.10 The substances on the right because they are the weaker acid and base.

15.11 (a) $HF < HBr < HI$

(b) $PH_3 < H_2S < HCl$

(c) $H_2O < H_2Se < H_2Te$

(d) $AsH_3 < H_2Se < HBr$

(e) $PH_3 < H_2Se < HI$

15.12 (a) HBr is the stronger acid since binary acid strength increases from left to right within a period.

(b) H_2Te is the stronger acid since binary acid strength increases from top to bottom within a group.

(c) H_2S since acid strength increases from top to bottom within a group.

15.13 (a) $HClO_3$ (b) H_2SO_4

15.14 (a) H_3AsO_4 (b) H_2TeO_4

15.15 (a) HIO_4 (b) H_2TeO_4 (c) H_3AsO_4

15.16 (a) H_2SO_4 (b) H_3AsO_4

15.17 (a) NH_3 is the Lewis base since it has an unshared pair of electrons.
H^+ is the Lewis acid since it can accept a pair of electrons
(b) $(CH_3)_2O$ Lewis base
BCl_3 Lewis acid
(c) Ag^+ is the Lewis acid
NH_3 is the Lewis base

15.18 (a) Fluoride ions have a filled octet of electrons and are likely to behave as Lewis bases, i.e., electron pair donors.
(b) $BeCl_2$ is a likely Lewis acid since it has an incomplete shell. The Be atom has only two valence electrons and it can easily accept a pair of electrons.
(c) It could reasonably be considered a potential Lewis base since it contains three oxygens, each with lone pairs and partial negative charges. However, it is more effective as a Lewis acid, since the central sulfur bears a significant positive charge.

15.19 $K_w = 1.0 \times 10^{-14} = [H^+][OH^-]$

$$\left[OH^-\right] = \frac{1.0\times10^{-14}}{\left[H^+\right]} = \frac{1.0\times10^{-14}}{12} = 8.3 \times 10^{-16}\ M$$

15.20 $K_w = 1.0 \times 10^{-14} = [H^+][OH^-]$

$$\left[H^+\right] = \frac{1.0 \times 10^{-14}}{\left[OH^-\right]} = \frac{1.0 \times 10^{-14}}{7.8 \times 10^{-6}} = 1.3 \times 10^{-9}\ M$$

Since $[OH^-] > [H^+]$, the solution is basic.

15.21 $pOH = 14 - pH = 14 - 4.25 = 9.75$
$[H^+] = 10^{-4.25} = 5.62 \times 10^{-5}\ M$
$[OH^-] = 10^{-9.75} = 1.78 \times 10^{-10}\ M$

15.22 $pH = -\log[H^+] = -\log[3.67 \times 10^{-4}] = 3.44$
$pOH = 14.00 - pH = 14.00 - 3.44 = 10.56$

The solution is acidic since pH is below 7.0.

15.23 $pOH = -\log[OH^-] = -\log[1.47 \times 10^{-9}] = 8.83$
$pH = 14.00 - pOH = 14.00 - 8.83 = 5.17$

15.24 Since the soil is acidic, a base should be added, and choosing between $Al_2(SO_4)_3$ and CaO, the base is CaO.

15.25 $[H^+] = 12\ M$ $pH = -\log 12 = -1.1$
$[OH^-] = 8.3 \times 10^{-16}\ M$ $pOH = -\log (8.3 \times 10^{-16}) = 15.1$

15.26 In general, we have the following relationships between pH, $[H^+]$, and $[OH^-]$:
$[H^+] = 10^{-pH}$
$[H^+][OH^-] = 1.00 \times 10^{-14}$
(a) $[H^+] = 10^{-2.90} = 1.3 \times 10^{-3}\ M$
$[OH^-] = 1.00 \times 10^{-14}/1.3 \times 10^{-3}\ M = 7.7 \times 10^{-12}\ M$
The solution is acidic.

(b) $[H^+] = 10^{-3.85} = 1.4 \times 10^{-4}\ M$
$[OH^-] = 1.00 \times 10^{-14}/1.4 \times 10^{-4}\ M = 7.1 \times 10^{-11}\ M$
The solution is acidic.

(c) $[H^+] = 10^{-10.81} = 1.5 \times 10^{-11}\ M$
$[OH^-] = 1.00 \times 10^{-14}/1.5 \times 10^{-11}\ M = 6.7 \times 10^{-4}\ M$
The solution is basic.

(d) $[H^+] = 10^{-4.11} = 7.8 \times 10^{-5}\ M$
$[OH^-] = 1.00 \times 10^{-14}/7.8 \times 10^{-5}\ M = 1.3 \times 10^{-10}\ M$
The solution is acidic.

(e) $[H^+] = 10^{-11.61} = 2.5 \times 10^{-12}\ M$
$[OH^-] = 1.00 \times 10^{-14}/2.5 \times 10^{-12}\ M = 4.0 \times 10^{-3}\ M$
The solution is basic.

15.27 $[H^+] = 0.0050\ M$
$pH = -\log[H^+] = -\log[0.0050] = 2.30$
$pOH = 14.0 - pH = 14.00 - 2.30 = 11.70$

15.28 First determine the number of moles of KOH and then the molarity of the solution

$$\text{mol KOH} = (1.20\text{ g KOH})\left(\frac{1\text{ mol KOH}}{56.11\text{ g KOH}}\right) = 0.0214\text{ mol KOH}$$

$$\text{molarity} = \frac{0.0214\text{ mol KOH}}{0.250\text{ L solution}} = 0.0855\ M\text{ KOH}$$

$pOH = -\log 0.0855\ M = 1.068$
$pH = 14 - 1.068 = 12.932$
$[H^+] = 1.17 \times 10^{-13}\ M$

15.29 $[H^+] = 10^{-5.5} = 3.2 \times 10^{-6}\ M$

Review Problems

15.41 (a) HF (b) $N_2H_5^+$ (c) $C_5H_5NH^+$
(d) HO_2^- (e) H_2CrO_4

15.43 (a)

conjugate pair

$HNO_3 + N_2H_4 \rightleftharpoons N_2H_5^+ + NO_3^-$
acid base acid base

conjugate pair

(b)

conjugate pair

$NH_3 + N_2H_5^+ \rightleftharpoons NH_4^+ + N_2H_4$
base acid acid base

conjugate pair

(c) conjugate pair

$H_2PO_4^- + CO_3^{2-} \rightleftharpoons HCO_3^- + HPO_4^{2-}$

acid base acid base

conjugate pair

(d) conjugate pair

$HIO_3 + HC_2O_4^- \rightleftharpoons H_2C_2O_4 + IO_3^-$

acid base acid base

conjugate pair

15.45 (a) HBr, HBr bond is weaker
(b) HF, more electronegative F polarizes and weakens the bond
(c) HBr, larger Br forms a weaker bond with H

15.47 (a) $HClO_2$, because it has more oxygen atoms
(b) H_2SeO_4, because it has more lone oxygen atoms

15.49 (a) $HClO_3$, because Cl is more electronegative
(b) $HClO_3$, because the charge is more evenly distributed
(c) $HBrO_4$, because the negative charge is more evenly distributed

15.51

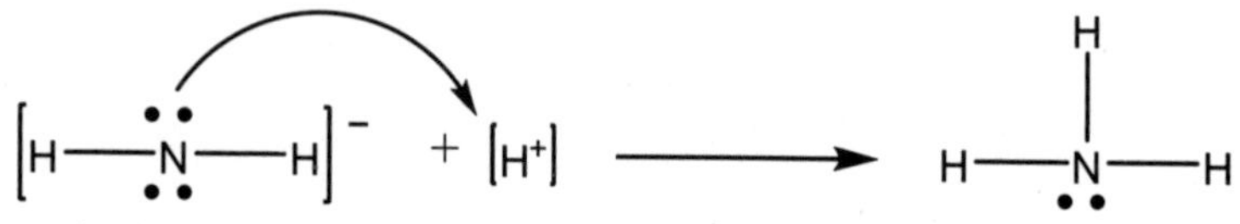

$[H—\ddot{N}—H]^- + [H^+] \longrightarrow NH_3$

Lewis acid: H^+ Lewis base: NH_2^-

15.53

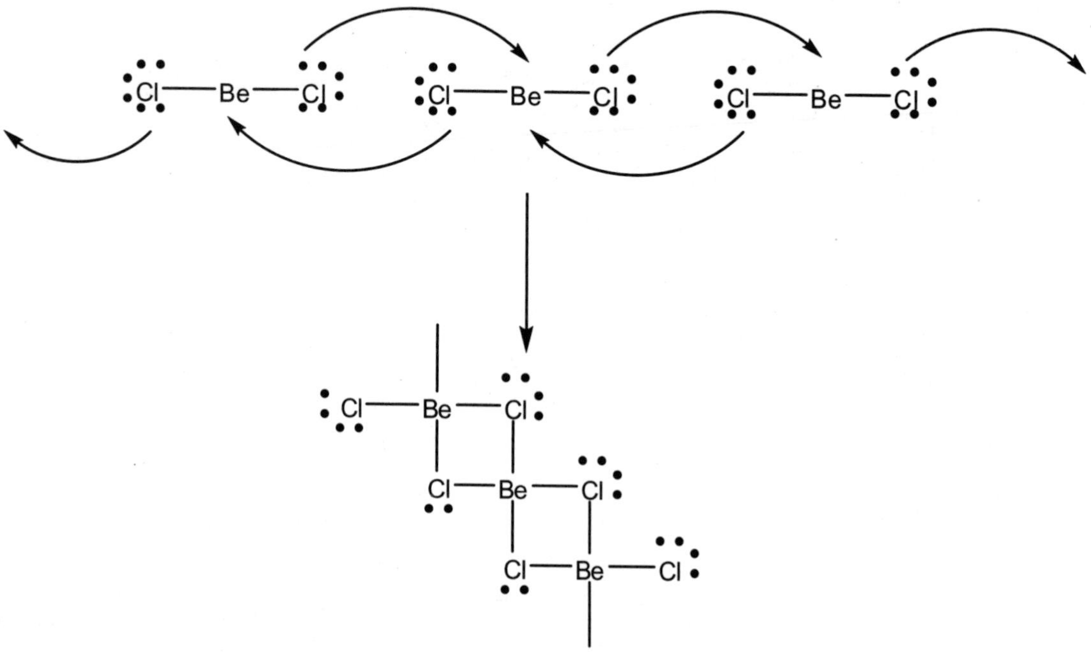

$BeCl_2 + BeCl_2 + BeCl_2 \longrightarrow (BeCl_2)_n$ chain

15.55

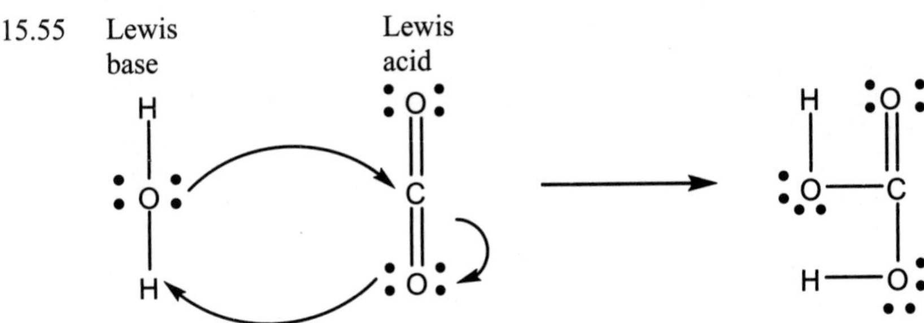

15.57

Lewis bases: NH_2^- and OH^- Lewis acid: H^+

15.59 $D_2O \rightleftharpoons D^+ + OD^-$, $K_w = [D^+] \times [OD^-] = 8.9 \times 10^{-16}$
Since $[D^+] = [OD^-]$, we can rewrite the above expression to give:
$8.9 \times 10^{-16} = ([D^+])^2$, $[D^+] = 3.0 \times 10^{-8}\ M = [OD^-]$
$pD = -\log[D^+] = -\log(3.0 \times 10^{-8}) = 7.52$
$pOD = -\log[OD^-] = -\log(3.0 \times 10^{-8}) = 7.52$
$pK_w = pD + pOD = 15.04$

Alternatively, we can calculate:
$pK_w = -\log(K_w) = -\log(8.9 \times 10^{-16}) = 15.05$

15.61 At 25 °C, $K_w = 1.0 \times 10^{-14} = [H^+] \times [OH^-]$. Let $x = [H^+]$, for each of the following:

(a) $x(0.0068) = 1.0 \times 10^{-14}$
$[H^+] = (1.0 \times 10^{-14}) \div (0.0068) = 1.5 \times 10^{-12}\ M$
$pH = -\log[H^+] = -\log(1.5 \times 10^{-12}) = 11.83$
$pOH = 14 - pH = 14 - 11.83 = 2.17$

(b) $x(6.4 \times 10^{-5}) = 1.0 \times 10^{-14}$
$[H^+] = (1.0 \times 10^{-14}) \div (6.4 \times 10^{-5}) = 1.6 \times 10^{-10}\ M$
$pH = -\log[H^+] = -\log(1.6 \times 10^{-10}) = 9.81$
$pOH = 14 - pH = 14 - 9.81 = 4.19$

(c) $x(1.6 \times 10^{-8}) = 1.0 \times 10^{-14}$
$[H^+] = (1.0 \times 10^{-14}) \div (1.6 \times 10^{-8}) = 6.3 \times 10^{-7}\ M$
$pH = -\log[H^+] = -\log(6.3 \times 10^{-7}) = 6.20$
$pOH = 14 - pH = 14 - 6.20 = 7.80$

(d) $x(8.2 \times 10^{-12}) = 1.0 \times 10^{-14}$
$[H^+] = (1.0 \times 10^{-14}) \div (8.2 \times 10^{-12}) = 1.2 \times 10^{-3}\ M$
$pH = -\log[H^+] = -\log(1.2 \times 10^{-3}) = 2.91$
$pOH = 14 - pH = 14 - 2.91 = 11.09$

15.63 $pH = -\log[H^+] = -\log(1.9 \times 10^{-5}) = 4.72$

15.65 $[H^+] = 10^{-pH}$ and $[OH^-] = 10^{-pOH}$
At 25 °C, pH + pOH = 14.00

(a) $[H^+] = 10^{-pH} = 10^{-8.14} = 7.2 \times 10^{-9}\ M$
$pOH = 14.00 - pH = 14.00 - 8.14 = 5.86$
$[OH^-] = 10^{-pOH} = 10^{-5.86} = 1.4 \times 10^{-6}\ M$

(b) $[H^+] = 10^{-pH} = 10^{-2.56} = 2.7 \times 10^{-3}\ M$
$pOH = 14.00 - pH = 14.00 - 2.56 = 11.44$
$[OH^-] = 10^{-pOH} = 10^{-11.44} = 3.6 \times 10^{-12}\ M$

(c) $[H^+] = 10^{-pH} = 10^{-11.25} = 5.6 \times 10^{-12}\ M$
$pOH = 14.00 - pH = 14.00 - 11.25 = 2.75$
$[OH^-] = 10^{-pOH} = 10^{-2.75} = 1.8 \times 10^{-3}\ M$

(d) $[H^+] = 10^{-pH} = 10^{-13.28} = 5.3 \times 10^{-14}\ M$
$pOH = 14.00 - pH = 14.00 - 13.28 = 0.72$
$[OH^-] = 10^{-pOH} = 10^{-0.76} = 1.9 \times 10^{-1}\ M$

(e) $[H^+] = 10^{-pH} = 10^{-6.70} = 2.0 \times 10^{-7}\ M$
$pOH = 14.00 - pH = 14.00 - 6.70 = 7.30$
$[OH^-] = 10^{-pOH} = 10^{-7.30} = 5.0 \times 10^{-8}\ M$

15.67 $[H^+] = 10^{-pH} = 10^{-5.7} = 2.0 \times 10^{-6}\ M$
$pOH = 14.00 - pH = 14.00 - 5.7 = 8.3$
$[OH^-] = 10^{-pOH} = 10^{-8.3} = 5.0 \times 10^{-9}\ M$

15.69 HNO_3 is a strong acid so $[H^+] = [HNO_3] = 0.00065\ M$
$pH = -\log[H^+] = -\log(0.00065) = 3.19$
$pOH = 14.00 - pH = 14.00 - 3.19 = 10.81$
$[OH^-] = 10^{-pOH} = 10^{-10.81} = 1.55 \times 10^{-11}\ M$

15.71

$$\text{M OH}^- = \frac{\text{moles OH}^-}{\text{L solution}} = \left(\frac{6.0\text{ g NaOH}}{1.00\text{ L solution}}\right)\left(\frac{1\text{ mole NaOH}}{40.0\text{ g NaOH}}\right)\left(\frac{1\text{ mole OH}^-}{1\text{ mole NaOH}}\right)$$

$$= 0.15\text{ M OH}^-$$

$pOH = -\log[OH^-] = -\log(0.15) = 0.82$
$pH = 14.00 - pOH = 14.00 - 0.82 = 13.18$
$[H^+] = 10^{-pH} = 10^{-13.18} = 6.61 \times 10^{-14}\ M$

15.73 $pOH = 14.00 - pH = 14.00 - 11.60 = 2.40$
$[OH^-] = 10^{-pOH} = 10^{-2.40} = 4.0 \times 10^{-3}\ M$

$$[Ca(OH)_2] = \left(\frac{4.0 \times 10^{-3}\text{ mol OH}^-}{1\text{ L solution}}\right)\left(\frac{1\text{ mol Ca(OH)}_2}{2\text{ mol OH}^-}\right)$$

$$= 2.0 \times 10^{-3}\text{ M Ca(OH)}_2$$

$pH = 10.60$
$pOH = 14 - 10.60 = 3.4$
$[OH^-] = 10^{-pOH} = 10^{-3.40} = 4.0 \times 10^{-4}\ M$

$$[Ca(OH)_2] = \left(\frac{4.0 \times 10^{-4}\text{ mol OH}^-}{1\text{ L solution}}\right)\left(\frac{1\text{ mol Ca(OH)}_2}{2\text{ mol OH}^-}\right)$$

$$= 2.0 \times 10^{-4}\text{ M Ca(OH)}_2$$

15.75 First, we must find the concentration of the HCl solution. Since HCl is a strong acid, we know that [HCl] = $[H^+]$. We can find $[H^+]$ from the pH:
$[H^+] = 10^{-pH} = 10^{-2.25} = 5.6 \times 10^{-3}\ M\ H^+$

Now we can solve the problem using the given conversion factors:

$$\text{volume of } 0.0100\ M\text{ KOH} = 300\text{ mL}\left(\frac{5.6\times10^{-3}\text{ mol H}^+}{1000\text{ mL solution}}\right)\left(\frac{1\text{ mol OH}^-}{1\text{ mol H}^+}\right)\left(\frac{1000\text{ mL OH}^-}{0.0100\text{ mol OH}^-}\right)$$

$$= 168\text{ mL } 0.0100\ M\text{ KOH}$$

15.77 Since NaOH is a strong base, $[NaOH] = [OH^-] = 0.0020\ M\ OH^-$. (Next, we make the simplifying assumption that the amount of hydroxide ion formed from the dissociation of water is so small that we can neglect it in calculating pOH for the solution.)

$[H^+] = 1 \times 10^{-14} \div 0.0020 = 5 \times 10^{-12}\ M\ H^+$
The only source of H^+ is the autoionization of water. Therefore, the molarity of OH^- from the ionization water is $5 \times 10^{-12}\ M$.

Practice Exercises

16.1 (a) $HC_2H_3O_2 + H_2O \rightleftharpoons H_3O^+ + C_2H_3O_2^-$

$$K_a = \frac{[H_3O^+][C_2H_3O_2^-]}{[HC_2H_3O_2]}$$

(b) $(CH_3)_3NH^+ + H_2O \rightleftharpoons H_3O^+ + (CH_3)_3N$

$$K_a = \frac{[H_3O^+][(CH_3)_3N]}{[(CH_3)_3NH^+]}$$

(c) $H_3PO_4 + H_2O \rightleftharpoons H_3O^+ + H_2PO_4^-$

$$K_a = \frac{[H_3O^+][H_2PO_4^-]}{[H_3PO_4]}$$

16.2 (a) $HCHO_2 + H_2O \rightleftharpoons H_3O^+ + CHO_2^-$

$$K_a = \frac{[H_3O^+][CHO_2^-]}{[HCHO_2]}$$

(b) $(CH_3)_2NH_2^+ + H_2O \rightleftharpoons H_3O^+ + (CH_3)_2NH$

$$K_a = \frac{[H_3O^+][CH_3NH]}{[(CH_3)_2NH_2^+]}$$

(c) $H_2PO_4^- + H_2O \rightleftharpoons H_3O^+ + HPO_4^{2-}$

$$K_a = \frac{[H_3O^+][HPO_4^{2-}]}{[H_2PO_4^-]}$$

16.3 The smaller the value of pK_a, the stronger the acid.
The acids stronger than acetic acid and weaker than formic acid from Table 16.1 are barbituric acid.

16.4 The acid with the smaller pK_a (H*A*) is the strongest acid.
Since $pK_a = -\log K_a$, $K_a = 10^{-pKa}$
For H_A: $K_a = 10^{-3.16} = 6.9 \times 10^{-4}$
For H_B: $K_a = 10^{-4.14} = 7.2 \times 10^{-5}$

16.5 (a) $(CH_3)_3N + H_2O \rightleftharpoons (CH_3)_3NH^+ + OH^-$

$$K_b = \frac{[(CH_3)_3NH^+][OH^-]}{[(CH_3)_3N]}$$

(b) $SO_3^{2-} + H_2O \rightleftharpoons HSO_3^- + OH^-$

$$K_b = \frac{[HSO_3^-][OH^-]}{[SO_3^{2-}]}$$

(c) $NH_2OH + H_2O \rightleftharpoons NH_3OH^+ + OH^-$

$$K_b = \frac{[NH_3OH^+][OH^-]}{[NH_2OH]}$$

16.6 (a) $HSO_4^- + H_2O \rightleftharpoons H_2SO_4 + OH^-$

$$K_b = \frac{[H_2SO_4][OH^-]}{[HSO_4^-]}$$

(b) $H_2PO_4^- + H_2O \rightleftharpoons H_3PO_4 + OH^-$

$$K_b = \frac{[H_3PO_4][OH^-]}{[H_2PO_4^-]}$$

(c) $HPO_4^{2-} + H_2O \rightleftharpoons H_2PO_4^- + OH^-$

$$K_b = \frac{[H_2PO_4^-][OH^-]}{[HPO_4^{2-}]}$$

(d) $HCO_3^- + H_2O \rightleftharpoons H_2CO_3 + OH^-$

$$K_b = \frac{[H_2CO_3][OH^-]}{[HCO_3^-]}$$

(e) $HSO_3^- + H_2O \rightleftharpoons H_2SO_3 + OH^-$

$$K_b = \frac{[H_2SO_3][OH^-]}{[HSO_3^-]}$$

16.7 For conjugate acid base pairs, $K_a \times K_b = K_w$:
$K_b = K_w \div K_a = 1.0 \times 10^{-14} \div 2.3 \times 10^{-11} = 4.3 \times 10^{-4}$

16.8 For conjugate acid base pairs, $K_a \times K_b = K_w$:
$K_b = K_w \div K_a = 1.0 \times 10^{-14} \div 1.8 \times 10^{-4} = 5.6 \times 10^{-11}$

16.9 $HSal \rightleftharpoons H^+ + Sal^-$

$$K_a = \frac{[H^+][Sal^-]}{[HSal]}$$

$[H^+] = 10^{-pH} = 10^{-1.836} = 0.0146\ M$

	[HSal]	$[H^+]$	$[Sal^-]$
I	0.200	–	–
C	–x	+x	+x
E	0.200–x	+x	+x

We know the $[H^+]$ is 0.0146 M
x = 0.0146
[HSal] = 0.200 – 0.0146 = 0.185 M
$[Sal^-]$ = 0.0146 M

$$K_a = \frac{[H^+][Sal^-]}{[HSal]} = \frac{(0.0146)(0.0146)}{(0.185)} = 1.15 \times 10^{-3}$$

$pK_a = -\log (K_a) = -\log (1.15 \times 10^{-3}) = 2.938$

16.10 $HBu \rightleftharpoons H^+ + Bu^-$

$$K_a = \frac{[H^+][Bu^-]}{[HBu]}$$

	[HBu]	[H+]	[Bu−]
I	0.01000	–	–
C	–x	+x	+x
E	0.01000–x	+x	+x

We know that the acid is 4.0% ionized so x = 0.01000 *M* × 0.040 = 0.00040 *M*. Therefore, our equilibrium concentrations are $[H^+] = [Bu^-] = 0.00040\ M$, and [HBu] = 0.01000 *M* – 0.00040 *M* = 0.00960 *M*. Substituting these values into the mass action expression gives:

$$K_a = \frac{(0.00040)(0.00040)}{0.00960} = 1.7 \times 10^{-5}$$

$pK_a = -\log(K_a) = -\log(1.7 \times 10^{-5}) = 4.78$

16.11 We will use the symbol Mor and $HMor^+$ for the base and its conjugate acid respectively:
$Mor + H_2O \rightleftharpoons HMor^+ + OH^-$

$$K_b = \frac{[HMor^+][OH^-]}{[Mor]}$$

	[Mor]	[HMor+]	[OH−]
I	0.010	–	–
C	–x	+x	+x
E	0.010–x	+x	+x

At equilibrium, the pOH = 3.90.
The $[OH^-] = 10^{-pOH} = 10^{-3.90} = 1.26 \times 10^{-4}\ M = x$.

Substituting these values into the mass action expression gives:

$$K_b = \frac{(1.26 \times 10^{-4})(1.26 \times 10^{-4})}{0.010 - 1.26 \times 10^{-4}} = 1.61 \times 10^{-6}$$

$pK_b = -\log(K_b) = -\log(1.61 \times 10^{-6}) = 5.79$

16.12 $H_3BO_3 \rightleftharpoons H^+ + H_2BO_3^-$

$$K_a = \frac{[H_2BO_3^-][H^+]}{[H_3BO_3]} = 5.8 \times 10^{-10}$$

	[H2BO3−]	[H+]	[H3BO3−]
I	0.050	–	–
C	–x	+x	+x
E	0.050–x	+x	+x

Assume that x << 0.050 and substitute the equilibrium values into the mass action expression to get:

$$K_a = \frac{[H_2BO_3^-][H^+]}{[H_3BO_3]} = 5.8 \times 10^{-10}$$

$$K_a = \frac{(x)(x)}{(0.050)} = 5.8 \times 10^{-10}$$

$x = 5.4 \times 10^{-6}$
$[H^+] = x = 5.4 \times 10^{-6}\ M$
$pH = -\log[H^+] = -\log(5.4 \times 10^{-6}) = 5.27$

16.13 $HC_2H_6NO_2 \rightleftharpoons H^+ + C_2H_6NO_2^-$

$$K_a = \frac{[H^+][C_2H_6NO_2^-]}{[HC_2H_6NO_2]} = 1.4 \times 10^{-5}$$

	$[HC_2H_6NO_2]$	$[H^+]$	$[C_2H_6NO_2^-]$
I	0.050	–	–
C	– x	+ x	+ x
E	0.050 – x	+ x	+ x

Assume that x << 0.050 and substitute the equilibrium values into the mass action expression to get:

$$K_a = \frac{[x][x]}{[0.050]} = 1.4 \times 10^{-5}$$

Solving for x we determine that $x = 8.4 \times 10^{-4}\ M = [H^+]$.
$pH = -\log[H^+] = -\log(8.4 \times 10^{-4}) = 3.08$

16.14 $C_6H_5NH_2 + H_2O \rightleftharpoons C_6H_5NH_3^+ + OH^-$

$$K_b = \frac{[C_6H_5NH_3^+][OH^-]}{[C_6H_5NH_2]} = 4.1 \times 10^{-10}$$

	$[C_6H_5NH_2]$	$[C_6H_5NH_3^+]$	$[OH^-]$
I	0.025	–	–
C	– x	+ x	+ x
E	0.025 – x	+ x	+ x

Assume that x << 0.025 and substitute the equilibrium values into the mass action expression to get:

$$K_b = \frac{(x)(x)}{(0.025)} = 4.1 \times 10^{-10}$$

$x = 3.2 \times 10^{-6}$
$[OH^-] = 3.2 \times 10^{-6}$
$pOH = -\log[OH^-] = -\log(3.2 \times 10^{-6}) = 5.49$

16.15 $C_5H_5N + H_2O \rightleftharpoons C_5H_5NH^+ + OH^-$

$$K_b = \frac{[C_5H_5NH^+][OH^-]}{[C_5H_5N]} = 1.5 \times 10^{-9}$$

	$[C_5H_5N]$	$[C_5H_5NH^+]$	$[OH^-]$
I	0.010	–	–
C	– x	+ x	+ x
E	0.010 – x	+ x	+ x

Assume that x << 0.010 and substitute the equilibrium values into the mass action expression to get:

$$K_b = \frac{[x][x]}{[0.010]} = 1.5 \times 10^{-9}$$

Solving for x we determine that $x = 3.9 \times 10^{-6}\ M = [OH^-]$.
$pOH = -\log[OH^-] = -\log(3.9 \times 10^{-6}) = 5.41$
$pH = 14.00 - pOH = 8.59$

16.16 We will use the notation Hphenol and phenol for the acid and its conjugate base:

$$\text{Hphenol} \rightleftharpoons \text{H}^+ + \text{phenol}$$

$$K_a = \frac{[\text{H}^+][\text{phenol}]}{[\text{Hphenol}]} = 1.3 \times 10^{-10}$$

	[Hphenol]	$[H^+]$	[phenol]
I	0.15	–	–
C	– x	+ x	+ x
E	0.15 – x	+ x	+ x

If we assume that x << 0.15, a good assumption based upon the size of K_a, we can substitute the equilibrium values in to the mass action expression to get:

$$K_a = \frac{(x)(x)}{0.15} = 1.3 \times 10^{-10}$$

Solving gives $x = 4.4 \times 10^{-6}\ M = [H^+]$
$pH = -\log[H^+] = -\log(4.4 \times 10^{-6}) = 5.36$

16.17 Examine the ions in solution one at a time: The cation for acidity and the anion for basicity.

(a) $NaNO_2$
Na^+ is an ion of a Group I metal and is not acidic
NO_3^- is the conjugate base of HNO_2, a weak acid, therefore it is weak base.
The solution should be basic.

(b) KCl
K^+ is an ion of a Group I metal and is not acidic
Cl^- is the conjugate base of HCl, a strong acid, therefore it is not a strong base.
The solution should be neutral.

(c) NH_4Br
NH_4^+ is the conjugate acid of NH_3, a weak base, therefore it is a weak acid
Br^- is the conjugate base of HBr, a strong acid, therefore it is not a strong base.
The solution should be acidic.

16.18 Examine the ions in solution one at a time: The cation for acidity and the anion for basicity.

(a) $NaNO_3$
Na^+ is an ion of a Group I metal and is not acidic
NO_3^- is the conjugate base of HNO_3, a strong acid, therefore it is not a strong base.
The solution should be neutral.

(b) KOCl
K^+ is an ion of a Group I metal and is not acidic
OCl^- is the conjugate base of HOCl, a weak acid, therefore it is a weak base.
The solution should be basic.

(c) NH_4NO_3
NH_4^+ is the conjugate acid of NH_3, a weak base, therefore it is a weak acid
NO_3^- is the conjugate base of HNO_3, a strong acid, therefore it is not a strong base.
The solution should be acidic.

16.19 The chloride ion is neutral since it is the conjugate base of a strong acid, HCl. The ion $CH_3NH_3^+$ forms from the reaction of CH_3NH_2 with water and will act as a weak acid.

$$CH_3NH_3^+ + H_2O \rightleftharpoons H_3O^+ + CH_3NH_2$$

$$K_a = \frac{K_w}{K_b} = \frac{1.0 \times 10^{-14}}{4.4 \times 10^{-4}} = 2.3 \times 10^{-11}$$

$$K_a = \frac{[CH_3NH_2][H_3O^+]}{[CH_3NH_3^+]}$$

The concentration of the salt, CH_3NH_3Cl, in water is

$$M\ CH_3NH_3C = (25.0 \text{ g } CH_3NH_3Cl)\left(\frac{1 \text{ mol } CH_3NH_3Cl}{67.53 \text{ g } CH_3NH_3Cl}\right)\left(\frac{1}{0.500 \text{ L}}\right) = 0.740\ M\ CH_3NH_3Cl$$

	$[CH_3NH_3^+]$	$[H_3O^+]$	$[CH_3NH_2]$
I	0.740	–	–
C	– x	+ x	+ x
E	0.740 – x	+ x	+ x

Assume that x is small relative to 0.740 M CH_3NH_3Cl.

$$K_a = \frac{(x)(x)}{(0.740)} = 2.3 \times 10^{-11}$$

$x = 4.13 \times 10^{-6} = [H^+]$
$pH = -\log[H^+] = -\log(4.13 \times 10^{-6}) = 5.39$

16.20 The sodium ion is neutral since it is the salt of the strong base, NaOH. The nitrite ion is basic since it is the salt of nitrous acid, HNO_2, a weak acid. The equilibrium we are interested in for this problem is:
$NO_2^- + H_2O \rightleftharpoons HNO_2 + OH^-$.

$$K_b = \frac{[HNO_2][OH^-]}{[NO_2^-]}$$

In order to determine the value for K_b recall that $K_a - K_b = K_w$. We can look for the value of K_a for HNO_2
$K_b = 1.0 \times 10^{-14} \div 7.1 \times 10^{-4} = 1.4 \times 10^{-11}$.

	$[NO_2^-]$	$[HNO_2]$	$[OH^-]$
I	0.10	–	–
C	– x	+ x	+ x
E	0.10 – x	+ x	+ x

Assume that x << 0.10 and substitute the equilibrium values into the mass action expression to get:

$$K_b = \frac{[x][x]}{[0.10]} = 1.4 \times 10^{-11}$$

Solving we determine that $x = 1.2 \times 10^{-6}\ M = [OH^-]$.
$pOH = -\log[OH^-] = -\log(1.2 \times 10^{-6}) = 5.93$
$pH = 14.00 - pOH = 8.07$

16.21 Upon mixing, the NH_3 will react with HBr to form NH_4Br. The initial concentration of the NH_4Br will be:

$$\text{mol } NH_3 = (500 \text{ mL solution})\left(\frac{0.20 \text{ mol } NH_3}{1000 \text{ mL solution}}\right) = 0.10 \text{ mol } NH_3$$

$$\text{mol HBr} = (500 \text{ mL solution})\left(\frac{0.20 \text{ mol HBr}}{1000 \text{ mL solution}}\right) = 0.10 \text{ mol HBr}$$

The NH_3 and HBr are in a 1:1 ratio, therefore the number of moles of NH_4Br is 0.10. The volume of the solution is 500 mL + 500 mL = 1000 mL = 1.0 L
The concentration of NH_4Br is:

$$M\ NH_4Br = \frac{0.10 \text{ mol } NH_4Br}{1.00 \text{ L solution}} = 0.10\ M\ NH_4Br$$

As previously determined, a solution of NH_4Br will be acidic since NH_4^+ is the salt of a weak base and Br^- is the salt of a strong acid. As in the previous Practice Exercise, we need to determine the value for the dissociation constant using the relationship $K_a \times K_b = K_w$ and the value of K_b for NH_3 as listed in Table 16.5.
$K_a = 1.0 \times 10^{-14} \div 1.8 \times 10^{-5} = 5.6 \times 10^{-10}$. The equilibrium reaction is:
$NH_4^+ \rightleftharpoons NH_3 + H^+$

	$[NH_4^+]$	$[NH_3]$	$[H^+]$
I	0.10	–	–
C	– x	+ x	+ x
E	0.10 – x	+ x	+ x

Assume that x << 0.10 and substitute the equilibrium values into the mass action expression to get:

$$K_a = \frac{[x][x]}{[0.10]} = 5.6 \times 10^{-10}$$

Solving we determine that $x = 7.5 \times 10^{-6}\ M = [H^+]$.
$pH = -\log[H^+] = -\log(7.5 \times 10^{-6}) = 5.13$

16.22 Since the ammonium ion is the salt of a weak base, NH_3, it is acidic. The cyanide ion is the salt of a weak acid, HCN, so it is basic. In order to determine if the solution is acidic or basic, we need to determine the relative strength of the two components. Use the relationship $K_a \times K_b = K_w$ in order to determine the dissociation constants for the cyanide ion and the ammonium ion.

$K_a(NH_4^+) = K_w \div K_b(NH_3) = 1.0 \times 10^{-14} \div 1.8 \times 10^{-5} = 5.6 \times 10^{-10}$
$K_b(CN^-) = K_w \div K_a(HCN) = 1.0 \times 10^{-14} \div 6.2 \times 10^{-14} = 1.6 \times 10^{-5}$

Since the $K_b(CN^-)$ is larger than the $K_a(NH_4^+)$ the NH_4CN solution will be basic.

16.23 Using the same logic as the previous question:
NH_4^+ is the salt of a weak base, and is acidic
$C_2H_3O_2^-$ is the salt of a weak acid, and is basic
The relative strengths of the two components must be compared.
$K_a(NH_4^+) = K_w \div K_b(NH_3) = 1.0 \times 10^{-14} \div 1.8 \times 10^{-5} = 5.6 \times 10^{-10}$
$K_b(C_2H_3O_2^-) = K_w \div K_a(HC_2H_3O_2) = 1.0 \times 10^{-14} \div 1.8 \times 10^{-5} = 5.6 \times 10^{-10}$

The $K_a\ (NH_4^+)$ equals the $K_b(C_2H_3O_2^-)$, thus the solution will be neutral.

16.24 $(CH_3)_2NH + H_2O \rightleftharpoons (CH_3)_2NH_2^+ + OH^-$

$$K_b = \frac{[(CH_3)_2NH_2^+][OH^-]}{[(CH_3)_2NH]} = 9.6 \times 10^{-4}$$

	$[(CH_3)_2NH]$	$[(CH_3)_2NH_2^+]$	$[OH^-]$
I	0.0010	–	–
C	– x	+ x	+ x
E	0.0010 – x	+ x	+ x

Check the assumption:
If $[(CH_3)_2NH] \geq 400 \times K_a$ then the simplification will work
$400 \times (9.6 \times 10^{-4}) = 0.384$
$[(CH_3)_2NH] = 0.0010 < 0.384$
The simplification will not work.

$$K_b = \frac{[x][x]}{[0.0010 - x]} = 9.6 \times 10^{-4}$$

$9.6 \times 10^{-4}(0.0010 - x) = x^2$
$x^2 + (9.6 \times 10^{-4})x - (9.6 \times 10^{-7}) = 0$
$x = 6.1 \times 10^{-4}$
$x = [OH^-]$
$pOH = -\log[OH^-] = -\log(6.1 \times 10^{-4}) = 3.21$
$pH = 14.00 - pOH = 14.00 - 3.21 = 10.79$

16.25 $HF \rightleftharpoons H^+ + F^-$

$$K_a = \frac{[H^+][F^-]}{[HF]} = 6.8 \times 10^{-4}$$

	[HF]	$[H^+]$	$[F^-]$
I	0.030	–	–
C	–x	+x	+x
E	0.030–x	+x	+x

$$K_a = \frac{[x][x]}{[0.030 - x]} = 6.8 \times 10^{-4}$$

Solving using the quadratic formula:

$6.8 \times 10^{-4}(0.030 - x) = x^2$

$x^2 + (6.8 \times 10^{-4})x - (2.04 \times 10^{-5}) = 0$

$$x = \frac{-\left(6.8\times10^{-4}\right) \pm \sqrt{\left(6.8\times10^{-4}\right)^2 - 4(1)\left(-2.04\times10^{-5}\right)}}{2(1)}$$

$x = 4.18 \times 10^{-3}$

$x = [H^+]$

$pH = -\log[H^+] = -\log(4.18 \times 10^{-3}) = 2.38$

Solving by simplification

$$K_a = \frac{[x][x]}{[0.030]} = 6.8 \times 10^{-4}$$

$x = 4.52 \times 10^{-3}$

$x = [H^+]$

$pH = -\log[H^+] = -\log(4.52 \times 10^{-3}) = 2.35$

The difference is a difference of 0.03 pH units.

16.26 $CH_3NH_2 + H_2O \rightleftharpoons CH_3NH_3^+ + OH^-$

$$K_b = \frac{[CH_3NH_3^+][OH^-]}{[CH_3NH_2]} = 4.4 \times 10^{-4}$$

	$[CH_3NH_2]$	$[CH_3NH_3^+]$	$[OH^-]$
I	0.0010	–	–
C	–x	+x	+x
E	0.0010–x	+x	+x

Check the assumption:

If $[(CH_3)_2NH] \geq 400 \times K_a$ then the simplification will work

$400 \times (4.4 \times 10^{-4}) = 0.176$

$[(CH_3)_2NH] = 0.0010 < 0.176$

The simplification will not work.

$$K_b = \frac{[x][x]}{[0.0010 - x]} = 4.4 \times 10^{-4}$$

$4.4 \times 10^{-4}(0.0010 - x) = x^2$

$x^2 + (4.4 \times 10^{-4})x - (4.4 \times 10^{-7}) = 0$

$$x = \frac{-\left(4.4\times10^{-4}\right) \pm \sqrt{\left(4.4\times10^{-4}\right)^2 - 4(1)\left(-4.4\times10^{-7}\right)}}{2(1)} = 4.8 \times 10^{-4}$$

$x = [OH^-]$

$pOH = -\log[OH^-] = -\log(4.8 \times 10^{-4}) = 3.32$

$pH = 14.00 - pOH = 14.00 - 3.32 = 10.68$

16.27 In the acetate buffer, there are $HC_2H_3O_2$ and $C_2H_3O_2^-$ present.
Upon addition of a strong acid, the concentration of $HC_2H_3O_2$ will increase:

$$H^+ + C_2H_3O_2^- \rightarrow HC_2H_3O_2$$

When a strong base is added, it reacts with the acid to form more of the acetate ion; therefore the concentration of the acetic acid will decrease:

$$HC_2H_3O_2 + OH^- \rightarrow C_2H_3O_2^- + H_2O$$

16.28 (a) $H^+ + NH_3 \rightarrow NH_4^+$
(b) $OH^- + NH_4^+ \rightarrow H_2O + NH_3$

16.29 The equation is: $C_2H_3O_2^- + H_2O \rightleftharpoons HC_2H_3O_2 + OH^-$
Start by determining K_b for acetate ion using $K_w = K_aK_b$

$K_b = K_w/K_a = 1.0 \times 10^{-14}/1.8 \times 10^{-5} = 5.6 \times 10^{-10}$

$$K_b = \frac{[HC_2H_3O_2][OH^-]}{[C_2H_3O_2^-]}$$

	$[C_2H_3O_2^-]$	$[HC_2H_3O_2]$	$[OH^-]$
I	0.11	0.090	–
C	–x	+x	+x
E	0.11 – x	0.090 +x	+x

$$K_b = \frac{[x][0.090 + x]}{[0.11 - x]}$$

assume x << 0.090 and solve for x

$x = [OH^-] = 6.8 \times 10^{-10}$

pOH = 9.16

pH = 14.00 – 9.16 = 4.84, the difference is due to rounding errors

16.30 Find the concentrations of the acetic acid and the acetate ion:

$$M\ HC_2H_3O_2 = (100.0\text{ g } HC_2H_3O_2)\left(\frac{1\text{ mol } HC_2H_3O_2}{60.053\text{ g } HC_2H_3O_2}\right)\left(\frac{1}{1\text{ L solution}}\right) = 1.665\ M$$

$$M\ NaC_2H_3O_2 = (100.0\text{ g } HC_2H_3O_2)\left(\frac{1\text{ mol } HC_2H_3O_2}{82.035\text{ g } HC_2H_3O_2}\right)\left(\frac{1}{1\text{ L solution}}\right) = 1.219\ M$$

$$HC_2H_3O_2 \rightleftharpoons C_2H_3O_2^- + H^+$$

$$K_a = \frac{[C_2H_3O_2^-][H^+]}{[HC_2H_3O_2]} = 1.8 \times 10^{-5}$$

	$[HC_2H_3O_2]$	$[C_2H_3O_2^-]$	$[H^+]$
I	1.665	1.219	–
C	–x	+x	+x
E	1.665 – x	1.219 +x	+x

$$K_a = \frac{(1.219 + x)(x)}{(1.665 - x)} = 1.8 \times 10^{-5}$$

Assume that x is small and solve for x

$x = 2.5 \times 10^{-5}$

$pH = -\log[H^+] = -\log(2.5 \times 10^{-5}) = 4.61$

16.31 Use propanoic acid. It has a pK_a of 4.87 which is within the range of $pH = pK_a \pm 1$
You may assume no volume change when adding the sodium salt of the acid to the solution.

$$pH = pK_a + \log\frac{[salt]}{[acid]}$$

[acid] = 0.200 *M*

$$5.25 = 4.87 + \log\frac{[salt]}{[0.200]}$$

[salt] = 0.480 *M*

$$\text{g salt} = (0.5000\ L)\left(\frac{0.480\ \text{mol}\ NaC_3H_5O_2}{1\ L}\right)\left(\frac{96.06\ \text{g}\ NaC_3H_5O_2}{1\ \text{mol}\ NaC_3H_5O_2}\right) = 23.1\ \text{g}\ NaC_3H_5O_2$$

Propanoic acid buffer, 23.1 g $NaC_3H_5O_2$

Using the same calculations, the following buffers can also be used:
Acetic acid buffer, 26.2 g $NaC_2H_3O_2$
Hydrazoic acid buffer, 21.0 NaN_3,
Butanoic acid buffer, 29.6 g $NaC_4H_7O_2$,

16.32 Yes, formic acid and sodium formate would make a good buffer solution since $pK_a = 3.74$ and the desired pH is within one pH unit of this value.

Using Equation 16.10; $[H^+] = K_a \times \frac{(\text{mol}\ HCHO_2)_{initial}}{(\text{mol}\ CHO_2^-)_{initial}}$

Rearranging and substituting the known values we get;

$$\frac{\text{mol}\ HCHO_{2\,initial}}{\text{mol}\ CHO_2^-{}_{initial}} = \frac{[H^+]}{K_a} = \frac{1.3 \times 10^{-4}}{1.8 \times 10^{-4}} = 0.72\ \text{mol}\ HCHO_2 \text{ for every mol } CHO_2^-.$$

For 0.10 mol $HCHO_2$, 0.14 mol $NaCHO_2$ would be needed. Converting to grams, this is (0.14 mol $NaCHO_2$)(68.02 g/1 mol) = 9.5 g $NaCHO_2$.

16.33 The NaOH added to a buffer solution will react with the $HC_2H_3O_2$.
$NaOH + HC_2H_3O_2 \rightarrow NaC_2H_3O_2 + H_2O$
First, calculate the amount of acid and base after the addition of the NaOH:
mol $NaC_2H_3O_2$ = 1.00 mol $NaC_2H_3O_2$ + 0.15 mol $NaC_2H_3O_2$ = 1.15 mol $NaC_2H_3O_2$
mol $HC_2H_3O_2$ = 1.00 mol $HC_2H_3O_2$ – 0.15 mol $NaC_2H_3O_2$ = 0.85 mol $HC_2H_3O_2$

$$M\ NaC_2H_3O_2 = \frac{1.15\ \text{mol}\ NaC_2H_3O_2}{1\ \text{L solution}}$$

$$M\ HC_2H_3O_2 = \frac{0.85\ \text{mol}\ HC_2H_3O_2}{1\ \text{L solution}}$$

$$pH = pK_a + \log\frac{[salt]}{[acid]} = 4.74 + \log\frac{(1.15)}{(0.85)} = 4.87$$

The pH of the initial solution was 4.74 since the concentration of the acid equals the concentration of the salt.
The pH change:
4.74 – 4.87 = 0.13 pH units

16.34 Calculate the moles of the salt and the acid, then calculate the pH of the solution.

$$\text{mol } NH_3 = (50.0 \text{ g } NH_3)\left(\frac{1 \text{ mol } NH_3}{17.03 \text{ g } NH_3}\right) = 2.94 \text{ mol } NH_3$$

$$\text{mol } NH_4Cl = (50.0 \text{ g } NH_4Cl)\left(\frac{1 \text{ mol } NH_4Cl}{53.49 \text{ g } NH_4Cl}\right) = 0.935 \text{ mol } NH_4Cl$$

$NH_4^+ \rightarrow NH_3 + H^+$

$$K_a = \frac{K_w}{K_b} = \frac{1.0 \times 10^{-14}}{1.8 \times 10^{-5}} = 5.6 \times 10^{-10}$$

$pK_a = -\log(K_a) = -\log(5.6 \times 10^{-10}) = 9.26$

$$pH = pK_a + \log\frac{[\text{salt}]}{[\text{acid}]} = 9.26 + \log\frac{(2.94)}{(0.935)} = 9.76$$

If 5.00 g of HCl is add, the solution will become more acidic:
$NH_3 + H^+ \rightarrow NH_4^+$

$$\text{mol HCl} = (5.00 \text{ g HCl})\left(\frac{1 \text{ mol HCl}}{36.46 \text{ g HCl}}\right) = 0.137 \text{ mol HCl}$$

The number of moles of NH_4^+ will increase by this amount and the number of moles of NH_3 will decrease by this amount.
$\text{mol } NH_3 = 2.94 \text{ mol } NH_3 - 0.137 \text{ mol} = 2.80 \text{ mol } NH_3$
$\text{mol } NH_4^+ = 0.935 \text{ mol} + 0.137 \text{ mol} = 1.07 \text{ mol } NH_4^+$

$$pH = pK_a + \log\frac{[\text{salt}]}{[\text{acid}]} = 9.26 + \log\frac{(2.80)}{(1.07)} = 9.67$$

16.35 $H_3PO_4 \rightleftharpoons H^+ + H_2PO_4^- \quad K_a = \dfrac{[H^+][H_2PO_4^-]}{[H_3PO_4]}$

$H_2PO_4^- \rightleftharpoons H^+ + HPO_4^{2-} \quad K_a = \dfrac{[H^+][HPO_4^{2-}]}{[H_2PO_4^-]}$

$HPO_4^{2-} \rightleftharpoons H^+ + PO_4^{3-} \quad K_a = \dfrac{[H^+][PO_4^{3-}]}{[HPO_4^{2-}]}$

16.36 $[H^+]$ is determined by the first protic equilibrium:
$H_2C_6H_6O_6 \rightleftharpoons H^+ + HC_6H_6O_6^-$

The mass action expression is: $K_{a_1} = 6.8 \times 10^{-5} = x^2/0.10$
$x = [H^+] = 2.6 \times 10^{-3}\ M$ $\qquad pH = -\log(2.6 \times 10^{-3}) = 2.58$

The concentration of the anion, $[HC_6H_6O_6^-]$, is given almost entirely by the second ionization equilibrium:
$HC_6H_6O_6^- \rightleftharpoons H^+ + C_6H_6O_6^{2-}$ for which the mass action expression is:

$$K_{a2} = \frac{[H^+][C_6H_6O_6^{2-}]}{[HC_6H_6O_6^-]} = 2.7 \times 10^{-12}$$

We have used the value for K_{a_2} from Table 16.3. Using the value of x from the first step above gives:

$$2.7 \times 10^{-12} = \frac{(2.6 \times 10^{-3})[C_6H_6O_6^{2-}]}{(2.6 \times 10^{-3})}$$

$[HC_6H_6O_6^-] = 2.7 \times 10^{-12}$

16.37 $CO_3^{2-}(aq) + H_2O \rightleftharpoons HCO_3^-(aq) + OH^-(aq)$

$$K_b = \frac{K_w}{K_{a_2}} = \frac{1.0\times10^{-14}}{4.7\times10^{-11}} = 2.1 \times 10^{-4}$$

$$K_b = 2.1 \times 10^{-4} = \frac{[HCO_3^-][OH^-]}{[CO_3^{2-}]}$$

	$[CO_3^{2-}]$	$[HCO_3^-]$	$[OH^-]$
I	0.10	–	–
C	–x	+x	+x
E	0.10–x	+x	+x

$$K_b = 2.1 \times 10^{-4} = \frac{(x)(x)}{(0.10-x)}$$

$x = [OH^-] = 4.5 \times 10^{-3}\ M$
pOH = –log(4.5 × 10^{-3}) = 2.35
pH = 14.00 – 2.35 = 11.66
It is basic, but it may not be a substitute for $NaHCO_3$.

16.38 The equilibrium we are interested in for this problem is:
$SO_3^{2-}(aq) + H_2O \rightleftharpoons HSO_3^-(aq) + OH^-(aq)$

$K_b = K_w/K_{a_2} = 1.0 \times 10^{-14} / 6.6 \times 10^{-8} = 1.5 \times 10^{-7}$

$$K_b = 1.5 \times 10^{-7} = \frac{[HSO_3^-][OH^-]}{[SO_3^{2-}]}$$

	$[SO_3^{2-}]$	$[HSO_3^-]$	$[OH^-]$
I	0.20	–	–
C	–x	+x	+x
E	0.20–x	+x	+x

Substituting these values into the mass action expression gives:

$$K_b = 1.5 \times 10^{-7} = \frac{(x)(x)}{0.20-x}$$

Assume that x << 0.20 and solving gives x = 1.7 × 10^{-4}.
$x = [OH^-] = 1.7 \times 10^{-4}\ M$
pOH = –log(1.7 × 10^{-4}) = 3.76
pH = 14.00 – pOH = 14.00 – 3.76 = 10.24

16.39 For weak polyprotic acids, the concentration of the polyvalent ions is equal to the volume of K_{an} where n is the valency. By analogy, the concentration of H_2SO_3 in 0.010 *M* Na_2SO_3 will be equal to K_{b2} for SO_3^{2-}.

16.40 (a) H_2O, K^+, $HC_2H_3O_2$, H^+, $C_2H_3O_2^-$, and OH^-
$[H_2O] > [K^+] > [C_2H_3O_2^-] > [OH^-] > [HC_2H_3O_2] > [H^+]$

(b) H_2O, $HC_2H_3O_2$, H^+, $C_2H_3O_2^-$, and OH^-
$[H_2O] > [HC_2H_3O_2] > [H^+] = [C_2H_3O_2^-] > [OH^-]$

(c) H_2O, K^+, $HC_2H_3O_2$, H^+, $C_2H_3O_2^-$, and OH^-
$[H_2O] > [K^+] > [OH^-] > [C_2H_3O_2^-] > [HC_2H_3O_2] > [H^+]$

(d) H_2O, K^+, $HC_2H_3O_2$, H^+, $C_2H_3O_2^-$, and OH^-
$[H_2O] > [HC_2H_3O_2] > [K^+] > [C_2H_3O_2^-] > [OH^-] > [H^+]$

16.41 $HCHO_2 + H_2O \rightleftharpoons H_3O^+ + CHO_2^-$

$$Ka = \frac{[H_3O^+][CHO_2^-]}{[HCHO_2]} = 1.8 \times 10^{-4}$$

(a)

	$[HCHO_2]$	$[H_3O^+]$	$[CHO_2^-]$
I	0.100	–	–
C	–x	+x	+x
E	0.100–x	+x	+x

Assume x << 0.100. Solving we get $x = [H_3O^+] = 4.2 \times 10^{-3}$. The pH is 2.37.

(b) $[HCHO_2] = [CHO_2^-]$ so $[H_3O^+] = Ka = 1.8 \times 10^{-4}$ and the pH = 3.74.

(c) $$\text{mol base added} = (15.0\text{ mL})\left(\frac{0.100\text{ mol}}{1000\text{ mL}}\right) = 1.50 \times 10^{-3}$$

$$\text{mol acid initially present} = (20.0\text{ mL})\left(\frac{0.10\text{ mol}}{1000\text{ mL}}\right) = 2.00 \times 10^{-3}$$

$$\text{excess acid} = 2.00 \times 10^{-3} - 1.50 \times 10^{-3} = 0.50 \times 10^{-3}\text{ moles acid}$$

$$[\text{acid}] = \frac{0.50 \times 10^{-3}\text{ moles}}{(35\text{ mL})\left(\frac{1\text{ L}}{1000\text{ mL}}\right)} = 1.43 \times 10^{-2}\ M$$

$$[\text{base}] = \frac{1.50 \times 10^{-3}\text{ moles}}{(35\text{ mL})\left(\frac{1\text{ L}}{1000\text{ mL}}\right)} = 4.29 \times 10^{-2}\ M$$

Substituting into the equilibrium expression and solving we get $[H_3O^+] = 6.00 \times 10^{-5}$ and the pH = 4.22.

(d) We now have a solution of formate ion with a concentration of 0.0500 *M*. We need K_b for formate ion: $K_b = K_w/K_a = 5.6 \times 10^{-11}$. If we set up the equilibrium problem and solve we get: $[OH^-] = 1.7 \times 10^{-6}$.

The pOH = 5.78 and the pH = 8.22.

16.42 $$\text{mol base added} = (30.0\text{ mL})\left(\frac{0.15\text{ mol}}{1000\text{ mL}}\right) = 4.50 \times 10^{-3}$$

$$\text{mol acid initially present} = (50.0\text{ mL})\left(\frac{0.20\text{ mol}}{1000\text{ mL}}\right) = 1.00 \times 10^{-2}$$

excess acid = $(1.00 \times 10^{-3}) - (4.50 \times 10^{-3}) = 5.50 \times 10^{-3}$ mol acid

$$[\text{acid}] = \frac{5.50 \times 10^{-3}\text{ moles}}{(80\text{ mL})\left(\frac{1\text{ L}}{1000\text{ mL}}\right)} = 6.88 \times 10^{-2}\ M$$

$$[\text{base}] = \frac{4.50 \times 10^{-3}\text{ moles}}{(80\text{ mL})\left(\frac{1\text{ L}}{1000\text{ mL}}\right)} = 5.63 \times 10^{-2}\ M$$

$HCHO_2 + H_2O \rightleftharpoons H_3O^+ + CHO_2^-$

$$Ka = \frac{[H_3O^+][CHO_2^-]}{[HCHO_2]}$$

	$[HCHO_2]$	$[H_3O^+]$	$[CHO_2^-]$
I	6.88×10^{-2}	–	5.63×10^{-2}
C	–x	+x	+x
E	6.88×10^{-2}–x	+x	5.63×10^{-2}+x

Assume $x << 5.63 \times 10^{-2}$

$$K_a = \frac{[x][5.63 \times 10^{-2}]}{[6.88 \times 10^{-2}]}$$

$$x = 2.20 \times 10^{-2} = [H_3O^+]$$

pH = 3.66

Review Problems

16.32 At 25 °C, $K_a \times K_b = K_w$
$K_b = K_w/K_a = 1.0 \times 10^{-14} \div 6.8 \times 10^{-4} = 1.5 \times 10^{-11}$

16.34 At 25 °C, $K_a \times K_b = K_w$
$K_b = K_w/K_a = 1.0 \times 10^{-14} \div 1.4 \times 10^{-4} = 7.1 \times 10^{-11}$

16.36 pH = 3.22 $[H^+] = 10^{-3.22} = 6.03 \times 10^{-4}\ M$

$$\text{Percentage ionization} = \frac{\text{moles ionized per liter}}{\text{moles available per liter}} \times 100\%$$

$$\text{Percentage ionization} = \frac{6.03 \times 10^{-4}\ M}{0.20\ M} \times 100\% = 0.30\%$$

$HA \rightleftharpoons H^+ + A^-$

	[HA]	$[H^+]$	$[A^-]$
I	0.20	–	–
C	-6.03×10^{-4}	$+6.03 \times 10^{-4}$	$+6.03 \times 10^{-4}$
E	0.20	6.03×10^{-4}	6.03×10^{-4}

$$K_a = \frac{[H^+][A^-]}{[HA]} = \frac{[6.03 \times 10^{-4}][6.03 \times 10^{-4}]}{[0.20]} = 1.82 \times 10^{-6}$$

16.38 $HIO_4 \rightleftharpoons H^+ + IO_4^-$

$$K_a = \frac{[H^+][IO_4^-]}{[HIO_4]}$$

	$[HIO_4]$	$[H^+]$	$[IO_4^-]$
I	0.10	–	–
C	– x	+ x	+ x
E	0.10 – x	+ x	+ x

We know that at equilibrium $[H^+] = 0.038\ M = x$. The equilibrium concentrations of the other components of the mixture are:

$[HIO_4] = 0.10 - x = 0.062\ M$ and $[IO_4^-] = x = 0.038\ M$.

Substituting the above values for equilibrium concentrations into the mass action expression gives:

$$K_a = \frac{(0.038)(0.038)}{0.062} = 2.3 \times 10^{-2}$$

$pK_a = -\log(K_a) = -\log(2.3 \times 10^{-2}) = 1.6$

16.40 $pOH = 14.00 - pH = 14.00 - 11.86 = 2.14$
$[OH^-] = 10^{-pOH} = 10^{-2.14} = 7.2 \times 10^{-3}\ M$

$CH_3CH_2NH_2 + H_2O \rightleftharpoons CH_3CH_2NH_3^+ + OH^-$

$$K_b = \frac{[CH_3CH_2NH_3^+][OH^-]}{[CH_3CH_2NH_2]}$$

	$[CH_3CH_2NH_2]$	$[CH_3CH_2NH_3^+]$	$[OH^-]$
I	0.10	–	–
C	–x	+x	+x
E	0.10–x	+x	+x

In the equilibrium analysis, the value of x is, therefore, equal to $7.2 \times 10^{-3}\ M$. Therefore, our equilibrium concentrations are $[CH_3CH_2NH_3^+] = [OH^-] = 7.2 \times 10^{-3}\ M$, and $[CH_3CH_2NH_2] = 0.10\ M - 7.2 \times 10^{-3}\ M = 0.09\ M$.

Substituting these values into the mass action expression gives:

$$K_b = \frac{(7.2 \times 10^{-3})(7.2 \times 10^{-3})}{0.093} = 5.6 \times 10^{-4}$$

$pK_b = -\log(K_b) = -\log(5.6 \times 10^{-4}) = 3.25$

$$\text{Percentage ionization} = \frac{\text{moles ionized per liter}}{\text{moles available per liter}} \times 100\%$$

$$\text{Percentage ionization} = \frac{7.2 \times 10^{-3}}{0.10} \times 100\% = 7.2\%$$

16.42 $HC_3H_5O_2 + H_2O \rightleftharpoons H_3O^+ + C_3H_5O_2^-$

$$K_a = \frac{[H_3O^+][C_3H_5O_2^-]}{[HC_3H_5O_2]} = 1.4 \times 10^{-4}$$

	$[HC_3H_5O_2]$	$[H_3O^+]$	$[C_3H_5O_2^-]$
I	0.15	–	–
C	– x	+ x	+ x
E	0.150 – x	+ x	+ x

Assume x << 0.15

$$K_a = \frac{[x][x]}{[0.150]} = 1.4 \times 10^{-4} \quad x = 4.6 \times 10^{-3} = [H_3O^+]$$

pH = 2.34

$[HC_3H_5O_2] = 0.145$
$[H^+] = 4.6 \times 10^{-3}$
$[C_3H_5O_2^-] = 4.6 \times 10^{-3}$

16.44 $K_b = 10^{-pKb} = 10^{-5.79} = 1.6 \times 10^{-6}$
$Cod + H_2O \rightleftharpoons HCod^+ + OH^-$

$$K_b = \frac{[HCod^+][OH^-]}{[Cod]} = 1.6 \times 10^{-6}$$

	[Cod]	[HCod$^+$]	[OH$^-$]
I	0.020	–	–
C	– x	+ x	+ x
E	0.020 – x	+ x	+ x

Substituting these values into the mass action expression gives:

$$K_b = \frac{(x)(x)}{0.020 - x} = 1.6 \times 10^{-6}$$

If we assume that x << 0.020 we get; $x^2 = 3.2 \times 10^{-8}$,
$x = 1.8 \times 10^{-4}\ M = [OH^-]$
$pOH = -\log[OH^-] = -\log(1.8 \times 10^{-4}) = 3.74$
$pH = 14.00 - pOH = 14.00 - 3.74 = 10.26$

16.46 $[H^+] = 10^{-pH} = 10^{-2.54} = 2.9 \times 10^{-3}\ M$
$HC_2H_3O_2 \rightleftharpoons H^+ + C_2H_3O_2^-$

$$K_a = \frac{[H^+][C_2H_3O_2^-]}{[HC_2H_3O_2]} = 1.8 \times 10^{-5}$$

	[$HC_2H_3O_2$]	[H^+]	[$C_2H_3O_2^-$]
I	Z	–	–
C	– x	+ x	+ x
E	Z – x	+ x	+ x

Substituting these values into the mass action expression gives:

$$K_a = \frac{(x)(x)}{Z - x} = 1.8 \times 10^{-5}$$

Assuming x << Z and knowing that $x = 2.9 \times 10^{-3}\ M$, we can solve for Z and find Z = 0.47. The initial concentration of $HC_2H_3O_2$ is 0.47 *M*.

16.48 $CN^- + H_2O \rightleftharpoons HCN + OH^-$

$$K_b = \frac{[HCN][OH^-]}{[CN^-]} = \frac{K_w}{K_a} = \frac{1.0 \times 10^{-14}}{6.2 \times 10^{-10}} = 1.6 \times 10^{-5}$$

	[CN^-]	[HCN]	[OH^-]
I	0.0050	–	–
C	– x	+ x	+ x
E	0.0050 – x	+ x	+ x

Assume that x << 0.0050

$$K_b = \frac{(x)(x)}{0.0050} = 1.6 \times 10^{-5} \qquad x = 2.8 \times 10^{-4} = [OH^-]$$

Wait!!!! 2.8×10^{-4} is not << 0.0050
So, use the method of successive approximations.

$$K_b = \frac{(x)(x)}{0.0050 - 0.00028} = 1.6 \times 10^{-5} \qquad x = 2.7 \times 10^{-4}$$

$$Kb = \frac{(x)(x)}{0.0050 - 0.00027} = 1.6 \times 10^{-5} \qquad x = 2.7 \times 10^{-4}$$

$x = 2.7 \times 10^{-4} = [OH^-]$
pOH = 3.56
pH = 10.44

16.50 $K_a = 10^{-pKa} = 10^{-4.92} = 1.2 \times 10^{-5}$
H–Paba $\rightleftharpoons$ H^+ + $Paba^-$

$$K_a = \frac{[H^+][Paba^-]}{[H-Paba]} = 1.2 \times 10^{-5}$$

	[H–Paba]	$[H^+]$	$[Paba^-]$
I	0.030	–	–
C	– x	+ x	+ x
E	0.030 – x	+ x	+ x

Substituting the above values for equilibrium concentrations into the mass action expression and assuming that x << 0.030 gives:

$$K_a = \frac{[x][x]}{[0.030]} = 1.2 \times 10^{-5}$$

$x^2 = 3.6 \times 10^{-7}$, $x = 6.0 \times 10^{-4}\ M = [H^+]$
$pH = -\log[H^+] = -\log(6.0 \times 10^{-4}) = 3.22$

16.52 NaCN will be basic in solution since CN^- is a basic ion and Na^+ is a neutral ion.
$CN^- + H_2O \rightleftharpoons HCN + OH^-$
For HCN, $K_a = 6.2 \times 10^{-10}$, we need K_b for CN^-;
$K_b = K_w/K_a = (1.0 \times 10^{-14}) \div (6.2 \times 10^{-10}) = 1.6 \times 10^{-5}$

$$K_b = \frac{[HCN][OH^-]}{[CN^-]} = 1.6 \times 10^{-5}$$

	$[CN^-]$	[HCN]	$[OH^-]$
I	0.20	–	–
C	– x	+ x	+ x
E	0.20 – x	+ x	+ x

Substituting these values into the mass action expression gives:

$$K_b = \frac{(x)(x)}{0.20 - x} = 1.6 \times 10^{-5}$$

Assuming that x << 0.20 we can solve for x an determine;
$x = 1.8 \times 10^{-3}\ M = [OH^-]$
$pOH = -\log[OH^-] = -\log(1.8 \times 10^{-3}) = 2.74$
pH = 14.00 – pOH = 14.00 – 2.74 = 11.26
Concentration of HCN is equal to that of hydroxide ion: $1.8 \times 10^{-3}\ M$

16.54 A solution of CH_3NH_3Cl will be acidic since the Cl^- ion is neutral and the $CH_3NH_3^+$ ion is acidic.
$CH_3NH_3^+ \rightleftharpoons H^+ + CH_3NH_2$
For CH_3NH_2, $K_b = 4.4 \times 10^{-4}$. We need K_a for $CH_3NH_3^+$;
$K_a = K_w/K_b = (1.0 \times 10^{-14}) \div (4.4 \times 10^{-4}) = 2.3 \times 10^{-11}$

$$K_a = \frac{[H^+][CH_3NH_2]}{[CH_3NH_3^+]} = 2.3 \times 10^{-11}$$

	$[CH_3NH_3^+]$	$[H^+]$	$[CH_3NH_2]$
I	0.15	–	–
C	– x	+ x	+ x
E	0.15 – x	+ x	+ x

Substituting these values into the mass action expression gives:

$$K_a = \frac{(x)(x)}{0.15 - x} = 2.3 \times 10^{-11}$$

Assuming that x << 0.15 we can solve for x an determine;
$x = 1.9 \times 10^{-6}\ M = [H_3O^+]$
$pH = -\log[H_3O^+] = -\log(1.9 \times 10^{-6}) = 5.72$

16.56 Let HNic symbolize the nicotinic acid
$Nic^- + H_2O \rightleftharpoons HNic + OH^-$

$$K_b = \frac{[HNic][OH^-]}{[Nic^-]}$$

	$[Nic^-]$	[HNic]	$[OH^-]$
I	0.18	–	–
C	– x	+ x	+ x
E	0.18 – x	+ x	+ x

Assume x << 0.18

$$K_b = \frac{(x)(x)}{0.18}$$

We know $x = [OH^-]$
pH = 9.05 and pOH = 4.95
and $[OH^-] = 10^{-pOH} = 1.1 \times 10^{-5} = x$

$$K_b = \frac{(1.1 \times 10^{-5})^2}{0.18} = 7.0 \times 10^{-10}$$

$$K_a = \frac{1 \times 10^{-14}}{7.0 \times 10^{-10}} = 1.4 \times 10^{-5}$$

16.58 $OCl^- + H_2O \rightleftharpoons HOCl + OH^-$

$$K_b = \frac{[HOCl][OH^-]}{[OCl^-]} = \frac{K_w}{K_a} = \frac{1.0 \times 10^{-14}}{3.0 \times 10^{-8}} = 3.3 \times 10^{-7}$$

$$[OCl^-] = \left(\frac{5.0\text{ g NaOCl}}{100\text{ g solution}}\right)\left(\frac{1\text{ mol NaOCl}}{74.5\text{ g NaOCl}}\right)\left(\frac{1.0\text{ g}}{1\text{ mL}}\right)\left(\frac{1000\text{ mL}}{1\text{ L}}\right)$$
$= 0.67\ M$

	$[OCl^-]$	$[HOCl]$	$[OH^-]$
I	0.67	–	–
C	– x	+ x	+ x
E	0.67 – x	+ x	+ x

Assume that x << 0.67

$$K_b = \frac{(x)(x)}{0.67} = 3.3 \times 10^{-7} \qquad x = 4.7 \times 10^{-4} = [OH^-]$$

$pOH = 3.33$

$pH = 10.67$

16.60 $HC_2H_3O_2 \rightleftharpoons H^+ + C_2H_3O_2^-$

$$K_a = \frac{[H^+][C_2H_3O_2^-]}{[HC_2H_3O_2]} = 1.8 \times 10^{-5}$$

	$[HC_2H_3O_2]$	$[H^+]$	$[C_2H_3O_2^-]$
I	0.15	–	0.25
C	– x	+ x	+ x
E	0.15 – x	+ x	0.25 + x

Substituting these values into the mass action expression gives:

$$K_a = \frac{(x)(0.25+x)}{0.15-x} = 1.8 \times 10^{-5}$$

Assume that x << 0.15 *M* and x << 0.25 *M*, then;

$$x \times \left(\frac{0.25}{0.15}\right) \approx 1.8 \times 10^{-5}$$

$$x \approx \left(\frac{0.15}{0.25}\right) \times 1.8 \times 10^{-5}$$

$$x \approx 1.1 \times 10^{-5}\ M = [H^+]$$

$pH = -\log [H^+] = 4.97$

16.62 The equilibrium we will consider in this problem is: $NH_3 + H_2O \rightleftharpoons NH_4^+ + OH^-$

$$K_b = \frac{[NH_4^+][OH^-]}{[NH_3]} = 1.8 \times 10^{-5}$$

$$= \frac{(0.45)[OH^-]}{0.25} = 1.8 \times 10^{-5}$$

$[OH^-] = 1.0 \times 10^{-5}\ M$

$pOH = -\log[OH^-] = -\log(1.0 \times 10^{-5}) = 5.00$

$pH = 14.00 - pOH = 14.00 - 5.00 = 9.00$

16.64 The answer is not intuitive due to the dilution effect of the added acid.
For a conjugate acid-base pair, $K_a \times K_b = 1.00 \times 10^{-14}$. The K_b of ammonia is given in Table 16.2, so we can find the K_a for the ammonium ion:

$K_a \times K_b = 1.00 \times 10^{-14}$

$K_a \times 1.8 \times 10^{-5} = 1.00 \times 10^{-14}$

$K_a = 5.56 \times 10^{-10}$

Using equation 16.10:

$$[H^+] = K_a \frac{[HA]}{[A^-]}$$

$$[H^+] = K_a \frac{[NH_4^+]}{[NH_3]}$$

$$[H^+] = 5.56\times10^{-10} \frac{[0.20]}{[0.25]} = 4.45\times10^{-10}$$

Therefore, the initial pH of the buffer = $-\log(4.45 \times 10^{-10}) = 9.35$

Initial amounts in the solution are:
mol NH_3 = (0.25 mol/L)0.25 L) = 0.0625 mol NH_3
mol NH_4^+ = (0.20 mol/L)(0.25 L) = 0.050 mol NH_4^+

The added acid (0.0250 L)(0.10 mol/L) = 0.00250 mol HCl will react with the ammonia present in the buffer solution. Assume the added acid reacts completely. For each mole of acid added, one mole of NH_3 is converted to NH_4^+. Since 0.00250 mol of acid is added;

mol $NH_4^+{}_{final}$ = (0.050 + 0.00250) mol = 0.053 mol
mol NH_{3final} = (0.0625 – 0.00250) mol = 0.060 mol

The final volume of solution is 250 mL + 25.0 mL = 275 mL.

The final concentrations are:

$[NH_4^+]_{final}$ = 0.053 mol/0.275 L = 0.19 M NH_4^+
$[NH_3]_{final}$ = 0.060 mol/0.275 L = 0.22 M NH_3

So the *changes* in concentrations are:

$\Delta[NH_4^+] = [NH_4^+]_{final} - [NH_4^+]_{initial} = 0.19\ M - 0.20\ M = -0.01\ M$
$\Delta[NH_3] = [NH_3]_{final} - [NH_3]_{initial} = 0.22\ M - 0.25\ M = -0.03\ M$

16.66 The initial pH of the buffer is 4.97 as determined in Exercise 16.60. The added acid, 0.025 mol, will react with the acetate ion present in the buffer solution. Assume the added acid reacts completely. For each mole of acid added, one mole of $C_2H_3O_2^-$ is converted to $HC_2H_3O_2$. Since 0.025 mol of acid is added;

$[HC_2H_3O_2]_{final} = (0.15 + 0.025)\ M = 0.175\ M$
$[C_2H_3O_2^-]_{final} = (0.25 - 0.025)\ M = 0.225\ M$

Now, substitute these values into the mass action expression to calculate the final $[H^+]$ in solution;

$$\frac{[H^+](0.225)}{(0.175)} = 1.8 \times 10^{-5}$$

$[H^+] = 1.4 \times 10^{-5}$ mol L^{-1} and the pH = 4.85

The pH of the solution changes by 4.97 – 4.85 = –0.12 pH units upon addition of the acid.

16.68 The initial pH is 9.00 as calculated in Exercise 16.62. For every mole of H^+ added, one mol of NH_3 will be changed to one mol of NH_4^+. Since we added 0.040 mol H^+:

$$\left[NH_4^+\right]_{final} = 0.45\ M + 0.040\ M = 0.49\ M$$

$$\left[NH_3\right]_{final} = 0.25\ M - 0.040\ M = 0.21\ M$$

Using these new concentrations, we can calculate a new pH:

$$K_b = \frac{\left[NH_4^+\right]\left[OH^-\right]}{\left[NH_3\right]} = \frac{(0.49)\left[OH^-\right]}{0.21} = 1.8 \times 10^{-5}$$

$[OH^-] = 7.7 \times 10^{-6}\ M$, the pOH = 5.11 and the pH = 8.89

As expected, when an acid is added, the pH decreases. In this problem, the pH decreases by 0.11 pH units from 9.00 to 8.89.

16.70 $$pH = pK_a + \log\frac{[anion]}{[acid]} = pK_a + \log\frac{[A^-]}{[HA]}$$

$4.00 = 4.74 + \log([NaC_2H_3O_2]/[HC_2H_3O_2])$
$-0.74 = \log[NaC_2H_3O_2]/[HC_2H_3O_2]$
$[NaC_2H_3O_2]/[HC_2H_3O_2] = 0.18$
$[NaC_2H_3O_2] = 0.18 \times [HC_2H_3O_2] = 0.18 \times 0.15 = 0.027\ M$
Thus to the 1 L of acetic acid solution we add: 0.027 mol $NaC_2H_3O_2 \times 82.0$ g/mol = 2.2 g $NaC_2H_3O_2$.

16.72 The equilibrium is; $HC_2H_3O_2 \rightleftharpoons H^+ + C_2H_3O_2^-$

$$K_a = \frac{[H^+][C_2H_3O_2^-]}{[HC_2H_3O_2]} = 1.8 \times 10^{-5}$$

The initial pH is; $\frac{\left[H^+\right](0.110)}{0.100} = 1.8 \times 10^{-5}$, $[H^+] = 1.64 \times 10^{-5}\ M$ and pH = 4.78. In this calculation we are able to use either the molar concentration or the number of moles since the volume is constant in this portion of the problem.

In order to calculate the change in pH, we need to determine the concentrations of $HC_2H_3O_2$ and $C_2H_3O_2^-$ after the complete reaction of the added acid. One mole of $C_2H_3O_2^-$ will be consumed for every mole of acid added and one mole of $HC_2H_3O_2$ will be produced. The number of moles of acid added is:

$$\text{mol } H^+ = (30.00\ \text{mL HCl})\left(\frac{0.100\ \text{mol HCl}}{1000\ \text{mL HCl}}\right)\left(\frac{1\ \text{mol } H^+}{1\ \text{mol HCl}}\right) = 3.00 \times 10^{-3}\ \text{mol } H^+$$

Since the volume of the new concentrations of $HC_2H_3O_2$ and $C_2H_3O_2^-$ are:

$$\left[HC_2H_3O_2\right]_{final} = \frac{(0.100\ \text{mol} + 0.00300\ \text{mol})}{0.130\ \text{L}} = 0.792\ M$$

$$\left[C_2H_3O_2^-\right]_{final} = \frac{(0.110\ \text{mol} - 0.00300\ \text{mol})}{0.130\ \text{L}} = 0.823\ M$$

Note: The new volume has been used in these calculations.

$$K_a = \frac{[H^+][C_2H_3O_2^-]}{\left[HC_2H_3O_2\right]} = \frac{[H^+](0.792)}{(0.823)} = 1.8 \times 10^{-5}$$

$[H^+] = 1.87 \times 10^{-5}\ M$ and pH = 4.73

Notice that the change in pH is very small in spite of adding a strong acid. If the same amount of HCl were added to water, a completely different effect would be observed.

Since HCl is a strong acid, the $[H^+]$ in a water solution will be the result of the strong acid dissociation. We do, of course, need to account for the dilution. Using the dilution equation, $M_1V_1 = M_2V_2$, we determine the $[H^+] = 0.0231$ mol L^{-1} and the pH = 1.64. The change in pH in this case is 7.00 – 1.64 = 5.36 pH units. A significantly larger change!!

16.74 $H_2C_6H_6O_6 + H_2O \rightleftharpoons H_3O^+ + HC_6H_6O_6^-$ $\qquad K_{a1} = 6.7 \times 10^{-5}$

$HC_6H_6O_6^- + H_2O \rightleftharpoons H_3O^+ + C_6H_6O_6^{2-}$ $\qquad K_{a2} = 2.7 \times 10^{-12}$

	$[H_2C_6H_6O_6]$	$[H_3O^+]$	$[HC_6H_6O_6^-]$
I	0.15	–	–
C	– x	+ x	+ x
E	0.15 – x	+ x	+ x

Assume x << 0.15

$$K_a = \frac{[x][x]}{[0.15]} = 6.7 \times 10^{-5} \qquad x = 3.2 \times 10^{-3}$$

$[H_2C_6H_6O_6] \cong 0.15\ M$

$[H_3O^+] = [HC_6H_6O_6^-] = 3.2 \times 10^{-3}\ M$

$[C_6H_6O_6^{2-}] = 2.7 \times 10^{-12}\ M$

pH = 2.50

pOH = 11.50

$[OH^-] = 3.2 \times 10^{-12}\ M$

16.76 $H_3PO_4 \rightleftharpoons H_2PO_4^- + H^+$ $\qquad K_{a1} = \dfrac{[H_2PO_4^-][H^+]}{[H_3PO_4]} = 7.1 \times 10^{-3}$

$H_2PO_4^- \rightleftharpoons HPO_4^{2-} + H^+$ $\qquad K_{a2} = \dfrac{[HPO_4^{2-}][H^+]}{[H_2PO_4^-]} = 6.3 \times 10^{-8}$

$HPO_4^{2-} \rightleftharpoons PO_4^{3-} + H^+$ $\qquad K_{a3} = \dfrac{[PO_4^{3-}][H^+]}{[HPO_4^{2-}]} = 4.5 \times 10^{-13}$

The following assumptions are made:

$[H^+]_{total} \approx [H^+]_{first\ step}$

$[H_2PO_4^-]_{total} \approx [H_2PO_4^-]_{first\ step}$

$[HPO_4^{2-}]_{total} \approx [HPO_4^{2-}]_{second\ step}$

The first dissociation:

	$[H_3PO_4]$	$[H_2PO_4^-]$	$[H^+]$
I	2.0	–	–
C	– x	+ x	+ x
E	2.0 – x	+ x	+ x

$$K_{a1} = \frac{[H_2PO_4^-][H^+]}{[H_3PO_4]} = \frac{(x)(x)}{(2.0 - x)} = 7.1 \times 10^{-3}$$

Solve for x by successive approximations or by the quadratic equation:

$x = 0.12\ M = [H^+] = [H_2PO_4^-]$
$[H_3PO_4] = 2.0 - 0.12 = 1.9\ M$
$pH = -\log(0.12) = 0.92$

The second dissociation:

	$[H_2PO_4^-]$	$[HPO_4^{2-}]$	$[H^+]$
I	0.12	–	0.14
C	– x	+ x	+ x
E	0.12 – x	+ x	0.14 + x

Assume that x << 0.14, therefore $0.14 - x \approx 0.14$ and $0.14 + x \approx 0.14$

$$K_{a2} = \frac{(x)(0.12)}{(0.12)} = 6.3 \times 10^{-8}$$

$x = 6.3 \times 10^{-8} = [HPO_4^{2-}]$

The third dissociation:

	$[HPO_4^{2-}]$	$[PO_4^{3-}]$	$[H^+]$
I	6.3×10^{-8}	–	0.14
C	– x	+ x	+ x
E	$6.3 \times 10^{-8} - x$	+ x	0.14 + x

Assume that $x << 6.3 \times 10^{-8}$, therefore $6.3 \times 10^{-8} - x \approx 6.3 \times 10^{-8}$ and $0.14 + x \approx 0.14$

$$K_{a3} = \frac{(x)(0.12)}{(6.3 \times 10^{-8})} = 4.5 \times 10^{-13}$$

Solving for x we get, $x = 2.4 \times 10^{-19} = [PO_4^{3-}]$

16.78 $H_3PO_3 \rightleftharpoons H_2PO_3^- + H^+$ $\quad K_{a_1} = \frac{[H_2PO_3^-][H^+]}{[H_3PO_3]} = 3.0 \times 10^{-2}$

$H_2PO_3^- \rightleftharpoons HPO_3^{2-} + H^+$ $\quad K_{a_2} = \frac{[HPO_3^{2-}][H^+]}{[H_2PO_3^-]} = 2.6 \times 10^{-7}$

To simplify the calculation, assume that the second dissociation does not contribute a significant amount of H^+ to the final solution. Solving the equilibrium problem for the first dissociation gives:

	$[H_3PO_3]$	$[H_2PO_3^-]$	$[H^+]$
I	1.0	–	–
C	– x	+ x	+ x
E	1.0 – x	+ x	+ x

$$K_{a_1} = \frac{[H_2PO_3^-][H^+]}{[H_3PO_3]} = \frac{(x)(x)}{(1.0 - x)} = 3.0 \times 10^{-2}$$

Because K_{a_1} is so large, a quadratic equation must be solved. On doing so we learn that $x = 0.16\ M = [H^+] = [H_2PO_3^-]$.
$pH = -\log[H^+] = -\log(0.16) = 0.80$

The $[HPO_3^{2-}]$ may be determined from the second ionization constant.

	$[H_2PO_3^-]$	$[HPO_3^{2-}]$	$[H^+]$
I	0.16	–	0.16
C	– x	+ x	+ x
E	0.16 – x	+ x	0.16 + x

$$K_{a_2} = \frac{[HPO_3^{2-}][H^+]}{[H_2PO_3^-]} = \frac{(x)(0.16 + x)}{(0.16 - x)} = 1.6 \times 10^{-7}$$

If we assume that x is small then $0.16 \pm x \approx 0.16$. Then, $x = [HPO_3^{2-}] = 1.6 \times 10^{-7}\ M$.

16.80 The hydrolysis equation is:

$$SO_3^{2-} + H_2O \rightleftharpoons HSO_3^- + OH^- \qquad K_b = \frac{[HSO_3^-][OH^-]}{[SO_3^{2-}]}$$

In order to obtain K_b we will use the relationship $K_w = K_a \times K_b$

$$K_b = \frac{K_w}{K_a} = \frac{1.0 \times 10^{-14}}{6.6 \times 10^{-8}} = 1.5 \times 10^{-7}$$

	$[SO_3^{2-}]$	$[HSO_3^-]$	$[OH^-]$
I	0.24	–	–
C	– x	+ x	+ x
E	0.24 – x	+ x	+ x

Since K_b is so small, assume that $x << 0.24$ and we determine $x = 1.9 \times 10^{-4}\ M = [OH^-]$
$pOH = -\log(1.9 \times 10^{-4}) = 3.72$, $pH = 14.00 - pOH = 10.28$

For the concentration of H_2SO_3

$$HSO_3^- + H_2O \rightleftharpoons H_2SO_3 + OH^- \qquad K_{b_2} = \frac{[H_2SO_3][OH^-]}{[HSO_3^-]}$$

$$K_{b_2} = \frac{K_w}{K_{a_1}} = \frac{1.0 \times 10^{-14}}{1.2 \times 10^{-2}} = 8.3 \times 10^{-13}$$

$$K_{b_2} = \frac{[H_2SO_3][1.9 \times 10^{-4}]}{[1.9 \times 10^{-4}]} = 8.3 \times 10^{-13}$$

$[H_2SO_3] = 8.3 \times 10^{-13}$

16.82 $C_6H_5O_7^{3-} + H_2O \rightleftharpoons HC_6H_5O_7^{2-} + OH^-$

$$K_b = \frac{[HC_6H_5O_7^{2-}][OH^-]}{[C_6H_5O_7^{3-}]} = \frac{K_w}{K_{a_3}} = \frac{1.0 \times 10^{-14}}{6.3 \times 10^{-6}} = 1.6 \times 10^{-9}$$

	$[C_6H_5O_7^{3-}]$	$[HC_6H_5O_7^{2-}]$	$[OH^-]$
I	0.10	–	–
C	– x	+ x	+ x
E	0.10 – x	+ x	+ x

Assume x << 0.10

$$K_b = \frac{(x)(x)}{0.10} = 1.6 \times 10^{-9} \qquad x = 1.3 \times 10^{-5} = [OH^-]$$

pOH = 4.90 pH = 9.10

16.84 $PO_4^{3-} + H_2O \rightleftharpoons HPO_4^{2-} + OH^-$ $K_{b1} = \frac{K_w}{K_{a3}} = 2.2 \times 10^{-2}$

$HPO_4^{2-} + H_2O \rightleftharpoons H_2PO_4^- + OH^-$ $K_{b2} = \frac{K_w}{K_{a2}} = 1.6 \times 10^{-7}$

$H_2PO_4^- + H_2O \rightleftharpoons H_3PO_4 + OH^-$ $K_{b3} = \frac{K_w}{K_{a1}} = 1.4 \times 10^{-12}$

By analogy with polyprotic acids, we know that
$[H_2PO_4^-] = 1.6 \times 10^{-7}$
We need to solve the first equilibrium expression to determine $[HPO_4^{2-}]$.

$$K_{b1} = \frac{[HPO_4^{2-}][OH^-]}{[PO_4^{3-}]} = 2.2 \times 10^{-2}$$

	$[PO_4^{3-}]$	$[HPO_4^{2-}]$	$[OH^-]$
I	0.50	–	–
C	– x	+ x	+ x
E	0.50 – x	+ x	+ x

Assume x << 0.50

$$K_{b1} = \frac{(x)(x)}{0.50 - x} = 2.2 \times 10^{-2}$$

Use the quadratic equation to solve for x since K_b is so large
$x = 9.4 \times 10^{-2}\ M = [OH^-] = [HPO_4^{2-}]$
pOH = 1.027 pH = 12.97

$[PO_4^{3-}] = 0.50 - 9.4 \times 10^{-2} = 0.41\ M$
Solve the third equilibrium expression to determine $[H_3PO_4]$:

$$K_{b3} = \frac{[H_3PO_4][OH^-]}{[H_2PO_4^-]} = 1.4 \times 10^{-12}$$

Substitute the calculated values of $[H_2PO_4^-]$ and $[OH^-]$ and solve for x:

$$K_{b3} = \frac{(9.4 \times 10^{-2})x}{(1.6 \times 10^{-7})} = 1.4 \times 10^{-12}$$

$$x = [H_3PO_4] = 2.4 \times 10^{-18}$$

16.86 Since HCO_2H and NaOH react in a 1:1 ratio:

$$HCO_2H + NaOH \rightarrow NaHCO_2 + H_2O$$

we can use the equation $V_a \times M_a = V_b \times M_b$ to determine the volume of NaOH that is required to reach the equivalence point, i.e. the point at which the number of moles of NaOH is equal to the number of moles of HCO_2H:
V_{NaOH} = 50 mL × 0.050/0.050 = 50 mL
Thus the final volume at the equivalence point will be 50 + 50 = 100 mL.
The concentration of $NaHCO_2$ would then be:
0.050 mol/L × 0.050 L = 2.5×10^{-3} mol HCO_2H = 2.5×10^{-3} mol $NaHCO_2$
2.5×10^{-3} mol/0.100 L = 2.5×10^{-2} M $NaHCO_2$

The hydrolysis of this salt at the equivalence point proceeds according to the following equilibrium:

$$HCO_2^- + H_2O \rightleftharpoons HCO_2H + OH^-$$

$$K_b = \frac{[HCO_2H][OH^-]}{[HCO_2^-]} = 5.6 \times 10^{-11}$$

	$[HCO_2^-]$	$[HCO_2H]$	$[OH^-]$
I	0.025	–	–
C	– x	+x	+ x
E	0.025 – x	+ x	+ x

Substituting the above values for equilibrium concentrations into the mass action expression and assuming x << 0.050 gives:

$$K_b = \frac{(x)(x)}{0.025} = 5.6 \times 10^{-11}$$

$x^2 = 1.4 \times 10^{-12}$ $\quad$ $x = 1.2 \times 10^{-6}$ M = $[OH^-]$ = $[HCO_2H]$
pOH = $-\log[OH^-]$ = $-\log(1.2 \times 10^{-6})$ = 5.93
pH = 14.00 – pOH = 14.00 – 5.93 = 8.07

Cresol red would be a good indicator, since it has a color change near the pH at the equivalence point.

16.88 mol $HC_2H_3O_2$ = (25.0 mL $HC_2H_3O_2$)$\left(\frac{0.180 \text{ mol } HC_2H_3O_2}{1000 \text{ mL } HC_2H_3O_2}\right)$ = 4.50×10^{-3} mol $HC_2H_3O_2$

mol OH^- = (40.0 mL OH^-)$\left(\frac{0.250 \text{ mol } OH^-}{1000 \text{ mL } OH^-}\right)$ = 1.00×10^{-2} mol OH^-

excess OH^- = $(1.00 \times 10^{-2}) - (4.5 \times 10^{-3}) = 5.5 \times 10^{-3}$ mol

$$[OH^-] = \frac{5.5 \times 10^{-3} \text{ moles}}{(25.0 + 40.0 \text{ mL})\left(\frac{1 \text{ L}}{1000 \text{ mL}}\right)} = 8.46 \times 10^{-2}\ M$$

pOH = 1.07
pH = 12.93

16.90 (a) $HC_2H_3O_2 \rightleftharpoons H^+ + C_2H_3O_2^-$ $K_a = \dfrac{[H^+][C_2H_3O_2^-]}{[HC_2H_3O_2]} = 1.8 \times 10^{-5}$

	$[HC_2H_3O_2]$	$[H^+]$	$[C_2H_3O_2^-]$
I	0.1000	–	–
C	– x	+ x	+ x
E	0.1000 – x	+ x	+ x

Substituting the above values for equilibrium concentrations into the mass action expression and assuming that x << 0.1000 gives:

$x = [H^+] = 1.342 \times 10^{-3}\ M$.

$pH = -\log [H^+] = -\log (1.342 \times 10^{-3}) = 2.8724$.

(b) When NaOH is added, it will react with the acetic acid present decreasing the amount in solution and producing additional acetate ion. Since this a one–to–one reaction, the number of moles of acetic acid will decrease by the same amount as the number of moles of NaOH added and the number of moles of acetate ion will increase by an identical amount. We must determine the number of moles of all ions present and calculate new concentrations accounting for dilution.

$$\text{mol } HC_2H_3O_2 = (0.07500 \text{ L solution})\left(\frac{0.1000 \text{ moles } HC_2H_3O_2}{1 \text{ L solution}}\right) = 7.500 \times 10^{-3} \text{ mol } HC_2H_3O_2$$

$$\text{mol } OH^- = (0.02500 \text{ L solution})\left(\frac{0.1000 \text{ moles } OH^-}{1 \text{ L solution}}\right) = 2.500 \times 10^{-3} \text{ mol } OH^-$$

$$[HC_2H_3O_2] = \frac{7.500 \times 10^{-3} \text{ moles} - 2.500 \times 10^{-3} \text{ moles}}{0.07500 \text{ L} + 0.02500 \text{ L}} = 5.000 \times 10^{-2}\ M$$

$$[C_2H_3O_2^-] = \frac{0 \text{ moles} + 2.500 \times 10^{-3} \text{ moles}}{0.07500 \text{ L} + 0.02500 \text{ L}} = 2.500 \times 10^{-2}\ M$$

$$pH = pK_a + \log \frac{[C_2H_3O_2^-]}{[HC_2H_3O_2]}$$

$$= 4.7447 + \log \frac{(2.500 \times 10^{-2})}{(5.000 \times 10^{-2})} = 4.444$$

(c) When half the acetic acid has been neutralized, there will be equal amounts of acetic acid and acetate ion present in the solution. At this point, $pH = pK_a = 4.7447$.

(d) At the equivalence point, all of the acetic acid will have been converted to acetate ion. The concentration of the acetate ion will be half the original concentration of acetic acid since we have doubled the volume of the solution. We then need to solve the equilibrium problem that results when we have a solution that possesses a $[C_2H_3O_2^-] = 0.05000\ M$.

$C_2H_3O_2^- + H_2O \rightleftharpoons HC_2H_3O_2 + OH^-$

$$K_b = \frac{[HC_2H_3O_2][OH^-]}{[C_2H_3O_2^-]} = 5.6 \times 10^{-10}$$

	$[C_2H_3O_2^-]$	$[HC_2H_3O_2]$	$[OH^-]$
I	0.05000	–	–
C	– x	+ x	+ x
E	0.05000 – x	+ x	+ x

Substituting the above values for equilibrium concentrations into the mass action expression and assuming that x << 0.05000 gives: $x = [OH^-] = 5.292 \times 10^{-6}\ M$.

$pOH = -\log [OH^-] = -\log (5.292 \times 10^{-6}) = 5.2764$.

$pH = 14.0000 - pOH = 14.0000 - 5.2764 = 8.7236$.

Practice Exercises

17.1 $Ba_3(PO_4)_2(s) \rightleftharpoons 3Ba^{2+}(aq) + 2PO_4^{3+}(aq)$
$K_{sp} = [Ba^{2+}]^3[PO_4^{3+}]^2$

17.2 (a) $K_{sp} = [Ba^{2+}][C_2O_4^{2-}]$ (b) $K_{sp} = [Ag^+]^2[SO_4^{2-}]$

17.3 $TlI(s) \rightleftharpoons Tl^+(aq) + I^-(aq)$
$K_{sp} = [Tl^+][I^-]$

$$\text{mol TlI} = 5.9 \times 10^{-3}\text{ g}\left(\frac{1\text{ mol TlI}}{331.3\text{ g TlI}}\right) = 1.78 \times 10^{-5}\text{ mol TlI}$$

$$[Tl^+] = [I^-] = \frac{1.78\times10^{-5}\text{ mol}}{1\text{ L}} = 1.78 \times 10^{-5}\ M$$

$K_{sp} = (1.8 \times 10^{-5})(1.8 \times 10^{-5}) = 3.2 \times 10^{-10}$

17.4 $PbF_2(s) \rightleftharpoons Pb^{2+}(aq) + 2F^-(aq)$ $K_{sp} = \left[Pb^{2+}\right]\left[F^-\right]^2$

$$K_{sp} = \left(2.15 \times 10^{-3}\right)\left(2\left(2.15 \times 10^{-3}\right)\right)^2 = 3.98 \times 10^{-8}$$

17.5 $Ag_2SO_4(s) \rightleftharpoons 2Ag^+(aq) + SO_4^{2-}(aq)$

$$\text{Mol } Na_2SO_4 \text{ added} = (28.4\text{ g } Na_2SO_4)\left(\frac{1\text{ mol } Na_2SO_4}{142.04\text{ g } Na_2SO_4}\right) = 0.200\text{ mol } Na_2SO_4$$

$$\text{Concentration of } SO_4^{2-} = \frac{0.200\text{ mol } SO_4^{2-}}{1\text{ L solution}} = 0.200\ M\ SO_4^{2-}$$

	$[Ag^+]$	$[SO_4^{2-}]$
I	–	0.200
C	$+ (4.3 \times 10^{-3}) \times 2$	$+ 4.3 \times 10^{-3}$
E	$+ 8.6 \times 10^{-3}$	$0.200 + 4.3 \times 10^{-3}$

$K_{sp} = [Ag^+]^2[SO_4^{2-}]$
$K_{sp}\ (8.6 \times 10^{-3})^2(0.204) = 1.5 \times 10^{-5}$

17.6 $CoCO_3(s) \rightleftharpoons Co^{2+}(aq) + CO_3^{2-}(aq)$

	$[Co^{2+}]$	$[CO_3^{2-}]$
I	–	0.10
C	$+1.0 \times 10^{-9}$	$+ 1.0 \times 10^{-9}$
E	$+1.0 \times 10^{-9}$	$0.10 + 1.0 \times 10^{-9}$

Substituting the above values for equilibrium concentrations into the expression for K_{sp} gives:

$K_{sp} = [Co^{2+}][CO_3^{2-}] = (1.0 \times 10^{-9})(0.10 + 1.0 \times 10^{-9}) = 1.0 \times 10^{-10}$

17.7 $PbF_2(s) \rightleftharpoons Pb^{2+}(aq) + 2F^-(aq)$

	$[Pb^{2+}]$	$[F^-]$
I	0.10	–
C	$+ 3.1 \times 10^{-4}$	$+ 2(3.1 \times 10^{-4})$
E	$0.10 + (3.1 \times 10^{-4})$	$+ 6.2 \times 10^{-4}$

Substituting the above values for equilibrium concentrations into the expression for K_{sp} gives:

$K_{sp} = [Pb^{2+}][F^-]^2 = [0.10 + 3.1 \times 10^{-4}][6.2 \times 10^{-4}]^2$
Now $(0.10 + 3.1 \times 10^{-4})$ is also ≈ 0.10:
Hence, $K_{sp} = (0.10)(6.2 \times 10^{-4})^2 = 3.9 \times 10^{-8}$

17.8 $Ag_3PO_4(s) \rightleftharpoons 3Ag^+(aq) + PO_4^{3+}(aq)$
$K_{sp} = [Ag^+]^3[PO_4^{3+}] = 2.8 \times 10^{-18}$

	$[Ag^+]$	$[PO_4^{3+}]$
I	–	–
C	+ 3x	+ x
E	+ 3x	+ x

$K_{sp} = [3x]^3[x] = 2.8 \times 10^{-18}$
$2.8 \times 10^{-18} = 27x^4$
$x^4 = 1.04 \times 10^{-19}$
$x = \sqrt[4]{1.04 \times 10^{-19}}$
$x = 1.79 \times 10^{-5}$
Thus the solubility is 1.8×10^{-5} *M* Ag_3PO_4.

17.9 (a) $AgBr(s) \rightleftharpoons Ag^+(aq) + Br^-(aq)$ $K_{sp} = [Ag^+][Br^-] = 5.0 \times 10^{-13}$

	$[Ag^+]$	$[Br^-]$
I	–	–
C	+ x	+ x
E	+ x	+ x

Substituting the above values for equilibrium concentrations into the expression for K_{sp} gives:

$K_{sp} = 5.0 \times 10^{-13} = [Ag^+][Br^-] = (x)(x)$
$x = \sqrt{5.0 \times 10^{-13}} = 7.1 \times 10^{-7}$
Thus the solubility is 7.1×10^{-7} *M* AgBr.

(b) $Ag_2CO_3(s) \rightleftharpoons 2Ag^+(aq) + CO_3^{2-}(aq)$ $K_{sp} = [Ag^+]^2[CO_3^{2-}] = 8.1 \times 10^{-12}$

	$[Ag^+]$	$[CO_3^{2-}]$
I	–	–
C	+ 2x	+ x
E	+ 2x	+ x

Substituting the above values for equilibrium concentrations into the expression for K_{sp} gives:

$K_{sp} = 8.1 \text{ x } 10^{-12} = [Ag^+]^2[CO_3^{2-}] = (2x)^2(x)$ and $4x^3 = 8.1 \times 10^{-12}$

$x = \sqrt[3]{(8.1 \times 10^{-12})/4} = 1.3 \times 10^{-4}$
Thus the molar solubility is 1.3×10^{-4} *M* Ag_2CO_3.

17.10 $AgI(s) \rightleftharpoons Ag^+(aq) + I^-(aq)$ $K_{sp} = [Ag^+][I^-] = 8.3 \times 10^{-17}$

The initial concentration of I^- is 2×0.20 *M* from the CaI_2.

	$[Ag^+]$	$[I^-]$
I	–	0.40
C	+ x	+ x
E	+ x	0.40 + x

Substituting the above values for equilibrium concentrations into the expression for K_{sp} gives:

$K_{sp} = 8.3 \times 10^{-17} = [Ag^+][I^-] = (x)(0.40 + x)$

We know that the value of K_{sp} is very small, and it suggests the simplifying assumption that $(0.40 + x) \approx 0.40$:

Hence, $8.3 \times 10^{-17} = (0.40)x$, and $x = 2.1 \times 10^{-16}$.
The assumption that
$(0.40 + x) \approx 0.40$ is seen to be valid indeed.

Thus 2.1×10^{-16} *M* of AgI will dissolve in a 0.20 *M* CaI_2 solution.

In pure water,
$K_{sp} = 8.3 \times 10^{-17} = [Ag^+][I^-] = (x)(x)$
$x = [AgI(aq)] = 9.1 \times 10^{-9}$ *M* (much more soluble)

17.11 $Fe(OH)_3(s) \rightleftharpoons Fe^{3+}(aq) + 3OH^-(aq)$ $K_{sp} = [Fe^{3+}][OH^-]^3 = 1.6 \times 10^{-39}$

	$[Fe^{3+}]$	$[OH^-]$
I	–	0.050
C	+ x	+ 3x
E	+ x	0.050 + 3x

Substituting the above values for equilibrium concentrations into the expression for K_{sp} gives:

$$K_{sp} = 1.6 \times 10^{-39} = [Fe^{3+}][OH^-]^3 = (x)[0.050 + 3x]^3$$

We try to simplify by making the approximation that $(0.050 + 3x) \approx 0.050$:

$$1.6 \times 10^{-39} = (x)(0.050)^3 \quad \text{or } x = 1.3 \times 10^{-35}$$

Clearly the assumption that $(0.050 + 3x) \approx 0.050$ is justified.
Thus 1.3×10^{-35} *M* of $Fe(OH)_3$ will dissolve in a 0.050 *M* sodium hydroxide solution.

17.12 The expression for K_{sp} is $K_{sp} = [Ca^{2+}][SO_4^{2-}] = 2.4 \times 10^{-5}$ and the ion product for this solution would be:
$[Ca^{2+}][SO_4^{2-}] = (2.5 \times 10^{-3})(3.0 \times 10^{-2}) = 7.5 \times 10^{-5}$

Since the ion product is larger than the value of K_{sp}, a precipitate will form.

17.13 The solubility product constant is $K_{sp} = [Ag^+]^2[CrO_4^{2-}] = 1.2 \times 10^{-12}$ and the ion product for this solution would be:
$[Ag^+]^2[CrO_4^{2-}] = (4.8 \times 10^{-5})^2(3.4 \times 10^{-4}) = 7.8 \times 10^{-13}$

Since the ion product is smaller than the value of K_{sp}, no precipitate will form.

17.14 We expect $PbSO_4(s)$ since nitrates are soluble.

Because two solutions are to be mixed together, there will be a dilution of the concentrations of the various ions, and the diluted ion concentrations must be used. In general, on dilution, the following relationship is found for the concentrations of the initial solution (M_i) and the concentration of the final solution (M_f): $M_iV_i = M_fV_f$

Thus the final or diluted concentrations are:

$$\left[Pb^{2+}\right] = \left(1.0 \times 10^{-3}\ M\right)\left(\frac{100.0\text{ mL}}{200.0\text{ mL}}\right) = 5.0 \times 10^{-4}\ M$$

$$\left[SO_4^{\ 2-}\right] = \left(2.0 \times 10^{-3}\ M\right)\left(\frac{100.0\text{ mL}}{200.0\text{ mL}}\right) = 1.0 \times 10^{-3}\ M$$

The value of the ion product for the final (diluted) solution is:
$[Pb^{2+}][SO_4^{\ 2-}] = (5.0 \times 10^{-4})(1.0 \times 10^{-3}) = 5.0 \times 10^{-7}$

Since this is smaller than the value of K_{sp} (6.3×10^{-7}), a precipitate of $PbSO_4$ is not expected.

17.15 We expect a precipitate of $PbCl_2$ since nitrates are soluble.
We proceed as in Practice Exercise 14. $M_iV_i = M_fV_f$

$$\left[Pb^{2+}\right] = \left(0.10\ M\right)\left(\frac{50.0\text{ mL}}{70.0\text{ mL}}\right) = 0.071\ M$$

$$\left[Cl^{-}\right] = \left(0.040\ M\right)\left(\frac{20.0\text{ mL}}{70.0\text{ mL}}\right) = 0.011\ M$$

The value of the ion product for such a solution would be:
$[Pb^{2+}][Cl^-]^2 = (7.1 \times 10^{-2})(1.1 \times 10^{-2})^2 = 8.6 \times 10^{-6}$

Since the ion product is smaller than K_{sp} (1.7×10^{-5}), a precipitate of $PbCl_2$ is not expected.

17.16 CoS will precipitate if the H^+ concentration is too low. Solving for Q and then comparing Q to K_{spa}, we can determine whether or not CoS will precipitate.

$$K_{spa} = \frac{[Co^{2+}][H_2S]}{[H^+]^2} = 5 \times 10^{-1}$$

$$Q = \frac{[Co^{2+}][H_2S]}{[H^+]^2} = \frac{[0.005][0.10]}{[3.16 \times 10^{-4}]^2} = 5 \times 10^3$$

$Q > K_{spa}$
Since Q is greater than K_{spa}, then the reaction will move to reactants and CoS solid will form.

17.17 Consulting Table 17.2, we find that Fe^{2+} is much more soluble in acid than Hg^{2+}. We want to make the H^+ concentration large enough to prevent FeS from precipitating, but small enough that HgS *does* precipitate. First, we calculate the highest pH at which FeS will remain soluble, by using K_{spa} for FeS. (Recall that a saturated solution of $H_2S = 0.10\ M$.)

$$K_{spa} = \frac{[Fe^{2+}][H_2S]}{[H^+]^2} = \frac{[0.010][0.10]}{[H^+]^2} = 6 \times 10^2$$

$[H^+] = 0.0013\ M$
$pH = -\log [H^+] = 2.9$

Since Fe^{2+} is much more soluble in acid than Hg^{2+} we already know that this pH will precipitate HgS, but we can check it by using K_{spa} for HgS:

$$K_{spa} = \frac{[Hg^{2+}][H_2S]}{[H^+]^2} = \frac{[0.010][0.10]}{[H^+]^2} = 2 \times 10^{-32}$$

$[H^+] = 2.2 \times 10^{14}\ M$

(This concentration is impossibly high, but it tells us that this much acid would be required to dissolve HgS at these concentrations.)

17.18 $BaC_2O_4(s) \rightleftharpoons Ba^{2+}(aq) + C_2O_4^{2-}(aq)$ $\qquad K_{sp} = 1.2 \times 10^{-7} = [Ba^{2+}][C_2O_4^{2-}]$

$[Ba^{2+}] = 0.050\ M$

$1.2 \times 10^{-7} = (0.050)[C_2O_4^{2-}]$

$[C_2O_4^{2-}] = 2.4 \times 10^{-6}\ M$

$H_2C_2O_4 \rightleftharpoons H^+ + HC_2O_4^-$ $\qquad K_{a1} = 5.6 \times 10^{-2}$

$HC_2O_4^- \rightleftharpoons H^+ + C_2O_4^{2-}$ $\qquad K_{a2} = 5.4 \times 10^{-5}$

$H_2C_2O_4 \rightleftharpoons 2H^+ + C_2O_4^{2-}$ $\qquad K_a = (5.4 \times 10^{-5}) \times (5.6 \times 10^{-2}) = 3.0 \times 10^{-6}$

$[H_2C_2O_4] = 0.10$

$[C_2O_4^{2-}] = 2.4 \times 10^{-6}\ M$

$$K_a = 3.0 \times 10^{-6} = \frac{[H^+]^2[C_2O_4^{2-}]}{[H_2C_2O_4]} = \frac{[H^+]^2(2.4\times10^{-6})}{(0.10)}$$

Since the amount of oxalate formed is so small, the concentration of oxalic acid is essentially unchanged.

$[H^+] = 3.55 \times 10^{-1}$

This is the minimum concentration of H^+ that will prevent the formation of BaC_2O_4 precipitate.

17.19 Follow the exact procedure outlined in Example 17.10.

$K_{sp} = [Ca^{2+}][CO_3^{2-}] = 4.5 \times 10^{-9}$

$K_{sp} = [Ni^{2+}][CO_3^{2-}] = 1.3 \times 10^{-7}$

$NiCO_3$ is more soluble and will precipitate when:

$$[CO_3^{2-}] = \frac{K_{sp}}{[Ni^{2+}]} = \frac{1.3 \times 10^{-7}}{0.10} = 1.3 \times 10^{-6}$$

$CaCO_3$ will precipitate when:

$$[CO_3^{2-}] = \frac{K_{sp}}{[Ca^{2+}]} = \frac{4.5 \times 10^{-9}}{0.10} = 4.5 \times 10^{-8}$$

$CaCO_3$ will precipitate and $NiCO_3$ will not precipitate if $[CO_3^{2-}] > 4.5 \times 10^{-8}$ and $[CO_3^{2-}] < 1.3 \times 10^{-6}$.
Now, using the equation in example 17.10 we get:

$$[H^+]^2 = (2.4 \times 10^{-17})\left(\frac{0.030}{[CO_3^{2-}]}\right) \quad \text{NiCO}_3 \text{ will precipitate if}$$

$$[H^+]^2 = (2.4 \times 10^{-17})\left(\frac{0.030}{1.3 \times 10^{-6}}\right) = 5.5 \times 10^{-13}$$

$[H^+] = 7.4 \times 10^{-7}$ $\qquad$ pH = 6.13

$CaCO_3$ will precipitate if:

$$[H^+]^2 = (2.4 \times 10^{-17})\left(\frac{0.030}{4.5 \times 10^{-8}}\right) = 1.6 \times 10^{-11}$$

$$[H^+] = 4.0 \times 10^{-6} \qquad pH = 5.40$$

So $CaCO_3$ will precipitate and $NiCO_3$ will not if the pH is maintained between pH = 5.40 and pH = 6.13

17.20 The overall equilibrium is $AgCl(s) + 2NH_3(aq) \rightleftharpoons Ag(NH_3)_2^+(aq) + Cl^-(aq)$

$$K_c = \frac{[Ag(NH_3)_2^+][Cl^-]}{[NH_3]^2}$$

In order to obtain a value for K_c for this reaction, we need to use the expressions for K_{sp} of AgCl(*s*) and the K_{form} of $Ag(NH_3)_2^+$:

$$K_{sp} = [Ag^+][Cl^-] = 1.8 \times 10^{-10}$$

$$K_{form} = \frac{[Ag(NH_3)_2^+]}{[Ag^+][NH_3]^2} = 1.6 \times 10^7$$

$$K_c = K_{sp} \times K_{form} = \frac{[Ag(NH_3)_2^+][Cl^-]}{[NH_3]^2} = 2.9 \times 10^{-3}$$

Now we may use an equilibrium table for the reaction in question:

	$[NH_3]$	$[Ag(NH_3)_2^+]$	$[Cl^-]$
I	0.10	–	–
C	– 2x	+ x	+ x
E	0.10 – 2x	x	x

Substituting these values into the mass action expression gives:

$$K_c = 2.9 \times 10^{-3} = \frac{(x)(x)}{(0.10 - 2x)^2}$$

Take the square root of both sides to get $0.054 = \dfrac{(x)}{(0.10 - 2x)}$

Solving for x we get, x = 4.9×10^{-3} *M*. The molar solubility of AgCl in 0.10 *M* NH_3 is therefore 4.9×10^{-3} *M*.

In order to determine the solubility in pure water, we simply look at K_{sp}

$$AgCl(s) \rightleftharpoons Ag^+(aq) + Cl^-(aq) \quad K_{sp} = [Ag^+][Cl^-] = 1.8 \times 10^{-10}$$

At equilibrium; $[Ag^+] = [Cl^-] = 1.3 \times 10^{-5}$. Hence the molar solubility of AgCl in 0.10 *M* NH_3 is about 380 times greater than in pure water.

17.21 We will use the information gathered for the last problem. Specifically,

$$AgCl(s) + 2NH_3(aq) \rightleftharpoons Ag(NH_3)_2^+(aq) + Cl^-(aq)$$

$$K_c = \frac{[Ag(NH_3)_2^+][Cl^-]}{[NH_3]^2} = 2.9 \times 10^{-3}$$

If we completely dissolve 0.20 mol of AgCl, the equilibrium [Cl^-] and [$Ag(NH_3)_2^+$] will be 0.20 *M* in a one liter container. The question asks, therefore, what amount of NH_3 must be initially present so that the equilibrium concentration of Cl^- is 0.20 *M*?

	[NH_3]	[$Ag(NH_3)_2^+$]	[Cl^-]
I	Z	–	–
C	– 2x	+ x	+ x
E	Z – 2x	x	x

$$K_c = 2.9 \times 10^{-3} = \frac{(x)(x)}{(Z-2x)^2}$$

Take the square root of both sides to get;

$$0.054 = \frac{x}{Z-2x} = \frac{0.20}{Z-0.40}$$

We have substituted the known value of x. Solving for Z we get, Z = 4.1 *M*

Consequently, we would need to add 4.1 moles of NH_3 to a one liter container of 0.20 *M* AgCl in order to completely dissolve the AgCl.

Review Problems

17.16 (a) $CaF_2(s) \rightleftharpoons Ca^{2+} + 2F^-$ $K_{sp} = [Ca^{2+}][F^-]^2$

(b) $Ag_2CO_3(s) \rightleftharpoons 2Ag^+ + CO_3^{2-}$ $K_{sp} = [Ag^+]^2[CO_3^{2-}]$

(c) $PbSO_4(s) \rightleftharpoons Pb^{2+} + SO_4^{2-}$ $K_{sp} = [Pb^{2+}][SO_4^{2-}]$

(d) $Fe(OH)_3(s) \rightleftharpoons Fe^{3+} + 3OH^-$ $K_{sp} = [Fe^{3+}][OH^-]^3$

(e) $PbI_2(s) \rightleftharpoons Pb^{2+} + 2I^-$ $K_{sp} = [Pb^{2+}][I^-]^2$

(f) $Cu(OH)_2(s) \rightleftharpoons Cu^{2+} + 2OH^-$ $K_{sp} = [Cu^{2+}][OH^-]^2$

17.18 $PbCl_2 \rightleftharpoons Pb^{2+} + 2Cl^-$ $K_{sp} = [Pb^{2+}][Cl^-]^2$

At equilibrium [Pb^{2+}] = 0.016 *M* and [Cl^-] = 0.032 *M*

so $K_{sp} = (0.016)(0.032)^2 = 1.6 \times 10^{-5}$

17.20 $\text{mol BaSO}_4 = (0.00245 \text{ g BaSO}_4)\left(\frac{1 \text{ mole BaSO}_4}{233.3906 \text{ g BaSO}_4}\right) = 1.05 \times 10^{-5} \text{ mol}$

$[Ba^{2+}] = [SO_4^{2-}] = 1.05 \times 10^{-5}$ *M*

$K_{sp} = [Ba^{2+}][SO_4^{2-}] = (1.05 \times 10^{-5})^2 = 1.10 \times 10^{-10}$

17.22 $BaSO_3(s) \rightleftharpoons Ba^{2+} + SO_3^{2-}$ $K_{sp} = [Ba^{2+}][SO_3^{2-}]$

$K_{sp} = (0.10)(8.0 \times 10^{-6}) = 8.0 \times 10^{-7}$

In this problem, all of the Ba^{2+} comes from the $BaCl_2$.

17.24 $Ag_3PO_4(s) \rightleftharpoons 3Ag^+ + PO_4^{3-}$ $K_{sp} = [Ag^+]^3[PO_4^{3-}]$

$K_{sp} = [3(1.8 \times 10^{-5})]^3[1.8 \times 10^{-5}] = 2.8 \times 10^{-18}$

17.26 $PbBr_2(s) \rightleftharpoons Pb^{2+} + 2Br^-$ $K_{sp} = [Pb^{2+}][Br^-]^2$

	$[Pb^{2+}]$	$[Br^-]$
I	–	–
C	+ x	+ 2x
E	+ x	+ 2x

$K_{sp} = (x)(2x)^2 = 4x^3 = 2.1 \times 10^{-6}$, $x = \sqrt[3]{\dfrac{2.1 \times 10^{-6}}{4}} = 8.1 \times 10^{-3}\ M$

17.28 For every mole of CO_3^{2-} produced, 2 moles of Ag^+ will be produced. Let $x = [CO_3^{2-}]$ at equilibrium and $[Ag^+] = 2x$ at equilibrium. $K_{sp} = [Ag^+]^2[CO_3^{2-}] = 8.1 \times 10^{-12} = (2x)^2(x) = 4x^3$. Solving we find $x = 1.3 \times 10^{-4}$. Thus, the molar solubility of Ag_2CO_3 is 1.3×10^{-4} moles/L.

17.30 To solve this problem, determine the molar solubility for each compound.

LiF: let $x = [Li^+] = [F^-]$ $K_{sp} = [Li^+][F^-] = x^2 = 1.7 \times 10^{-3}$
$x = 4.1 \times 10^{-2}$ moles/L = molar solubility of LiF.

BaF_2: let $x = [Ba^{2+}]$; $[F^-]^2 = 2x$ $K_{sp} = [Ba^{2+}][F^-]^2 = (x)(2x)^2 = 1.7 \times 10^{-6}$
$4x^3 = 1.7 \times 10^{-6}$, and $x = 7.5 \times 10^{-3}\ M$ = molar solubility of BaF_2.

Because the molar solubility of LiF is greater than the molar solubility of BaF_2, LiF is more soluble.

17.32 First determine the molar solubility of the MX salt.
Let $x = [M^+] = [X^-]$, $K_{sp} = [M^+][X^-] = (x)(x) = 3.2 \times 10^{-10}$
$x = 1.8 \times 10^{-5}\ M$. This is the equilibrium concentration of the two ions.

For the MX_3 salt, let x = equilibrium concentration of M^{3+}, $[X^-] = 3x$.
$K_{sp} = [M^+][X^-]^3 = (x)(3x)^3 = 27x^4$. The value of x in this expression is the value determined in the first part of this problem.
So, $K_{sp} = (27)(1.8 \times 10^{-5})^4 = 2.8 \times 10^{-18}$

17.34 $CaSO_4(s) \rightleftharpoons Ca^{2+}(aq) + SO_4^{2-}(aq)$ $K_{sp} = [Ca^{2+}][SO_4^{2-}] = 2.4 \times 10^{-5}$
let $x = [Ca^{2+}] = [SO_4^{2-}]$ $K_{sp} = x^2 = 2.4 \times 10^{-5}$ and $x = 4.9 \times 10^{-3}\ M$.
The molar solubility of $CaSO_4$ is 4.9×10^{-3} moles/L.

17.36 (a) $CuCl(s) \rightleftharpoons Cu^+(aq) + Cl^-(aq)$ $K_{sp} = [Cu^+][Cl^-] = 1.9 \times 10^{-7}$

	$[Cu^+]$	$[Cl^-]$
I	–	–
C	+ x	+ x
E	+ x	+ x

$K_{sp} = x^2 = 1.9 \times 10^{-7}$ $\therefore$ x = molar solubility = $4.4 \times 10^{-4}\ M$

(b) $CuCl(s) \rightleftharpoons Cu^+(aq) + Cl^-(aq)$ $K_{sp} = [Cu^+][Cl^-] = 1.9 \times 10^{-7}$

	$[Cu^+]$	$[Cl^-]$
I	–	0.0200
C	+ x	+ x
E	+ x	0.0200 + x

$K_{sp} = (x)(0.0200+x) = 1.9 \times 10^{-7}$ Assume that x << 0.0200
$\therefore$ x = molar solubility = 9.5×10^{-6} *M*

(c) $CuCl(s) \rightleftharpoons Cu^+(aq) + Cl^-(aq)$ $K_{sp} = [Cu^+][Cl^-] = 1.9 \times 10^{-7}$

	$[Cu^+]$	$[Cl^-]$
I	–	0.200
C	+ x	+ x
E	+ x	0.200 + x

$K_{sp} = (x)(0.200+x) = 1.9 \times 10^{-7}$ Assume that x << 0.200
$\therefore$ x = molar solubility = 9.5×10^{-7} *M*

(d) $CuCl(s) \rightleftharpoons Cu^+(aq) + Cl^-(aq)$ $K_{sp} = [Cu^+][Cl^-] = 1.9 \times 10^{-7}$
Note that the Cl^- concentration equals (2)(0.150 *M*) since two moles of Cl^- are produced for every mole of $CaCl_2$.

	$[Cu^+]$	$[Cl^-]$
I	–	0.300
C	+ x	+ x
E	+ x	0.300 + x

$K_{sp} = (x)(0.300+x) = 1.9 \times 10^{-7}$ Assume that x << 0.300
$\therefore$ x = molar solubility = 6.3×10^{-7} *M*

17.38 $Mg(OH)_2 \rightleftharpoons Mg^{2+} + 2OH^-$ $K_{sp} = [Mg^{2+}][OH^-]^2 = 7.1 \times 10^{-12}$
The concentration of OH^- is determined from the pH:
pOH = 14 – 12.50 = 1.50
$[OH^-] = 0.032$ *M*
$[Mg^{2+}] = x$ $[OH^-] = 0.032$ *M*
$K_{sp} = x(0.032)^2 = 7.1 \times 10^{-12}$
$x = 6.9 \times 10^{-9}$ *M*
The molar solubility of $Mg(OH)_2$ is 6.9×10^{-9} *M* in a solution with a pH of 12.50.

17.40 $PbCl_2(s) \rightleftharpoons Pb^{2+} + 2Cl^-$ $K_{sp} = [Pb^{2+}][Cl^-]^2 = 1.7 \times 10^{-5}$
$[Cl^-] = 0.10$ *M*
$[Pb^{2+}][Cl^-]^2 = [Pb^{2+}][0.10]^2 = 1.7 \times 10^{-5}$
$[Pb^{2+}] = 1.7 \times 10^{-3}$ *M*

17.42 $Ag_2CrO_4(s) \rightleftharpoons 2Ag^+(aq) + CrO_4^{2-}(aq)$ $K_{sp} = [Ag^+]^2[CrO_4^{2-}] = 1.2 \times 10^{-12}$
(a)

	$[Ag^+]$	$[CrO_4^{2-}]$
I	0.200	–
C	0.200 + 2x	+ x
E	0.200 + 2x	+ x

$K_{sp} = (0.200+2x)^2(x)$ Assume that x << 0.200
$1.2 \times 10^{-12} = (0.200)^2(x)$ $x = 3.0 \times 10^{-11}$
The molar solubility is 3.0×10^{-11} moles/L

(b)

	$[Ag^+]$	$[CrO_4^{2-}]$
I	–	0.200
C	+2x	0.200 + x
E	+2x	0.200 + x

$K_{sp} = (2x)^2(0.200+x)$ Assume that $x << 0.200$
$1.2 \times 10^{-12} = (2x)^2(0.200)$ $x = 1.2 \times 10^{-6}$
The molar solubility is 1.2×10^{-6} moles/L.

17.44 $HC_2H_3O_2 + H_2O \rightleftharpoons H_3O^+ + C_2H_3O_2^-$

$$K_a = \frac{[H_3O^+][C_2H_3O_2^-]}{[HC_2H_3O_2]} = 1.8 \times 10^{-5}$$

First, we calculate the % ionization in water (no added sodium acetate):

	$[HC_2H_3O_2]$	$[H_3O^+]$	$[C_2H_3O_2^-]$
I	0.10	–	–
C	0.10–x	+x	+x
E	0.10 –x	x	+x

Assume $x << 0.10$

$$K_a = \frac{[x][x]}{[0.10]} = 1.8 \times 10^{-5} \quad x = 1.3 \times 10^{-3} = [H_3O^+]$$

% ionization = ($[H^+]$/[total acid]) × 100 = (1.3×10^{-3}/0.10) × 100 = 1.3 %

Now we calculate the % ionization using 0.050 mol/0.500 L = 0.10 *M* for the concentration of sodium acetate:

	$[HC_2H_3O_2]$	$[H_3O^+]$	$[C_2H_3O_2^-]$
I	0.10	–	0.10
C	0.10–x	+x	+x
E	0.10 –x	x	0.10 + x

Assume $x << 0.10$

$$K_a = \frac{[x][0.10]}{[0.10]} = 1.8 \times 10^{-5} \quad x = 1.8 \times 10^{-5} = [H_3O^+]$$

% ionization = ($[H^+]$/[total acid]) × 100 = (1.8×10^{-5}/0.10) × 100 = 0.018 %

So the % ionization decreased by (1.3 – 0.018) = 1.3 %, using correct significant figures. (This does not mean it has no dissociation, but that the dissociation is very small compared to 1.3%.)

Using the $[H^+]$ values above, the pH initially was:
$-\log [H^+] = -\log [1.3 \times 10^{-3}] = 2.89$

After addition of the sodium acetate it was:
$-\log [H^+] = -\log [1.8 \times 10^{-5}] = 4.74$

The pH changes by (4.7 – 2.9) = +1.85 pH units

17.46 $Fe(OH)_2(s) \rightleftharpoons Fe^{2+}(aq) + 2\,OH^-(aq)$ $\qquad K_{sp} = \left[Fe^{2+}\right]\left[OH^-\right]^2$

mol OH^- = 2.20 g NaOH(1 mol/40.01 g NaOH) = 0.0550 mol NaOH

$[OH^-]$ = mol OH^-/L solution = 0.0550 mol/0.250 L = 0.22 *M*

	$[Fe^{2+}]$	$[OH^-]$
I	–	0.22
C	+ x	+ 2x
E	x	0.22 + 2x

We assume that x << 0.22, so that 0.22 + 2x ≈ 0.22, then we enter the equilibrium values of the above table into the K_{sp} expression:

$K_{sp} = \left[Fe^{2+}\right]\left[OH^-\right]^2$
$7.9 \times 10^{-16} = x(0.22)^2$
x = molar solubility = 1.6×10^{-14} *M*

Next, we must determine how many moles of $Fe(OH)_2$ are formed in the reaction.
This is a limiting reactant problem.

The number of moles of OH^- is 0.0550 (see above).
The number of moles of Fe^{2+} is (0.250 L)(0.10 mol/L) = 0.025 mol

From the balanced equation at the top, we need two OH^- for every one Fe^{2+}.
This would be 2(0.025 mol) = 0.050 mol OH^-. Looking at the molar quantities above, we have more than enough OH^- so, Fe^{2+} is our limiting reactant:

0.025 mol $Fe(OH)_2$ will form in 0.25 L solution. If dissolved, this would be a concentration of 0.025 mol/0.25 L = 0.10 *M*. But from above, the maximum molar solubility of is 1.6×10^{-14} *M*.

This means that remainder of $Fe(OH)_2$ in excess of this value precipitates:
$0.10 - 1.6 \times 10^{-14} \approx 0.10$ *M*.

This works out to 0.25 L(0.10 mol/L) = 0.025 mol $Fe(OH)_2$(89.8 g/mol)
= 2.2 g solid $Fe(OH)_2$ (essentially all of it).

The remaining OH^-, 0.005 mol, gives a concentration of OH^- of

$$\frac{0.005\text{ mol OH}^-}{0.250\text{ L}} = 0.02\,M\,\text{OH}^-$$

$7.9 \times 10^{-16} = [Fe^{2+}][0.02]^2$
$[Fe^{2+}] = 2.0 \times 10^{-12}$ *M*

17.48 $Fe(OH)_2(s) \rightleftharpoons Fe^{2+}(aq) + 2\,OH^-(aq)$ $\qquad K_{sp} = \left[Fe^{2+}\right]\left[OH^-\right]^2 = 7.9 \times 10^{-16}$

pH = 9.50
pOH = 14.00 – pH = 4.50
$[OH^-] = 10^{-4.50} = 3.16 \times 10^{-5}$ *M*

	$[Fe^{2+}]$	$[OH^-]$
I	–	3.16×10^{-5}
C	+ x	+ 2x
E	x	$(3.16 \times 10^{-5}) + 2x$

Since K_{sp} for iron(II) hydroxide is so small, we can safely assume that $2x << 3.16 \times 10^{-5}$, so that $(3.16 \times 10^{-5}) + 2x \approx 3.16 \times 10^{-5}$, then we enter the equilibrium values of the above table into the K_{sp} expression:

$$K_{sp} = \left[Fe^{2+}\right]\left[OH^-\right]^2$$
$$7.9 \times 10^{-16} = x(3.16 \times 10^{-5})^2$$
$$x = \text{molar solubility} = 7.9 \times 10^{-7}\ M$$

17.50 In order for a precipitate to form, the value of the reaction quotient, Q, must be greater than the value of K_{sp}. For $PbCl_2$, $K_{sp} = 1.7 \times 10^{-5}$ (see Table 17.1).

$Q = \left[Pb^{2+}\right]\left[Cl^-\right]^2 = (0.0150)(0.0120)^2 = 2.16 \times 10^{-6}$. Since $Q < K_{sp}$, no precipitate will form.

17.52 To solve this problem, determine the value for Q and apply LeChâtelier's Principle.

(a) $\left[Pb^{2+}\right] = (50.0\text{ mL})(0.0100\text{ moles/L})/(100.0\text{ mL}) = 5.00 \times 10^{-3}$

$\left[Br^-\right] = (50.0\text{ mL})(0.0100\text{ moles/L})/(100.0\text{ mL}) = 5.00 \times 10^{-3}$

$Q = \left[Pb^{2+}\right]\left[Br^-\right]^2 = (5.00 \times 10^{-3})(5.00 \times 10^{-3})^2 = 1.25 \times 10^{-7}$

For $PbBr_2$, $K_{sp} = 2.1 \times 10^{-6}$

Since $Q < K_{sp}$, no precipitate will form.

(b) $\left[Pb^{2+}\right] = (50.0\text{ mL})(0.0100\text{ moles/L})/(100.0\text{ mL}) = 5.00 \times 10^{-3}$

$\left[Br^-\right] = (50.0\text{ mL})(0.100\text{ moles/L})/(100.0\text{ mL}) = 5.00 \times 10^{-2}$

$Q = \left[Pb^{2+}\right]\left[Br^-\right]^2 = (5.00 \times 10^{-3})(5.00 \times 10^{-2})^2 = 1.25 \times 10^{-5}$

For $PbBr_2$, $K_{sp} = 2.1 \times 10^{-6}$

Since $Q > K_{sp}$, a precipitate will form.

17.54 $AgCl(s) \rightleftharpoons Ag^+ + Cl^-$ $\quad K_{sp} = \left[Ag^+\right]\left[Cl^-\right] = 1.8 \times 10^{-10}$

$AgI(s) \rightleftharpoons Ag^+ + I^-$ $\quad K_{sp} = \left[Ag^+\right]\left[I^-\right] = 8.3 \times 10^{-17}$

When $AgNO_3$ is added to the solution, AgI will precipitate before any AgCl does due to the lower solubility of AgI. In order to answer the question, i.e., what is the $[I^-]$ when AgCl first precipitates, we need to find the minimum concentration of Ag^+ that must be added to precipitate AgCl.

Let $x = [Ag^+]$; $K_{sp} = (x)(0.050) = 1.8 \times 10^{-10}$; $x = 3.6 \times 10^{-9}\ M$

When the AgCl starts to precipitate, the solution will have a $[Ag^+]$ of $3.6 \times 10^{-9}\ M$. Now we ask, what is the $[I^-]$ if $[Ag^+] = 3.6 \times 10^{-9}\ M$?

So, $K_{sp} = \left[Ag^+\right]\left[I^-\right] = (3.6 \times 10^{-9})(x) = 8.3 \times 10^{-17}$; $x = 2.3 \times 10^{-8}\ M = [I^-]$

17.56 The precipitate that may form is $PbBr_2(s)$. To determine if a precipitate will form, a value for the reaction quotient, Q, must be calculated: $Q = [Pb^{2+}][Br^-]^2$. In performing this calculation, the dilution of the ions must be considered:

$[Pb^{2+}] = [Br^-] = 0.00500\ M.$ $Q = [0.00500][0.00500]^2 = 1.3 \times 10^{-7}$

If $Q > K_{sp}$, a precipitate will form. K_{sp} for $PbBr_2(s)$ is 2.1×10^{-6}. Therefore, a precipitate will not form. Hence, the concentrations calculated are the diluted concentrations. Since no precipitate forms, the concentrations are not equilibrium values.

17.58 The less soluble substance is PbS. We need to determine the minimum $[H^+]$ at which CoS will precipitate.

$$K_{spa} = \frac{[Co^{2+}][H_2S]}{[H^+]^2} = \frac{(0.010)(0.1)}{[H^+]^2} = 0.5 \quad \text{(from Table 17.2)}$$

$$[H^+] = \sqrt{\frac{(0.010)(0.1)}{0.5}} = 0.045$$

$pH = -\log[H^+] = 1.35$. At a pH lower than 1.35, PbS will precipitate and CoS will not. At larger values of pH, both PbS and CoS will precipitate.

17.60 $Cu(OH)_2(s) \rightleftharpoons Cu^{2+}(aq) + 2\ OH^-(aq)$

$$K_{sp} = [Cu^{2+}][OH^-]^2$$

$$4.8 \times 10^{-20} = [0.10][OH^-]^2$$

$[OH^-] = 6.9 \times 10^{-10}\ M$

$pOH = -\log[OH^-] = -\log[6.9 \times 10^{-10}] = 9.2$

$pH = 14.00 - pOH = 4.8$

4.8 is the pH *below* which all the $Cu(OH)_2$ will be soluble.

$Mn(OH)_2(s) \rightleftharpoons Mn^{2+}(aq) + 2\ OH^-(aq)$

$$K_{sp} = [Mn^{2+}][OH^-]^2$$

$$1.6 \times 10^{-13} = [0.10][OH^-]^2$$

$[OH^-] = 1.3 \times 10^{-6}\ M$

$pOH = -\log[OH^-] = -\log[1.3 \times 10^{-6}] = 5.9$

$pH = 14.00 - pOH = 8.1$

8.1 is the pH *below* which all the $Mn(OH)_2$ will be soluble.

Therefore, from pH = 4.8–8.1 $Mn(OH)_2$ will be soluble, but some $Cu(OH)_2$ will precipitate out of solution.

17.62 (a) $Cu^{2+}(aq) + 4Cl^-(aq) \rightleftharpoons CuCl_4^{2-}(aq)$ $$K_{form} = \frac{[CuCl_4^{2-}]}{[Cu^{2+}][Cl^-]^4}$$

(b) $Ag^+(aq) + 2I^-(aq) \rightleftharpoons AgI_2^-(aq)$ $$K_{form} = \frac{[AgI_2^-]}{[Ag^+][I^-]^2}$$

(c) $Cr^{3+}(aq) + 6NH_3(aq) \rightleftharpoons Cr(NH_3)_6^{3+}(aq)$ $$K_{form} = \frac{[Cr(NH_3)_6^{3+}]}{[Cr^{3+}][NH_3]^6}$$

17.64 (a) $Co^{3+}(aq) + 6NH_3(aq) \rightleftharpoons Co(NH_3)_6^{3+}(aq)$ $$K_{form} = \frac{[Co(NH_3)_6^{3+}]}{[Co^{3+}][NH_3]^6}$$

(b) $Hg^{2+}(aq) + 4I^-(aq) \rightleftharpoons HgI_4^{2-}(aq)$ $$K_{form} = \frac{[HgI_4^{2-}]}{[Hg^{2+}][I^-]^4}$$

(c) $Fe^{2+}(aq) + 6CN^-(aq) \rightleftharpoons Fe(CN)_6^{4-}(aq)$ $$K_{form} = \frac{[Fe(CN)_6^{4-}]}{[Fe^{2+}][CN^-]^6}$$

17.66 $K_c = K_{sp} \times K_{form} = (1.7 \times 10^{-5})(2.5 \times 10^1) = 4.3 \times 10^{-4}$

17.68 There are two events in this net process: one is the formation of a complex ion (an equilibrium which has an appropriate value for K_{form}), and the other is the dissolving of $Fe(OH)_3$, which is governed by K_{sp} for the solid.

$Fe(OH)_3(s) \rightleftharpoons Fe^{3+}(aq) + 3OH^-(aq)$ $$K_{sp} = [Fe^{3+}][OH^-]^3 = 1.6 \times 10^{-39}$$

$Fe^{3+}(aq) + 6CN^-(aq) \rightleftharpoons Fe(CN)_6^{3-}(aq)$ $$K_{form} = \frac{[Fe(CN)_6^{3-}]}{[Fe^{3+}][CN^-]^6} = 1.0 \times 10^{31}$$

The net process is:

$$Fe(OH)_3(s) + 6CN^-(aq) \rightleftharpoons Fe(CN)_6^{3-}(aq) + 3OH^-(aq)$$

The equilibrium constant for this process should be:

$$K_c = \frac{[Fe(CN)_6^{3-}][OH^-]^3}{\cdot[CN^-]^6}$$

The numerical value for the above K_c is equal to the product of K_{sp} for $Fe(OH)_3(s)$ and K_{form} for $Fe(CN)_6^{3-}$, as can be seen by multiplying the mass action expressions for these two equilibria:

$$K_c = K_{form} \times K_{sp} = 1.6 \times 10^{-8}$$

Because K_{form} is so very large, we can assume that all of the dissolved iron ion is present in solution as the complex, thus: $[Fe(CN)_6^{3-}] = 0.11 \text{ mol}/1.2 \text{ L} = 0.092\ M$. Also the reaction stoichiometry shows that each iron ion that dissolves gives 3 OH^- ions in solution, and we have: $[OH^-] = 0.092 \times 3 = 0.28\ M$. We substitute these values into the K_c expression and rearrange to get:

$$[CN^-] = \sqrt[6]{\frac{[Fe(CN)_6^{3-}][OH^-]^3}{K_c}}$$

$$= \sqrt[6]{\frac{(0.092)(0.28)^3}{1.6 \times 10^{-8}}}$$

Thus we arrive at the concentration of cyanide ion that is required in order to satisfy the mass action requirements of the equilibrium: $[CN^-] = 7.1$ mol L^{-1}. Since this concentration of CN^- must be present in 1.2 L, the number of moles of cyanide that are required is: 7.1 mol L^{-1} × 1.2 L = 8.5 mol CN^-.

Additionally, a certain amount of cyanide is needed to form the complex ion. The stoichiometry requires six times as much cyanide ion as iron ion. This is 0.11 moles × 6 = 0.66 mol. This brings the total required cyanide to (8.5 + 0.66) = 9.2 mol.

9.2 mol × 49.0 g/mol = 450 g NaCN are required.

17.70 The applicable equilibria are as follows:

$$AgI(s) \rightleftharpoons Ag^+(aq) + I^-(aq) \qquad K_{sp} = [Ag^+][I^-] = 8.3 \times 10^{-17}$$

$$Ag^+(aq) + 2I^-(aq) \rightleftharpoons AgI_2^-(aq) \qquad K_{form} = \frac{[AgI_2^-]}{[Ag^+][I^-]^2} = 1 \times 10^{11}$$

When a solution of AgI_2^- is diluted, all of the concentrations of the species in K_{form} above decrease. However, the decrease of $[I^-]$ has more effect on equilibrium because its expression is *squared*. Hence, the denominator is decreased more than the numerator in the reaction quotient, Q. The system reacts according to Le Châtelier's Principle, by moving to the left (toward reactants) to increase the value of $[I^-]$.

As the system moves to the left, more Ag^+ is created, which has an effect on the first equilibrium above. Again, Le Châtelier's Principle causes the reaction to move to the left to re-establish equilibrium, which produces AgI(*s*) precipitate.

The two equations above may be combined and K_c found as follows:

$$AgI(s) + I^-(aq) \rightleftharpoons AgI_2^-(aq) \qquad K_c = \frac{[AgI_2^-]}{[I^-]} = K_{sp} \times K_{form} = 8.3 \times 10^{-6}$$

To answer the second question, we make a table and fill in what we know. We begin with 1.0 *M* I^-. This is reduced by some amount (x) as it reacts with the silver ions, and $[AgI_2^-]$ is increased by the same amount:

	$[I^-]$	$[AgI_2^-]$
I	1.0	–
C	– x	+ x
E	1.0 – x	+ x

Now we insert the equilibrium values into the above equation:

$$K_c = \frac{[AgI_2^-]}{[I^-]} = 8.3 \times 10^{-6}$$

$$K_c = \frac{[x]}{[1.0 - x]} = 8.3 \times 10^{-6}$$

$$x = 8.3 \times 10^{-6}$$

This value represents the change in concentration of I^- which, from the balanced equation, equals the change in concentration of AgI(*s*). The given volume is 100 mL, which allows us to find the amount of AgI reacting:

$0.100\ L(8.3 \times 10^{-6}\ mol/L) = 8.3 \times 10^{-7}$ mol AgI

8.3×10^{-7} mol AgI(234.8 g/mol) = 1.9×10^{-4} g AgI

17.72 The applicable equilibria are as follows:

$AgI(s) \rightleftharpoons Ag^{+}(aq) + I^{-}(aq)$ $\qquad K_{sp} = [Ag^{+}][I^{-}] = 8.3 \times 10^{-17}$

$Ag^{+}(aq) + 2CN^{-}(aq) \rightleftharpoons Ag(CN)_2^{-}(aq)$ $\qquad K_{form} = \dfrac{[Ag(CN)_2^{-}]}{[Ag^{+}][CN^{-}]^2} = 5.3 \times 10^{18}$

The two equations above may be combined and K_c found as follows:

$AgI(s) + 2CN^{-}(aq) \rightleftharpoons Ag(CN)_2^{-}(aq) + I^{-}(aq)$ $\qquad K_c = \dfrac{[Ag(CN)_2^{-}][I^{-}]}{[CN^{-}]^2} = K_{sp} \times K_{form} = 4.4 \times 10^2$

We begin with 0.010 *M* CN^-. This is reduced by some amount (x) as it reacts with the silver ions, and $[AgI_2^-]$ is increased by the same amount:

	$[CN^-]$	$[Ag(CN)_2^-]$	I^-
I	0.010	–	–
C	– 2x	+ x	+ x
E	0.010 – 2x	x	x

Now we insert the equilibrium values into the above equation:

$$K_c = \frac{[Ag(CN)_2^{-}][I^{-}]}{[CN^{-}]^2} = 4.4 \times 10^2$$

$$K_c = \frac{[x][x]}{[0.010 - 2x]^2} = 4.4 \times 10^2$$

Take the square root of both sides and solve for x:

$$x = 4.9 \times 10^{-3}$$

This value represents the change in concentration of I^- which, from the balanced equation, equals the change in concentration of AgI(*s*).
The molar solubility of AgI in 0.010 *M* KCN is 4.9×10^{-3} *M*.

17.74 Recall that $K_{inst} = 1/K_{form}$.

$Zn(OH)_2(s) \rightleftharpoons Zn^{2+}(aq) + 2OH^{-}(aq)$ $\qquad K_{sp} = [Zn^{2+}][OH^{-}]^2 = 3.0 \times 10^{-16}$

$Zn^{2+}(aq) + 4NH_3(aq) \rightleftharpoons Zn(NH_3)_4^{2+}(aq)$ $\qquad K_{form} = \dfrac{[Zn(NH_3)_4^{2+}]}{[Zn^{2+}][NH_3]^4} = ?$

Combined, this is:

$Zn(OH)_2(s) + 4NH_3(aq) \rightleftharpoons Zn(NH_3)_4^{2+}(aq) + 2OH^{-}(aq)$

$$K_c = \frac{\left[Zn(NH_3)_4{}^{2+}\right]\left[OH^-\right]^2}{\left[NH_3\right]^4}$$

	$[NH_3]$	$[Zn(NH_3)_4{}^{2+}]$	$[OH^-]$
I	1.0	–	–
C	– 4x	+ x	+ 2x
E	1.0 – 4x	x	2x

$$K_c = \frac{[x][2x]^2}{[1.0 - 4x]^4}$$

The problem gives the molar solubility of $Zn(OH)_2$ as 5.7×10^{-3} *M*. This means in one liter of 1.0 *M* NH_3, $x = 5.7 \times 10^{-3}$ moles. Substituting this value in for x, we get $K_c = 8.1 \times 10^{-7}$.

$$K_c = K_{sp} \times K_{form}$$
$$8.1 \times 10^{-7} = 3.0 \times 10^{-16} \times K_{form}$$
$$K_{form} = 2.7 \times 10^9$$

$K_{inst} = 1/K_{form}$
$K_{inst} = 1/(2.7 \times 10^9) = 3.7 \times 10^{-10}$

Practice Exercises

18.1 $w = -P\Delta V = -(14.0 \text{ atm})(12.0 \text{ L} - 1.0 \text{ L}) = -154 \text{ L atm}$

$\Delta E = w + q = 0$
$0 = -154 + q$

Therefore, $q = +154$ L atm.

(The energy is converted into heat; since the heat does not leave the system the temperature increases.)

18.2 $\Delta E = q - P\Delta V$ since $q = 0$
$\Delta E = -P\Delta V$
but ΔV is negative for a compression so ΔE increases and T increases.
Energy is added to the system in the form of work.

18.3 $\Delta H = \Delta H°[NO_2(g)] - \{\Delta H°[N_2O\ (g)] + \Delta H°[O_2(g)]\}$
$\Delta H = \{4 \text{ mol } NO_2(g) \times (34 \text{ kJ/mol})\} - \{2 \text{ mol } N_2O \times (81.5 \text{ kJ/mol}) + 3 \text{ mol } O_2 \times (0 \text{ kJ/mol})\}$
$\Delta H = -27$ kJ

$\Delta E - \Delta H = \Delta nRT = (-1 \text{ mol})(8.314 \text{ J mol}^{-1} \text{ K}^{-1})(318 \text{ K}) = -2.64 \times 10^4 \text{ J}$
$\Delta E - \Delta H = -2.64$ kJ
ΔE is more exothermic.

18.4 $\Delta E° = \Delta H° - \Delta nRT = -217.1 \text{ kJ} - (-1 \text{ mol})(8.314 \text{ J mol}^{-1} \text{ K}^{-1})(298 \text{ K})$
$= -217.1 \text{ kJ} + 2.48 \text{ kJ}$
$= -214.6 \text{ kJ}$

% Difference = (2.48/217) × 100 = 1.14 %

18.5 ΔS should be negative since the reaction moves from great movement of ions in solution to less movement in the solid.

18.6 (a) ΔS is negative since the products have a lower entropy, i.e. a lower freedom of movement.
(b) ΔS is positive since the products have a higher entropy, i.e. a higher freedom of movement.

18.7 (a) ΔS is negative since there are less gas molecules. (The product is also more complex, indicating an increase in order.)
(b) ΔS is negative since there are less gas molecules. (The product is also more complex, indicating an increase in order.)

18.8 (a) ΔS is negative since there is a change from a gas phase to a liquid phase. (The product is also more complex, indicating an increase in order.)
(b) ΔS is negative since there are less gas molecules. (The product is also more complex, indicating an increase in order.)
(c) ΔS is positive since the particles go from an ordered, crystalline state to a more disordered, aqueous state.

18.9 $\frac{1}{2}N_2(g) + \frac{3}{2}H_2(g) \rightarrow NH_3(g)$

$$\Delta S_f^o = [S^o_{NH_3(g)}] - [S^o_{N_2(g)} + S^o_{H_2(g)}]$$

$$\Delta S_f^o = \left[(1 \text{ mol } NH_3)\times\left(\frac{192.5 \text{ J}}{\text{mol K}}\right)\right] - \left[\left(\frac{1}{2} \text{ mol } N_2\right)\times\left(\frac{191.5 \text{ J}}{\text{mol K}}\right) + \left(\frac{3}{2} \text{ mol } H_2\right)\times\left(\frac{130.6 \text{ J}}{\text{mol K}}\right)\right]$$

$$\Delta S_f^o = -99.1 \text{ J K}^{-1}$$

18.10 $\Delta S° = (\text{sum } S°[\text{products}]) - (\text{sum } S°[\text{reactants}])$

a) $\Delta S° = \{S°[H_2O(l)] + S°[CaCl_2(s)]\} - \{S°[CaO(s)] + 2S°[HCl(g)]\}$
$\Delta S° = \{1 \text{ mol} \times (69.96 \text{ J mol}^{-1} \text{ K}^{-1}) + 1 \text{ mol} \times (114 \text{ J mol}^{-1} \text{ K}^{-1})\}$
$- \{1 \text{ mol} \times (40 \text{ J mol}^{-1} \text{ K}^{-1}) + 2 \text{ mol} \times (186.7 \text{ J mol}^{-1} \text{ K}^{-1})\}$
$\Delta S° = -229 \text{ J/K}$

b) $\Delta S° = \{S°[C_2H_6(g)]\} - \{S°[H_2(g)] + S°[C_2H_4(g)]\}$
$\Delta S° = \{1 \text{ mol} \times (229.5 \text{ J mol}^{-1} \text{ K}^{-1})\}$
$- \{1 \text{ mol} \times (130.6 \text{ J mol}^{-1} \text{ K}^{-1}) + 1 \text{ mol} \times (219.8 \text{ J mol}^{-1} \text{ K}^{-1})\}$
$\Delta S° = -120.9 \text{ J/K}$

18.11 $N_2(g) + 2O_2(g) \rightarrow N_2O_4(g)$

$$\Delta H^o_{N_2O_4(g)} = 9.67 \text{ kJ mol}^{-1}$$

$$\Delta S^o_{N_2O_4(g)} = (1 \text{ mol } N_2O_4)(304 \text{ J mol}^{-1} \text{ K}^{-1}) - (1 \text{ mol } N_2)(191.5 \text{ J mol}^{-1} \text{ K}^{-1}) - (2 \text{ mol } O_2)(205.0 \text{ J mol}^{-1} \text{ K}^{-1}) = -297.5 \text{ J K}^{-1}$$

$$\Delta G_f^o = \Delta H_f^o - T\Delta S_f^o$$

$$\Delta G_f^o = 9.67 \text{ kJ mol}^{-1} - (298 \text{ K})(-0.2975 \text{ kJ mol}^{-1} \text{ K}^{-1})$$

$$\Delta G_f^o = 98.3 \text{ kJ mol}^{-1}$$

18.12 First, we calculate $\Delta S°$, using the data:

$\Delta S° = \{2S°[Fe_2O_3(s)]\} - \{3S°[O_2(g)] + 4S°[Fe(s)]\}$
$\Delta S° = \{2 \text{ mol} \times (90.0 \text{ J mol}^{-1} \text{ K}^{-1})\}$
$- \{3 \text{ mol} \times (205.0 \text{ J mol}^{-1} \text{ K}^{-1}) + 4 \text{ mol} \times (27 \text{ J mol}^{-1} \text{ K}^{-1})\}$
$\Delta S° = -543 \text{ J/K} = -0.543 \text{ kJ/mol}$

Next, we calculate $\Delta H°$ using the data :

$\Delta H° = (\text{sum } \Delta H_f°[\text{products}]) - (\text{sum } \Delta H_f°[\text{reactants}])$
$\Delta H° = \{2\Delta H_f°[Fe_2O_3(s)]\} - \{3\Delta H_f°[O_2(g)] + 4\Delta H_f°[Fe(s)]\}$
$\Delta H° = \{2 \text{ mol} \times (-822.2 \text{ kJ/mol})\} - \{3 \text{ mol} \times (0.0 \text{ kJ/mol}) + 4 \times (0.0 \text{ kJ/mol})\}$
$\Delta H° = -1644 \text{ kJ}$

The temperature is $25.0 + 273.15 = 298.15$ K, and the calculation of $\Delta G°$ is as follows: $\Delta G° = \Delta H° - T\Delta S° = -1644 \text{ kJ} - (298.15 \text{ K})(-0.543 \text{ kJ/K}) = -1482 \text{ kJ}$

18.13 $Fe_2O_3(s) + 3CO(g) \rightarrow 2Fe(s) + 3CO_2(g)$
$\Delta G° = \{2 \text{ mol } Fe(s) \times \Delta G_f°[Fe(s)] + 3 \text{ mol } CO_2(g) \times \Delta G_f°[CO_2(g)]\}$
$- \{1 \text{ mol } Fe_2O_3(s) \times \Delta G_f°[Fe_2O_3(s)] + 3 \text{ mol } CO(g) \times \Delta G_f°[CO(g)]\}$
$\Delta G° = \{2 \text{ mol } Fe(s) \times 0 \text{ kJ mol}^{-1} + 3 \text{ mol } CO_2(g) \times -394.4 \text{ kJ mol}^{-1}\}$
$- \{1 \text{ mol } Fe_2O_3(s) \times -741.0 \text{ kJ mol}^{-1} + 3 \text{ mol } CO(g) \times -137.3 \text{ kJ mol}^{-1}\}$
$\Delta G° = -30.3 \text{ kJ}$

18.14 We calculate $\Delta G°_{rxn}$, using the data from Table 20.2:

a) $\Delta G°_{rxn} = \{2\Delta G_f°[NO_2(g)]\} - \{2\Delta G_f°[NO(g)] + \Delta G_f°[O_2(g)]\}$

$\Delta G°_{rxn} = \{2 \text{ mol} \times (+51.84 \text{ kJ mol}^{-1})\}$
$- \{2 \text{ mol} \times (+86.69 \text{ kJ mol}^{-1}) + 1 \text{ mol} \times (0 \text{ kJ mol}^{-1})\}$

$\Delta G°_{rxn} = -69.7$ kJ/mol

b) $\Delta G°_{rxn} = \{\Delta G_f°[CaCl_2(s)] + 2\Delta G_f°[H_2O(g)]\} - \{\Delta G_f°[Ca(OH)_2(s)] + 2\Delta G_f°[HCl(g)]\}$

$\Delta G°_{rxn} = \{1 \text{ mol} \times (-750.2 \text{ kJ mol}^{-1}) + 2 \text{ mol} \times (-228.6 \text{ kJ mol}^{-1})\}$
$- \{1 \text{ mol} \times (-896.76 \text{ kJ mol}^{-1}) + 2 \text{ mol} \times (-95.27 \text{ kJ mol}^{-1})\}$

$\Delta G°_{rxn} = -120.1$ kJ/mol

18.15 $C_2H_5OH(l) + 3O_2(g) \rightarrow 2CO_2(g) + 3H_2O(l)$

$\Delta G°_{rxn} = \{2\Delta G_f°[CO_2(g)] + 3\Delta G_f°[H_2O(l)]\} - \{\Delta G_f°[C_2H_5OH(l)] + 3\Delta G_f°[O_2(g)]\}$

$\Delta G°_{rxn} = \{2 \text{ mol} \times (-394.4 \text{ kJ mol}^{-1}) + 3 \text{ mol} \times (-237.2 \text{ kJ mol}^{-1})\} - \{1 \text{ mol} \times (-174.8 \text{ kJ mol}^{-1})$
$+ 3 \text{ mol} \times (0 \text{ kJ mol}^{-1})\}$

$G°_{rxn} = -1325.6$ kJ

For 100 g C_2H_5OH:

$$\text{mol } C_2H_5OH = 100 \text{ g } C_2H_5OH\left(\frac{1 \text{ mol } C_2H_5OH}{46.08 \text{ g } C_2H_5OH}\right) = 2.17 \text{ mol } C_2H_5OH$$

Maximum work:

$(1325.6 \text{ kJ mol}^{-1})(2.17 \text{ mol } C_2H_5OH) = 2877$ kJ

$$C_8H_{18}(l) + \frac{25}{2}O_2(g) \rightarrow 8CO_2(g) + 9H_2O(l)$$

$$\Delta G°_{rxn} = \{8\Delta G_f°[CO_2(g)] + 9\Delta G_f°[H_2O(l)]\} - \{\Delta G_f°[C_8H_{18}(l)] + \frac{25}{2}\Delta G_f°[O_2(g)]\}$$

$$\Delta G°_{rxn} = \{8 \text{ mol} \times (-394.4 \text{ kJ mol}^{-1}) + 9 \text{ mol} \times (-237.2 \text{ kJ mol}^{-1})\} - \{1 \text{ mol} \times (+17.3 \text{ kJ mol}^{-1}) + \frac{25}{2} \text{ mol} \times (0 \text{ kJ mol}^{-1})\}$$

$G°_{rxn} = -5304.1$ kJ

For 100 g C_8H_{18}:

$$\text{mol } C_8H_{18} = 100 \text{ g } C_8H_{18}\left(\frac{1 \text{ mol } C_8H_{18}}{114.26 \text{ g } C_8H_{18}}\right) = 0.875 \text{ mol } C_8H_{18}$$

Maximum work:

$(5304.1 \text{ kJ mol}^{-1})(0.875 \text{ mol } C_8H_{18}) = 4640.$ kJ

The octane is a better fuel on both a per gram and per mole basis.

18.16 The maximum amount of work that is available is the free energy change for the process, in this case, the standard free energy change, $\Delta G°$, since the process occurs at 25 °C.

$4Al(s) + 3O_2(g) \rightarrow 2Al_2O_3(s)$

$\Delta G° = (\text{sum } \Delta G_f°[\text{products}]) - (\text{sum } \Delta G_f°[\text{reactants}])$

$\Delta G° = 2\Delta G_f°[Al_2O_3(s)] - \{3\Delta G_f°[O_2(g)] + 4\Delta G_f°[Al(s)]\}$

$\Delta G° = 2 \text{ mol} \times (-1576.4 \text{ kJ/mol}) - \{3 \text{ mol} \times (0.0 \text{ kJ/mol}) + 4 \text{ mol} \times (0.0 \text{ kJ/mol})\}$

$\Delta G° = -3152.8$ kJ, for the reaction as written.

This calculation conforms to the reaction *as written*. This means that the above value of $\Delta G°$ applies to the equation involving *4 mol* of Al. The conversion to give energy *per mole* of aluminum is then: –3152.8 kJ/4 mol Al = –788 kJ/mol

The maximum amount of energy that may be obtained is thus 788 kJ.

18.17 $T \approx \dfrac{\Delta H^\circ}{\Delta S^\circ}$

$$\Delta H^\circ = 21.7 \text{ kJ mol}^{-1}\left(\frac{1000 \text{ J}}{1 \text{ kJ}}\right) = 21.7 \times 10^3 \text{ J}$$

$$239.9 \text{ K} \approx \frac{21.7\times10^3 \text{ J}}{\Delta S^\circ}$$

$$\Delta S^\circ = 90.5 \text{ J K}^{-1}$$

18.18 For the vaporization process in particular, and for any process in general, we have:

$$\Delta G = \Delta H - T\Delta S$$

If the temperature is taken to be that at which equilibrium is obtained, that is the temperature of the boiling point (where liquid and vapor are in equilibrium with one another), then we also have the result that ΔG is equal to zero:

$$\Delta G = 0 = \Delta H - T\Delta S, \text{ or } T_{eq} = \Delta H/\Delta S$$

We know ΔH to be 60.7 kJ/mol; we need the value for ΔS in units kJ mol^{-1} K^{-1}:

ΔS° = (sum S°[products]) – (sum S°[reactants])
ΔS° = S°[Hg(g)] – S°[Hg(l)]
$\Delta S^\circ = (175 \times 10^{-3} \text{ kJ mol}^{-1} \text{ K}^{-1}) - (76.1 \times 10^{-3} \text{ kJ mol}^{-1} \text{ K}^{-1})$
$\Delta S^\circ = 98.9 \times 10^{-3} \text{ kJ mol}^{-1} \text{ K}^{-1}$

$T_{eq} = 60.7 \text{ kJ/mol} \div 98.9 \times 10^{-3} \text{ kJ/mol K} = 614 \text{ K } (341\ °\text{C})$

18.19 ΔG° = (sum ΔG_f°[products]) – (sum ΔG_f°[reactants])
$\Delta G^\circ = \Delta G_f^\circ[SO_3(g)] - \{\Delta G_f^\circ[SO_2(g)] + \Delta G_f^\circ[1/2O_2(g)]\}$
ΔG° = 1 mol × (–370.4 kJ/mol) – {1 mol × (–300.4 kJ/mol) + (0.0 kJ/mol)}
ΔG° = –70.0 kJ/mol

Since the sign of ΔG° is negative, the reaction should be spontaneous.

18.20 $\Delta G^\circ = \Delta H^\circ - T\Delta S^\circ$
$\Delta G^\circ = \{2\Delta G_f^\circ[HCl(g)] + \Delta G_f^\circ[CaCO_3(s)]\}$
$- \{\Delta G_f^\circ[CaCl_2(s)] + \Delta G_f^\circ[H_2O(g)] + \Delta G_f^\circ[CO_2(g)]\}$
ΔG° = {2 mol × (–95.27 kJ/mol) + 1 mol × (–1128.8 kJ/mol)}
– {1 mol × (–750.2 kJ/mol) + 1 mol × (–228.6 kJ/mol) +
1 mol × (–394.4 kJ/mol)}
ΔG° = +53.9 kJ

ΔG° is positive, the reaction is not spontaneous, and we do not expect to see products formed from reactants.

18.21 T = 75 °C + 273.15 K = 348 K
ΔG° = 119.2 kJ – (348 K)(0.3548 kJ/K)
–4.3 kJ

18.22 $\Delta G^\circ = \{\Delta G_f^\circ[Na_2CO_3(s)] + \Delta G_f^\circ[CO_2(g)] + \Delta G_f^\circ[H_2O(g)]\}$
$- \{2\Delta G_f^\circ[NaHCO_3(s)]\}$
ΔG° = {1 mol × (–1048 kJ/mol) + 1 mol × (–394.4 kJ/mol) + 1 mol × (–228.6 kJ/mol)}
– {2 mol × (–851.9 kJ/mol)}
ΔG° = +32.8 kJ

$\Delta H° = \{\Delta H_f°[Na_2CO_3(s)] + \Delta H_f°[CO_2(g)] + \Delta H_f°[H_2O(g)]\}$
$- \{2\Delta H_f°[NaHCO_3(s)]\}$
$\Delta H° = \{1 \text{ mol} \times (-1131 \text{ kJ/mol}) + 1 \text{ mol} \times (-393.5 \text{ kJ/mol}) + 1 \text{ mol} \times (-241.8 \text{ kJ/mol})\}$
$- \{2 \text{ mol} \times (-947.7 \text{ kJ/mol})\}$

$\Delta H° = +129.1 \text{ kJ}$

$\Delta S° = \{S_f°[Na_2CO_3(s)] + S_f°[CO_2(g)] + S_f°[H_2O(g)]\}$
$- \{2S_f°[NaHCO_3(s)]\}$
$\Delta S° = \{1 \text{ mol} \times (136 \text{ J/mol K}) + 1 \text{ mol} \times (213.6 \text{ J/mol K}) + 1 \text{ mol} \times (188.7 \text{ J/mol K})\}$
$- \{2 \text{ mol} \times (102 \text{ J/mol K})\}$

$\Delta S° = +334.3 \text{ J K}^{-1} = 0.3343 \text{ kJ K}^{-1}$

$\Delta G°_{473} = \Delta H° - T\Delta S°$
$\Delta G°_{473} = 129.1 \text{ kJ} - (473 \text{ K})(0.3343 \text{ kJ K}^{-1}) = -29.1 \text{ kJ}$
The equilibrium shifts to products.

18.23 $\Delta G = \Delta G°_{298} + RT \ln\left(\frac{P_{N_2O_4}}{P_{NO_2}{}^2}\right)$

$\Delta G = -5.40 \times 10^3 \text{ J mol}^{-1} + \left(8.314 \text{ J mol}^{-1} \text{ K}^{-1}\right)(298 \text{ K})\ln\left(\frac{0.598 \text{ atm}}{(0.260 \text{ atm})^2}\right)$

$\Delta G = 0 \text{ J mol}^{-1} = 0 \text{ kJ mol}^{-1}$
The system is at equilibrium, so that the reaction will not move.

18.24 Using the data provided we may write:

$$\Delta G = \Delta G° + RT \ln\left(\frac{P_{N_2O_4}}{P_{NO_2}{}^2}\right)$$

$$= -5.40 \times 10^3 \text{ J mol}^{-1} + \left(8.314 \text{ J mol}^{-1} \text{ K}^{-1}\right)\left(298 \text{ K}\right) \ln\left(\frac{0.25 \text{ atm}}{(0.60 \text{ atm})^2}\right)$$

$$= -5.40 \times 10^3 \text{ J mol}^{-1} + \left(-9.03 \times 10^2 \text{ J mol}^{-1}\right)$$

$$= -6.30 \times 10^3 \text{ J mol}^{-1}$$

Since ΔG is negative, the forward reaction is spontaneous and the reaction will proceed to the right.

18.25 $\Delta G° = -RT \ln K_p$
$\Delta G° = -(8.314 \text{ J K}^{-1} \text{ mol}^{-1})(25 + 273 \text{ K}) \times \ln(6.9 \times 10^5) = -33 \times 10^3 \text{ J}$
$\Delta G° = -33 \text{ kJ}$

18.26 $\Delta G° = -RT \ln K_p$
$3.3 \times 10^3 \text{ J mol}^{-1} = -(8.314 \text{ J K}^{-1} \text{ mol}^{-1})(25.0 + 273.15 \text{ K}) \times \ln(K_{-p})$
$K_p = 0.26$

Review Problems

18.46 $\Delta E = q + w = 0.300 \text{ kJ} + 0.700 \text{ kJ} = +1.000 \text{ kJ}$

The overall process is endothermic, meaning that the internal energy of the system increases. Notice that both terms, q and w, contribute to the increase in internal energy of the system; the system gains heat (+q) and has work done on it (+w).

18.48 work = $P \times \Delta V$
The total pressure is atmospheric pressure plus that caused by the hand pump:
$P = (30.0 + 14.7) \text{ lb/in}^2 = 44.7 \text{ lb/in}^2$

Converting to atmospheres we get:
$P = 44.7 \text{ lb/in}^2 \times 1 \text{ atm}/14.7 \text{ lb/in}^2 = 3.04 \text{ atm}$

Next we convert the volume change in units in^3 to units L:
$24.0 \text{ in}^3 \times (2.54 \text{ cm/in})^3 \times 1 \text{ L}/1000 \text{ cm}^3 = 0.393 \text{ L}$

Hence $P \times \Delta V = (3.04 \text{ atm})(0.393 \text{ L}) = 1.19 \text{ L·atm}$
$1.19 \text{ L·atm} \times 101.3 \text{ J/L·atm} = 121 \text{ J}$

18.50 We use the data supplied in Appendix.
(a) $3PbO(s) + 2NH_3(g) \rightarrow 3Pb(s) + N_2(g) + 3H_2O(g)$

$\Delta H° = \{\Delta H_f^\circ[Pb(s)] + \Delta H_f^\circ[N_2(g)] + 3\,\Delta H_f^\circ[H_2O(g)]\}$
$- \{3\,\Delta H_f^\circ[PbO(s)] + 2\,\Delta H_f^\circ NH_3(g)]\}$
$\Delta H° = \{3 \text{ mol} \times (0 \text{ kJ/mol}) + 1 \text{ mol} \times (0 \text{ kJ/mol}) + 3 \text{ mol} \times (-241.8 \text{ kJ/mol})\}$
$- \{3 \text{ mol} \times (-219.2 \text{ kJ/mol}) + 2 \text{ mol} \times (-46.19 \text{ kJ/mol})\}$
$\Delta H° = +24.58 \text{ kJ}$

$\Delta E = \Delta H° - \Delta nRT$
$\Delta E = 24.58 \text{ kJ} - (+2 \text{ mol})(8.314 \text{ J/mol K})(10^{-3} \text{ kJ/J})(298 \text{ K}) = 19.6 \text{ kJ}$

(b) $NaOH(s) + HCl(g) \rightarrow NaCl(s) + H_2O(l)$

$\Delta H° = \{\Delta H_f^\circ[NaCl(s)] + \Delta H_f^\circ[H_2O(l)]\} - \{\Delta H_f^\circ[NaOH(s)] + \Delta H_f^\circ[HCl(g)]\}$
$\Delta H° = \{1 \text{ mol} \times (-411.0 \text{ kJ/mol}) + 1 \text{ mol} \times (-285.9 \text{ kJ/mol})\}$
$- \{1 \text{ mol} \times (-426.8 \text{ kJ/mol}) + 1 \text{ mol} \times (-92.3)\}$
$\Delta H° = -178 \text{ kJ}$

$\Delta E = \Delta H° - \Delta nRT$
$\Delta E = -178 \text{ kJ} - (-1)(8.314 \text{ J/mol K})(10^{-3} \text{ kJ/J})(298 \text{ K}) = -175 \text{ kJ}$

(c) $Al_2O_3(s) + 2Fe(s) \rightarrow Fe_2O_3(s) + 2Al(s)$

$\Delta H° = \{\Delta H_f^\circ[Fe_2O_3(s)] + 2\,\Delta H_f^\circ[Al(s)]\} - \{\Delta H_f^\circ[Al_2O_3(s)] + 2\,\Delta H_f^\circ[Fe(s)]\}$
$\Delta H° = \{1 \text{ mol} \times (-822.2 \text{ kJ/mol}) + 2 \text{ mol} \times (0 \text{ kJ/mol})\}$
$- \{1 \text{ mol} \times (-1669.8 \text{ kJ/mol}) + 2 \text{ mol} \times (0 \text{ kJ/mol})\}$
$\Delta H° = 847.6 \text{ kJ}$

$\Delta E = \Delta H°$, since the value of Δn for this reaction is zero.

(d) $2CH_4(g) \rightarrow C_2H_6(g) + H_2(g)$

$\Delta H° = \{\Delta H_f^\circ[C_2H_6(g)] + \Delta H_f^\circ[H_2(g)]\} - \{2\,\Delta H_f^\circ[CH_4(g)]\}$

$\Delta H° = \{1 \text{ mol} \times (-84.667 \text{ kJ/mol}) + 1 \text{ mol} \times (0.0 \text{ kJ/mol})\}$
$- \{2 \text{ mol} \times (-74.848 \text{ kJ/mol})\}$

$\Delta H° = 65.029 \text{ kJ}$

$\Delta E = \Delta H°$, since the value of Δn for this reaction is zero.

18.52 $\Delta H = \Delta E + \Delta n_{gas}RT$
$\Delta E = \Delta H - \Delta n_{gas}RT$
$\Delta H = -163.14 \text{ kJ}$
$\Delta n_{gas} = 3 \text{ mol} - 2 \text{ mol} = 1 \text{ mol}$
$R = 8.314 \text{ J mol}^{-1} \text{ K}^{-1}$
$T = 25 \text{ °C} + 273 \text{ K} = 298 \text{ K}$

For 180 g N_2O, first calculate the number of moles of N_2O and then how many kJ of energy will be released for that many moles of N_2O
$\Delta E = -163.14 \text{ kJ} - (1 \text{ mol})(8.314 \times 10^{-3} \text{ kJ mol}^{-1} \text{ K}^{-1})(298 \text{ K})$
$\Delta E = -165.62 \text{ kJ}$

$$\text{mol } N_2O = 180 \text{ g}\left(\frac{1 \text{ mol } N_2O}{44.02 \text{ g } N_2O}\right) = 4.09 \text{ mol } N_2O$$

$$\text{Amount of Energy} = 4.09 \text{ mol } N_2O\left(\frac{-165.62 \text{ kJ}}{2 \text{ mol } N_2O}\right) = -338 \text{ kJ}$$

For $T = 200$ °C
$T = 200 \text{ °C} + 273 \text{ K} = 473 \text{ K}$
$\Delta E = -163.14 \text{ kJ} - (1 \text{ mol})(8.314 \times 10^{-3} \text{ kJ mol}^{-1} \text{ K}^{-1})(473 \text{ K})$
$\Delta E = -163.14 \text{ kJ} - 3.93 \text{ kJ}$
$\Delta E_{200\,°C} = -167.07 \text{ kJ}$ for 1 mole of N_2O

For 180 g N_2O, 4.09 mol N_2O:

$$\text{Amount of Energy at 200 °C} = 4.09 \text{ mol } N_2O\left(\frac{-167.07 \text{ kJ}}{2 \text{ mol } N_2O}\right) = -341 \text{ kJ}$$

18.54 In general, we have the equation: $\Delta H° = (\text{sum } \Delta H_f^\circ[\text{products}]) - (\text{sum } \Delta H_f^\circ[\text{reactants}])$

(a) $\Delta H° = \{\Delta H_f^\circ[CaCO_3(s)]\} - \{\Delta H_f^\circ[CO_2(g)] + \Delta H_f^\circ[CaO(s)]\}$
$\Delta H° = \{1 \text{ mol} \times (-1207 \text{ kJ/mol})\} - \{1 \text{ mol} \times (-394 \text{ kJ/mol}) + 1 \text{ mol} \times (-635.5 \text{ kJ/mol})\}$
$\Delta H° = -178 \text{ kJ}$ ∴ favored.

(b) $\Delta H° = \{\Delta H_f^\circ[C_2H_6(g)]\} - \{\Delta H_f^\circ[C_2H_2(g)] + 2\,\Delta H_f^\circ[H_2(g)]\}$
$\Delta H° = \{1 \text{ mol} \times (-84.5 \text{ kJ/mol})\} - \{1 \text{ mol} \times (227 \text{ kJ/mol}) + 2 \text{ mol} \times (0.0 \text{ kJ/mol})\}$
$\Delta H° = -311 \text{ kJ}$ ∴ favored.

(c) $\Delta H° = \{\Delta H_f^\circ[Fe_2O_3(s)] + 3\,\Delta H_f^\circ[Ca(s)]\} - \{2\,\Delta H_f^\circ[Fe(s)] + 3\,\Delta H_f^\circ[CaO(s)]\}$
$\Delta H° = \{1 \text{ mol} \times (-822.2 \text{ kJ/mol}) + 3 \text{ mol} \times (0.0 \text{ kJ/mol})\}$
$- \{2 \text{ mol} \times (0.0 \text{ kJ/mol}) + 3 \text{ mol} \times (-635.5 \text{ kJ/mol})\}$
$\Delta H° = +1084.3 \text{ kJ}$ ∴ not favorable from the standpoint of enthalpy alone.

18.56 $2N_2O(g) \rightarrow 2N_2(g) + O_2(g)$

The factors needed to be consider in order to determine the sign of ΔS are

(1) the number of moles of products versus reactants
in this case the number of moles increases

(2) the state of the products versus reactants
both products and reactants are gases

(3) the complexity of the molecules
N_2O is more complex than either N_2 or O_2

ΔS is expected to be positive.

18.58 (a) negative – since the number of moles of gaseous material decreases.
(b) negative – since the number of moles of gaseous material decreases.
(c) negative – since the number of moles of gas decreases.
(d) positive – since a gas appears where there formerly was none.

18.60 ΔS° = (sum S°[products]) – (sum S°[reactants])

(a) $\Delta S° = \{2S°[NH_3(g)]\} - \{3S°[H_2(g)] + S°[N_2(g)]\}$
$\Delta S° = \{2 \text{ mol} \times (192.5 \text{ J mol}^{-1} \text{ K}^{-1})\} - \{3 \text{ mol} \times (130.6 \text{ J mol}^{-1} \text{ K}^{-1}) + 1 \text{ mol} \times (191.5 \text{ J mol}^{-1} \text{ K}^{-1})\}$
$\Delta S° = -198.3$ J/K ∴ not spontaneous from the standpoint of entropy.

(b) $\Delta S° = \{S°[CH_3OH(l)]\} - \{2S°[H_2(g)] + S°[CO(g)]\}$
$\Delta S° = \{1 \text{ mol} \times (126.8 \text{ J mol}^{-1} \text{ K}^{-1})\} - \{2 \text{ mol} \times (130.6 \text{ J mol}^{-1} \text{ K}^{-1}) + 1 \text{ mol} \times (197.9 \text{ J mol}^{-1} \text{ K}^{-1})\}$
$\Delta S° = -332.3$ J/K ∴ not favored from the standpoint of entropy alone.

(c) $\Delta S° = \{6S°[H_2O(g)] + 4S°[CO_2(g)]\} - \{7S°[O_2(g)] + 2S°[C_2H_6(g)]\}$
$\Delta S° = \{6 \text{ mol} \times (188.7 \text{ J mol}^{-1} \text{ K}^{-1}) + 4 \text{ mol} \times (213.6 \text{ J mol}^{-1} \text{ K}^{-1})\} - \{7 \text{ mol} \times (205.0 \text{ J mol}^{-1} \text{ K}^{-1}) + 2 \text{ mol} \times (229.5 \text{ J mol}^{-1} \text{ K}^{-1})\}$
$\Delta S° = +92.6$ J/K ∴ favorable from the standpoint of entropy alone.

(d) $\Delta S° = \{2S°[H_2O(l)] + S°[CaSO_4(s)]\} - \{S°[H_2SO_4(l)] + S°[Ca(OH)_2(s)]\}$
$\Delta S° = \{2 \text{ mol} \times (69.96 \text{ J mol}^{-1} \text{ K}^{-1}) + 1 \text{ mol} \times (107 \text{ J mol}^{-1} \text{ K}^{-1})\} - \{1 \text{ mol} \times (157 \text{ J mol}^{-1} \text{ K}^{-1}) + 1 \text{ mol} \times (76.1 \text{ J mol}^{-1} \text{ K}^{-1})\}$
$\Delta S° = +14$ J/K ∴ favorable from the standpoint of entropy alone.

(e) $\Delta S° = \{2S°[N_2(g)] + S°[SO_2(g)]\} - \{2S°[N_2O(g)] + S°[S(s)]\}$
$\Delta S° = \{2 \text{ mol} \times (191.5 \text{ J mol}^{-1} \text{ K}^{-1}) + 1 \text{ mol} \times (248 \text{ J mol}^{-1} \text{ K}^{-1})\} - \{2 \text{ mol} \times (220.0 \text{ J mol}^{-1} \text{ K}^{-1}) + 1 \text{ mol} \times (31.9 \text{ J mol}^{-1} \text{ K}^{-1})\}$
$\Delta S° = +159$ J/K ∴ favorable from the standpoint of entropy alone.

18.62 The entropy change that is designated ΔS_f° is that which corresponds to the reaction in which one mole of a substance is formed from elements in their standard states. Since the value is understood to correspond to the reaction forming one mole of a single pure substance, the units may be written either J K^{-1} or $\text{J mol}^{-1} \text{ K}^{-1}$.

(a) $2C(s) + 2H_2(g) \rightarrow C_2H_4(g)$
$\Delta S° = \{S°[C_2H_4(g)]\} - \{2S°[C(s)] + 2S°[H_2(g)]\}$
$\Delta S° = \{1 \text{ mol} \times (219.8 \text{ J mol}^{-1} \text{ K}^{-1})\} - \{2 \text{ mol} \times (5.69 \text{ J mol}^{-1} \text{ K}^{-1}) + 2 \text{ mol} \times (130.6 \text{ J mol}^{-1} \text{ K}^{-1})\}$
$\Delta S° = -52.8$ J/K or $-52.8 \text{ J mol}^{-1} \text{ K}^{-1}$

(b) $Ca(s) + S(s) + 3O_2(g) + 2H_2(g) \rightarrow CaSO_4{\cdot}2H_2O(s)$

$\Delta S° = \{S°[CaSO_4{\cdot}2H_2O(s)]\} - \{2S°[H_2(g)] + 3S°[O_2(g)] + S°[S(s)] + S°[Ca(s)]\}$

$\Delta S° = \{1 \text{ mol} \times (194.0 \text{ J mol}^{-1} \text{ K}^{-1})\} - \{2 \text{ mol} \times (130.6 \text{ J mol}^{-1} \text{ K}^{-1})$
$+ 3 \text{ mol} \times (205.0 \text{ J mol}^{-1} \text{ K}^{-1}) + 1 \text{ mol} \times (31.9 \text{ J mol}^{-1} \text{ K}^{-1}) + 1 \text{ mol} \times (41.4 \text{ J mol}^{-1} \text{ K}^{-1})\}$

$\Delta S° = -755.5$ J/K or $-755.5 \text{ J mol}^{-1} \text{ K}^{-1}$

(c) $2H_2(g) + 2C(s) + O_2(g) \rightarrow HC_2H_3O_2(l)$

$\Delta S° = \{S°[HC_2H_3O_2(l)]\} - \{2S°[H_2(g)] + 2S°[C(s)] + S°[O_2(g)]\}$

$\Delta S° = \{1 \text{ mol} \times (160 \text{ J mol}^{-1} \text{ K}^{-1})\} - \{2 \text{ mol} \times (130.6 \text{ J mol}^{-1} \text{ K}^{-1})$
$+ 2 \text{ mol} \times (5.69 \text{ J mol}^{-1} \text{ K}^{-1}) + 1 \text{ mol} \times (205.0 \text{ J mol}^{-1} \text{ K}^{-1})\}$

$\Delta S° = -318$ J/K or $-318 \text{ J mol}^{-1} \text{ K}^{-1}$

18.64 $\Delta S° = (\text{sum } S°[\text{products}]) - (\text{sum } S°[\text{reactants}])$

$\Delta S° = \{2S°[HNO_3(l)] + S°[NO(g)]\} - \{3S°[NO_2(g)] + S°[H_2O(l)]\}$

$\Delta S° = \{2 \text{ mol} \times (155.6 \text{ J mol}^{-1} \text{ K}^{-1}) + 1 \text{ mol} \times (210.6 \text{ J mol}^{-1} \text{ K}^{-1})\}$
$- \{3 \text{ mol} \times (240.5 \text{ J mol}^{-1} \text{ K}^{-1}) + 1 \text{ mol} \times (69.96 \text{ J mol}^{-1} \text{ K}^{-1})\}$

$\Delta S° = -269.7$ J/K

18.66 The quantity ΔG_f° applies to the equation in which one mole of pure phosgene is produced from the naturally occurring forms of the elements:

$$C(s) + 1/2O_2(g) + Cl_2(g) \rightarrow COCl_2(g), \ \Delta G_f^\circ = ?$$

We can determine ΔG_f° if we can find values for ΔH_f° and ΔS_f°, because:

$\Delta G° = \Delta H° - T\Delta S°$

The value of ΔS_f° is determined using S° for phosgene in the following way:

$\Delta S_f^\circ = \{S°[COCl_2(g)]\} - \{S°[C(s)] + 1/2S°[O_2(g)] + S°[Cl_2(g)]\}$

$\Delta S_f^\circ = \{1 \text{ mol} \times (284 \text{ J mol}^{-1} \text{ K}^{-1})\} - \{1 \text{ mol} \times (5.69 \text{ J mol}^{-1} \text{ K}^{-1})$
$+ 1/2 \text{ mol} \times (205.0 \text{ J mol}^{-1} \text{ K}^{-1}) + 1 \text{ mol} \times (223.0 \text{ J mol}^{-1} \text{ K}^{-1})\}$

$\Delta S_f^\circ = -47 \text{ J mol}^{-1} \text{ K}^{-1}$ or -47 J/K

$\Delta G_f^\circ = \Delta H_f^\circ - T\Delta S_f^\circ = -223$ kJ/mol $-$ (298 K)(-0.047 kJ/mol K)
$= -209$ kJ/mol

18.68 $\Delta G° = (\text{sum } \Delta G_f^\circ[\text{products}]) - (\text{sum } \Delta G_f^\circ[\text{reactants}])$

(a) $\Delta G° = \{\Delta G_f^\circ[H_2SO_4(l)]\} - \{\Delta G_f^\circ[H_2O(l)] + \Delta G_f^\circ[SO_3(g)]\}$

$\Delta G° = \{1 \text{ mol} \times (-689.9 \text{ kJ/mol})\} -$
$\{1 \text{ mol} \times (-237.2 \text{ kJ/mol}) + 1 \text{ mol} \times (-370 \text{ kJ/mol})\}$

$\Delta G° = -82.3$ kJ

(b) $\Delta G° = \{2\Delta G_f^\circ[NH_3(g)] + \Delta G_f^\circ[H_2O(l)] + \Delta G_f^\circ[CaCl_2(s)]\}$
$- \{\Delta G_f^\circ[CaO(s)] + 2\Delta G_f^\circ[NH_4Cl(s)]\}$

$\Delta G° = \{2 \text{ mol} \times (-16.7 \text{ kJ/mol}) + 1 \text{ mol} \times (-237.2 \text{ kJ/mol})$
$+ 1 \text{ mol} \times (-750.2 \text{ kJ/mol})\} - \{1 \text{ mol} \times (-604.2 \text{ kJ/mol})$
$+ 2 \text{ mol} \times (-203.9 \text{ kJ/mol})\}$

$\Delta G° = -8.8$ kJ

(c) $\Delta G° = \{\Delta G_f^\circ[H_2SO_4(l)] + \Delta G_f^\circ[CaCl_2(s)]\} - \{\Delta G_f^\circ[CaSO_4(s)] + \Delta G_f^\circ[HCl(g)]\}$

$\Delta G° = \{1\text{ mol} \times (-689.9\text{ kJ/mol}) + 1\text{ mol} \times (-750.2\text{ kJ/mol})\}$
$\quad - \{1\text{ mol} \times (-1320.3\text{ kJ/mol}) + 2\text{ mol} \times (-95.27\text{ kJ/mol})\}$

$\Delta G° = +70.7\text{ kJ}$

18.70 Multiply the reverse of the second equation by 2 (remembering to multiply the associated free energy change by –2), and add the result to the first equation:

$4NO(g) \rightarrow 2N_2O(g) + O_2(g)$, $\quad \Delta G° = -139.56\text{ kJ}$
$4NO_2(g) \rightarrow 4NO(g) + 2O_2(g)$, $\quad \Delta G° = +139.40\text{ kJ}$

$4NO_2(g) \rightarrow 3O_2(g) + 2N_2O(g)$, $\quad \Delta G° = -0.16\text{ kJ}$

This result is the reverse of the desired reaction, which must then have $\Delta G° = +0.16\text{ kJ}$

18.72 The maximum work obtainable from a reaction is equal in magnitude to the value of ΔG for the reaction. Thus, we need only determine ΔG° for the process:

$\Delta G° = (\text{sum } \Delta G_f^\circ[\text{products}]) - (\text{sum } \Delta G_f^\circ[\text{reactants}])$

$\Delta G° = \{3\,\Delta G_f^\circ[H_2O(g)] + 2\,\Delta G_f^\circ[CO_2(g)]\} - \{3\,\Delta G_f^\circ[O_2(g)] + \Delta G_f^\circ[C_2H_5OH(l)]\}$

$\Delta G° = \{3\text{ mol} \times (-228.6\text{ kJ/mol}) + 2\text{ mol} \times (-394.4\text{ kJ/mol})\}$
$\quad - \{3\text{ mol} \times (0.0\text{ kJ/mol}) + 1\text{ mol} \times (-174.8\text{ kJ/mol})\}$

$\Delta G° = -1299.8\text{ kJ}$

18.74 At equilibrium, $\Delta G = 0 = \Delta H - T\Delta S$
$T_{eq} = \Delta H/\Delta S$, and assuming that ΔS is independent of temperature, we have:
$T_{eq} = (31.4 \times 10^3\text{ J mol}^{-1}) \div (94.2\text{ J mol}^{-1}\text{ K}^{-1}) = 333\text{ K}$

18.76 At equilibrium, $\Delta G = 0 = \Delta H - T\Delta S$
Thus $\Delta H = T\Delta S$, and if we assume that both ΔH and ΔS are independent of temperature, we have:

$\Delta S = \Delta H/T_{eq} = (37.7 \times 10^3\text{ J/mol}) \div (99.3 + 273.15\text{ K})$
$\Delta S = 101\text{ J mol}^{-1}\text{ K}^{-1}$

18.78 The reaction is spontaneous if its associated value for ΔG° is negative.
$\Delta G° = (\text{sum } \Delta G_f^\circ[\text{products}]) - (\text{sum } \Delta G_f^\circ[\text{reactants}])$

$\Delta G° = \{\Delta G_f^\circ[HC_2H_3O_2(l)] + \Delta G_f^\circ[H_2O(l)] + \Delta G_f^\circ[NO(g)] + \Delta G_f^\circ[NO_2(g)]\}$
$\quad - \{\Delta G_f^\circ[C_2H_4(g)] + \Delta G_f^\circ[HNO_3(l)]\}$

$\Delta G° = \{1\text{ mol} \times (-392.5\text{ kJ/mol}) + 1\text{ mol} \times (-237.2\text{ kJ/mol})$
$\quad + 1\text{ mol} \times (86.69\text{ kJ/mol}) + 1\text{ mol} \times (51.84\text{ kJ/mol})\}$
$\quad - \{1\text{ mol} \times (68.12\text{ kJ/mol}) + 1\text{ mol} \times (-79.91\text{ kJ/mol})\}$

$\Delta G° = -479.4\text{ kJ}$

Yes, the reaction is spontaneous.

18.80 (a) $\Delta G° = \{2 \times \Delta G_f^\circ[POCl_3(g)]\} - \{2 \times \Delta G_f^\circ[PCl_3(g)] + 1 \times \Delta G_f^\circ[O_2(g)]\}$

$\Delta G° = \{2\text{ mol} \times (-1019\text{ kJ/mol})\}$
$\quad - \{2\text{ mol} \times (-267.8\text{ kJ/mol}) + 1\text{ mol} \times (0\text{ kJ/mol})\}$

$\Delta G° = -1502\text{ kJ} = -1.502 \times 10^6\text{ J}$
$-1.502 \times 10^6\text{ J} = -RT\ln K_p = -(8.314\text{ J/K mol})(298\text{ K}) \times \ln K_p$
$\ln K_p = 606 \quad \therefore \log K_p = 263$, and $K_p = 10^{263}$.

(b) $\Delta G° = \{2 \times \Delta G^\circ_f[SO_2(g)] + 1 \times \Delta G^\circ_f[O_2(g)]\} - \{2 \times \Delta G^\circ_f[SO_3(g)]\}$
$\Delta G° = \{2\text{ mol} \times (-300.4\text{ kJ/mol}) + 1\text{ mol} \times (0\text{ kJ/mol})\} - \{2\text{ mol} \times (-370.4\text{ kJ/mol})\}$
$\Delta G° = 140\text{ kJ} = 1.40 \times 10^5\text{ J}$
$1.40 \times 10^5\text{ J} = -RT\ln K_p = -(8.314\text{ J/K mol})(298\text{ K}) \times \ln K_p$
$\ln K_p = -56.5$ and $K_p = 2.90 \times 10^{-25}$

18.82 $\Delta G° = -RT\ln K_p$
$-9.67 \times 10^3\text{ J} = -(8.314\text{ J/K mol})(1273\text{ K}) \times \ln K_p$
$\ln K_p = 0.914 \;\therefore\; K_p = 2.49$

$$Q = \frac{[N_2O][O_2]}{[NO_2][NO]} = \frac{(0.015)(0.0350)}{(0.0200)(0.040)} = 0.66$$

Since the value of Q is less than the value of K, the system is not at equilibrium and must shift to the right to reach equilibrium.

18.84 $\Delta G° = -RT\ln K_p$
$-50.79 \times 10^3\text{ J} = -(8.314\text{ J K}^{-1}\text{ mol}^{-1})(298\text{ K}) \times \ln K_p$
$\ln K_p = 20.50$
Taking the exponential of both sides of this equation gives: $K_p = 8.000 \times 10^8$
This is a favorable reaction, since the equilibrium lies far to the side favoring products and is worth studying as a method for methane production.

18.86 If $\Delta G° = 0$, $K_c = 1$. If we start with pure products, the value of Q will be infinite (there are zero reactants) and, since $Q > K_c$, the equilibrium will shift towards the reactants, i.e., the pure products will decompose to their elements.

18.88 This requires the breaking of three N–H single bonds:

$$NH_3 \rightarrow N + 3H$$

The enthalpy of atomization of NH_3 is thus three times the average N–H single bond energy:
$3 \times 388\text{ kJ/mol} = 1.16 \times 10^3\text{ kJ/mol}$

18.90 The heat of formation for ethanol vapor describes the following change:
$2C(s) + 3H_2(g) + 1/2O_2(g) \rightarrow C_2H_5OH(g)$

This can be arrived at by adding the following thermochemical equations, using data from Table 18.3:

$3H_2(g) \rightarrow 6H(g)$	$\Delta H_1° = (6)217.89\text{ kJ} = 1307.34\text{ kJ}$
$2C(s) \rightarrow 2C(g)$	$\Delta H_2° = (2)716.67\text{ kJ} = 1,433.34\text{ kJ}$
$1/2O_2(g) \rightarrow O(g)$	$\Delta H_3° = (1)249.17\text{ kJ} = 249.17\text{ kJ}$
$6H(g) + 2C(g) + O(g) \rightarrow C_2H_5OH(g)$	$\Delta H_4° = x$
$3H_2(g) + 2C(s) + 1/2O_2(g) \rightarrow C_2H_5OH(g)$	$\Delta H_f° = (2989.85 + x)\text{ kJ}$

Since $\Delta H_f°$ is given as –235.3 kJ…
$-235.3 = 2989.85 + x$
$x = -3225.2\text{ kJ}$

$\Delta H°_{atom}$ is the reverse reaction, so the sign will change:

$\Delta H°_{atom} = 3225.2\text{ kJ}$
The sum of all the bond energies in the molecule should be equal to the atomization energy:

ΔH°_{atom} = 1(C–C bond) + 5(C–H bonds) + 1(O–H bond) + 1(C–O bond)

We use values from Table 18.4:

3225.2 kJ = 1(348 kJ) + 5(412 kJ) + 1(463 kJ) + 1(C–O bond)

C–O bond energy = 354 kJ/mol

18.92 There are two C=S double bonds to be considered:

ΔH°_f = sum(ΔH°_f [gaseous atoms]) – sum(average bond energies in the molecule)

ΔH°_f [$CS_2(g)$] = 115.3 kJ/mol = [716.67 + 2 × 276.98] – [2 × C=S]

The C=S double bond energy is therefore given by the equation:

C=S = –(115.3 – 716.67 – 2 × 276.98) ÷ 2 = 577.7 kJ/mol

18.94 There are six S—F bonds in the molecule:

ΔH°_f = sum(ΔH°_f [gaseous atoms]) – sum(average bond energies in the molecule)

ΔH°_f [$SF_6(g)$] = –1096 kJ/mol = [277.0 + 6 × 79.14] – [6 × S—F]

S—F = (1096 + 277.0 + 6 × 79.14) ÷ 6 = 308.0 kJ/mol

18.96 See the method of review problems 18.88 through 18.95.

ΔH°_f = sum(ΔH°_f [gaseous atoms]) – sum(average bond energies in the molecule)

ΔH°_f [$C_2H_2(g)$] = [2 × 716.7 + 2 × 218.0] – [2 × 412 + 960]
= 85 kJ/mol

18.98 The heat of formation of CF_4 should be more exothermic than that of CCl_4 because more energy is released on formation of a C—F bond than on formation of a C—Cl bond. Also, less energy is needed to form gaseous F atoms than to form gaseous Cl atoms.

Practice Exercises

19.1 anode: $Mg(s) \rightarrow Mg^{2+}(aq) + 2e^-$
cathode: $Fe^{2+}(aq) + 2e^- \rightarrow Fe(s)$
cell notation: $Mg(s)|Mg^{2+}(aq)||Fe^{2+}(aq)|Fe(s)$

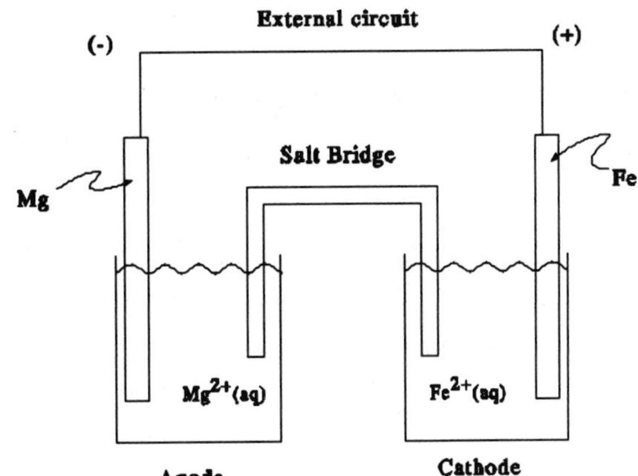

19.2 anode: $Al(s) \rightarrow Al^{3+}(aq) + 3e^-$
cathode: $Ni^{2+}(aq) + 2e^- \rightarrow Ni(s)$
overall: $3Ni^{2+}(aq) + 2Al(s) \rightarrow 2Al^{3+}(aq) + 3Ni(s)$

19.3 $E°_{cell} = E°_{substance\ reduced} - E°_{substance\ oxidized}$
$2Ag^+(aq) + Cu(s) \rightarrow 2Ag(s) + Cu^{2+}(aq)$
$E°_{cell} = E°_{Ag} - E°_{Cu}$
$E°_{cell} = 0.80\ V - 0.34\ V = 0.46\ V$

$2Ag^+(aq) + Zn(s) \rightarrow 2Ag(s) + Zn^{2+}(aq)$
$E°_{cell} = E°_{Ag} - E°_{Zn}$
$E°_{cell} = 0.80\ V - (-0.76\ V) = 1.56\ V$

Zinc will have the larger value for $E°_{cell}$.

19.4 $E°_{cell} = E°_{substance\ reduced} - E°_{substance\ oxidized}$
$Fe^{2+}(aq) + Mg(s) \rightarrow Mg^{2+}(aq) + Fe(s)$
$1.93\ V = E°_{Fe^{2+}} - (-2.37\ V)$
$E°_{Fe^{2+}} = 1.93\ V + (-2.37\ V) = -0.44\ V$

This agrees exactly with Table 19.1.

19.5 The half–reaction with the more positive value of E° (listed higher in Table 19.1) will occur as a reduction. The half–reaction having the less positive (more negative) value of E° (listed lower in Table 19.1) will be reversed and occur as an oxidation.

(a) $I_2(aq) + 2e^- \rightarrow 2I^-(aq)$ oxidation
$Fe^{3+}(aq) + 2e^- \rightarrow Fe^{2+}(aq)$ reduction
$2I^-(aq) + Fe^{3+}(aq) \rightarrow I_2(aq) + 2Fe^{2+}(aq)$

(b) $Mg^{2+}(aq) + 2e^- \rightarrow Mg(s)$ oxidation
$Cr^{3+}(aq) + 3e^- \rightarrow Cr(s)$ reduction
$Mg(s) + Cr(s) \rightarrow Mg^{2+}(aq) + Cr^{3+}(aq)$

(c) $Co^{2+}(aq) + 2e^- \rightarrow Co(s)$ oxidation
$SO_4^{2-}(aq) + 4H^+(aq) + 2e^- \rightarrow H_2SO_3(aq) + H_2O$ reduction
$SO_4^{2-}(aq) + 4H^+(aq) + Co(s) \rightarrow H_2SO_3(aq) + H_2O + Co^{2+}(aq)$

19.6 The half–reaction with the more positive value of E° (listed higher in Table 19.1) will occur as a reduction. The half–reaction having the less positive (more negative) value of E° (listed lower in Table 19.1) will be reversed and occur as an oxidation.

$Br_2(aq) + 2e^- \rightarrow 2Br^-(aq)$	reduction
$SO_4^{2-}(aq) + 4H^+(aq) + 2e^- \rightarrow H_2SO_3(aq) + H_2O$	oxidation
$Br_2(aq) + H_2SO_3(aq) + H_2O \rightarrow 2Br^-(aq) + SO_4^{2-}(aq) + 4H^+(aq)$	

19.7 Either nickel(II) or iron(III) will be reduced, depending on which way the reaction proceeds. Iron(III) is listed higher than nickel(II) in Table 19.1 (it has a greater reduction potential), so we would expect that the reaction would not be spontaneous in the direction shown.
The spontaneous reaction is:
$Ni(s) + 2Fe^{3+}(aq) \rightarrow Ni^{2+}(aq) + 2Fe^{2+}(aq)$

19.8 The half–reaction having the more positive value for E° will occur as a reduction. The other half–reaction should be reversed, so as to appear as an oxidation.

$NiO_2(s) + 2H_2O + 2e^- \rightarrow Ni(OH)_2(s) + 2OH^-(aq)$	reduction
$Fe(s) + 2OH^-(aq) \rightarrow 2e^- + Fe(OH)_2(s)$	oxidation
$NiO_2(s) + Fe(s) + 2H_2O \rightarrow Ni(OH)_2(s) + Fe(OH)_2(s)$	net reaction

$E°_{cell} = E°_{substance\ reduced} - E°_{substance\ oxidized}$
$E°_{cell} = E°_{NiO_2} - E°_{Fe}$
$E°_{cell} = 0.49 - (-0.88) = 1.37\ V$

19.9 The half–reaction having the more positive value for E° will occur as a reduction. The other half–reaction should be reversed, so as to appear as an oxidation.
$3Cu^{2+}(aq) + 2Cr(s) \rightarrow 3Cu(s) + 2Cr^{3+}(aq)$
$E°_{cell} = E°_{substance\ reduced} - E°_{substance\ oxidized}$
$E°_{cell} = E°_{Cu} - E°_{Cr}$
$E°_{cell} = 0.34 - (-0.74) = 1.08\ V$

19.10 The half–reaction having the more positive value for E° will occur as a reduction. The other half–reaction should be reversed, so as to appear as an oxidation.

Reduction:	$3 \times [MnO_4^-(aq) + 8H^+(aq) + 5e^- \rightarrow Mn^{2+}(aq) + 4H_2O]$
Oxidation:	$5 \times [Cr(s) \rightarrow Cr^{3+}(aq) + 3e^-]$
Net reaction:	$3MnO_4^-(aq) + 24H^+(aq) + 5Cr(s) \rightarrow 5Cr^{3+}(aq) + 3Mn^{2+}(aq) + 12H_2O$

$E°_{cell} = E°_{substance\ reduced} - E°_{substance\ oxidized}$
$E°_{cell} = E°_{MnO_4^-} - E°_{Cr}$
$E°_{cell} = 1.51\ V - (-0.74\ V) = 2.25\ V$

19.11 A reaction will occur spontaneously in the forward direction if the value of E° is positive. We therefore evaluate E° for each reaction using:

$$E°_{cell} = E°_{substance\ reduced} - E°_{substance\ oxidized}$$

(a) $Br_2(aq) + 2e^- \rightarrow 2Br^-(aq)$ reduction
$I_2(s) + 2e^- \rightarrow 2I^-(aq)$ oxidation

$E°_{cell} = E°_{Br_2} - E°_{I_2}$
$E°_{cell} = 1.07\ V - (0.54\ V) = 0.53\ V$, ∴ spontaneous

(b) $MnO_4^-(aq) + 8H^+ + 5e^- \rightarrow Mn^{2+}(aq) + 4H_2O$ reduction
$5Ag^+(aq) + 5e^- \rightarrow 5Ag(s)$ oxidation

$E^\circ_{cell} = E^\circ_{MnO_4} - E^\circ_{Ag}$
$E^\circ_{cell} = 1.51\ V - (0.80V) = +0.71\ V$, ∴ spontaneous

19.12 A reaction will occur spontaneously in the forward direction if the value of E° is positive. We therefore evaluate E° for each reaction using:

$$E^\circ_{cell} = E^\circ_{substance\ reduced} - E^\circ_{substance\ oxidized}$$

(a) $Br_2(aq) + 2e^- \rightarrow 2Br^-(aq)$ reduction
$Cl_2(aq) + 2H_2O \rightarrow 2HOCl(aq) + 2H^+(aq) + 2e^-$ oxidation

$E^\circ_{cell} = E^\circ_{Br_2} - E^\circ_{Cl_2}$
$E^\circ_{cell} = 1.07\ V - (1.36\ V) = -0.29\ V$, ∴ nonspontaneous

(b) $2Cr^{3+}(aq) + 6e^- \rightarrow 2Cr(s)$ reduction
$3Zn(s) \rightarrow 3Zn^{2+}(aq) + 6e^-$ oxidation
$E^\circ_{cell} = E^\circ_{Cr^{3+}} - E^\circ_{Zn}$
$E^\circ_{cell} = -0.74\ V - (-0.76\ V) = +0.02\ V$, ∴ spontaneous

19.13 From the equation $\Delta G^\circ = -nFE^\circ_{cell}$
$\Delta G^\circ = -30.9$ kJ or $-30{,}900$ J
$F = 96{,}500\ C\ mol^{-1}$
$E^\circ_{cell} = 0.107\ V$
$-30{,}900\ J = -n(96{,}500\ C\ mol^{-1})(0.107\ V)$
n = 2.99 which rounds to 2
Therefore, 3 moles of electrons are transferred in the reaction.

19.14 $\Delta G^\circ = -nFE^\circ_{cell}$
From Practice Exercise 11 (a): $n = 2\ e^-$, $E^\circ_{cell} = 0.53\ V$
$\Delta G^\circ = -nFE^\circ_{cell} = -(2\ e^-)(96{,}500\ F)(0.53\ V) = -102{,}000\ J = -102\ kJ$

From Practice Exercise 11 (b): $n = 5\ e^-$, $E^\circ_{cell} = 0.71\ V$
$\Delta G^\circ = -nFE^\circ_{cell} = -(5\ e^-)(96{,}500\ F)(0.71\ V) = -342{,}600\ J = -343\ kJ$

From Practice Exercise 12 (a): $n = 2\ e^-$, $E^\circ_{cell} = -0.29\ V$
$\Delta G^\circ = -nFE^\circ_{cell} = -(2\ e^-)(96{,}500\ F)(-0.29\ V) = 55{,}970\ J = 55.9\ kJ$

From Practice Exercise 12 (b): $n = 6\ e^-$, $E^\circ_{cell} = 0.02\ V$
$\Delta G^\circ = -nFE^\circ_{cell} = -(6\ e^-)(96{,}500\ F)(0.02\ V) = 11{,}600\ J = -11.6\ kJ$

19.15 Using Equation 19.7,

$$E^\circ_{cell} = \frac{RT}{nF} \ln K_c$$

$$-0.46\ V = \frac{(8.314\ J\ mol^{-1}K^{-1})(298\ K)}{2(96{,}500\ C\ mol^{-1})} \ln K_c$$

$$\ln K_c = -35.83$$

Taking the antilog (e^x) of both sides of the above equation gives $K_c = 2.7 \times 10^{-16}$.

This very small value for the equilibrium constant means that the products of the reaction are not formed spontaneously. The equilibrium lies far to the left, favoring reactants, and we do not expect much product to form.

The reverse reaction will be spontaneous, therefore, the value for K_c for the spontaneous reaction will be:
$Cu(s) + 2Ag^{+}(aq) \rightarrow Cu^{2+}(aq) + 2Ag(s)$

$$K_c' = \frac{1}{K_c} = \frac{1}{2.7\times10^{-16}} = 3.7 \times 10^{15}$$

19.16 $Ag^{+}(aq) + e^{-} \rightarrow Ag(s) \qquad E° = 0.80\text{ V}$
$AgBr(s) + e^{-} \rightarrow Ag(s) + Br^{-}(aq) \qquad E° = 0.07\text{ V}$
Equation for the spontaneous reaction:
$Ag^{+}(aq) + Br^{-}(aq) \rightarrow AgBr(s) \qquad E°_{cell} = 0.73\text{ V}$

$$K = \frac{1}{[Ag^{+}][Br^{-}]}$$

$$\ln K_c = \frac{E°_{cell}nF}{RT} = \frac{(0.73\text{ V})(1\text{ e}^{-})(96,500\text{ C mol}^{-1})}{(8.314\text{ J mol}^{-1}\text{ K}^{-1})(298\text{ K})} = 28.43$$

$K_c = 2.23 \times 10^{12}$

K_{sp} for AgBr is 5.0×10^{-13}

$$\frac{1}{K_c} = \frac{1}{2.23\times10^{12}} = 4.5 \times 10^{-13}$$

The K_c is the inverse of the K_{sp}.

19.17 $Cu^{2+}(aq) + Mg(s) \rightarrow Cu(s) + Mg^{2+}(aq) \qquad E°_{cell} = 0.34 - (-2.37\text{ V}) = 2.71\text{ V}$

$$E_{cell} = E°_{cell} - \frac{RT}{nF}\ln\frac{[Mg^{2+}]}{[Cu^{2+}]}$$

$$E_{cell} = 2.71\text{ V} - \frac{(8.314\text{ J mol}^{-1}\text{ K}^{-1})(298\text{ K})}{(2)(96,500\text{ C mol}^{-1})}\ln\frac{[2.2\times10^{-6}]}{[0.015]} = 2.82\text{ V}$$

19.18 $Zn(s) \rightarrow Zn^{2+}(aq) + 2e^{-}$ oxidation
$Cu^{2+}(aq) + 2e^{-} \rightarrow 2Cu(s)$ reduction

$E°_{cell} = E°_{Cu^{2+}} - E°_{Zn}$
$E°_{cell} = +0.34\text{ V} - (-0.76\text{ V}) = +1.10\text{ V}$

The Nernst equation for this cell is:

$$E_{cell} = E°_{cell} - \frac{RT}{nF}\ln\frac{[Zn^{2+}]}{[Cu^{2+}]}$$

$$E_{cell} = 1.10\text{ V} - \frac{(8.314\text{ J mol}^{-1}\text{K}^{-1})(298\text{ K})}{2(96,500\text{ C mol}^{-1})}\ln\frac{[1.0]}{[0.010]}$$

$$= 1.10\text{ V} - 0.01284(4.605) = 1.04\text{ V}$$

19.19 $Cu^{2+}(aq) + Mg(s) \rightarrow Cu(s) + Mg^{2+}(aq)$ $E°_{cell} = 0.34 - (-2.37 \text{ V}) = 2.71 \text{ V}$

$$E_{cell} = E^{\circ}_{cell} - \frac{RT}{nF} \ln \frac{[Mg^{2+}]}{[Cu^{2+}]}$$

$$2.79 = 2.71 \text{ V} - \frac{(8.314 \text{ J mol}^{-1} \text{ K}^{-1})(298 \text{ K})}{(2)(96{,}500 \text{ C mol}^{-1})} \ln \frac{[Mg^{2+}]}{[0.015]}$$

$$0.08 \text{ V} = -0.0128 \ln \frac{[Mg^{2+}]}{[0.015]} \ln \frac{[Mg^{2+}]}{[0.015]}$$

$$-6.25 = \ln \frac{[Mg^{2+}]}{[0.015]}$$

$$e^{-6.25} = \frac{[Mg^{2+}]}{[0.015]}$$

$[Mg^{2+}] = 2.95 \times 10^{-5}$ *M*

19.20 $Cu(s) \rightarrow Cu^{2+}(aq) + 2e^-$ oxidation
$Ag^+(aq) + e^- \rightarrow Ag(s)$ reduction

$E°_{cell} = E°_{Ag^+} - E°_{Cu}$
$E°_{cell} = +0.80 \text{ V} - (+0.34 \text{ V}) = +0.46 \text{ V}$

$$E_{cell} = E^{\circ}_{cell} - \frac{RT}{nF} \ln \frac{[Cu^{2+}]}{[Ag^+]^2}$$

$$\ln \frac{[Cu^{2+}]}{[Ag^+]^2} = \frac{(E^{\circ}_{cell} - E_{cell})}{\left(\frac{RT}{nF}\right)}$$

$$= \frac{(0.46 \text{ V} - 0.57 \text{ V})}{0.01284}$$

$$= -8.5670$$

$$\frac{[Cu^{2+}]}{[Ag^+]^2} = e^{-8.5670} = 1.9 \times 10^{-4}$$

Since the $[Ag^+] = 1.00$ *M*, $[Cu^{2+}] = 1.9 \times 10^{-4}$ *M*
Substituting the second value into the same expression gives
$[Cu^{2+}] = 6.6 \times 10^{-13}$ *M*

19.21 We are told that, in this galvanic cell, the chromium electrode is the anode, meaning that oxidation occurs at the chromium electrode.

Now in general, we have the equation:
$$E°_{cell} = E°_{reduction} - E°_{oxidation}$$
which becomes, in particular for this case:
$$E°_{cell} = E°_{Ni^{2+}} - E°_{Cr}$$

The net cell reaction is given by the sum of the reduction and the oxidation half–reactions, multiplied in each case so as to eliminate electrons from the result:

$3 \times [Ni^{2+}(aq) + 2e^- \rightarrow Ni(s)]$ reduction
$2 \times [Cr(s) \rightarrow Cr^{3+}(aq) + 3e^-]$ oxidation

$3Ni^{2+}(aq) + 2Cr(s) \rightarrow 2Cr^{3+}(aq) + 3Ni(s)$ net reaction

In this reaction, n = 6, and the Nernst equation becomes:

$$E_{cell} = E^{\circ}_{cell} - \frac{RT}{nF}\ln\frac{[Cr^{3+}]^2}{[Ni^{2+}]^3}$$

$$\ln\frac{[Cr^{3+}]^2}{[Ni^{2+}]^3} = \frac{(E^{\circ}_{cell} - E_{cell})}{\left(\frac{RT}{nF}\right)}$$

$$= \frac{(0.487\text{ V} - 0.552\text{ V})}{0.004279}$$

$$= -15.190$$

$$\frac{[Cr^{3+}]^2}{[Ni^{2+}]^3} = e^{-15.190} = 2.5 \times 10^{-7}$$

Substituting $[Ni^{2+}] = 1.20\ M$, we solve for $[Cr^{3+}]$ and get: $[Cr^{3+}] = 6.6 \times 10^{-4}\ M$.

19.22 $Fe^{3+}(aq) + e^- \rightarrow Fe^{2+}(aq)$ $E^{\circ} = 0.77$ V
$2I_2(s) + e^- \rightarrow 2I^-(aq)$ $E^{\circ} = +0.54$ V
$O_2(g) + 4H^+(aq) + 4e^- \rightarrow 2H_2O$ $E^{\circ} = +1.23$ V

The reaction with the least positive reduction potential will be the easiest to oxidize, and its product will be the product at the anode. I_2 will be produced.

19.23 The cathode is always where reduction occurs. We must consider which species could be candidates for reduction, then choose the species with the highest reduction potential from Table 19.1.

$Cd^{2+}(aq) + 2e^- \rightarrow 2Cd(s)$ $E^{\circ} = -0.40$ V
$Sn^{2+}(aq) + 2e^- \rightarrow 2Sn(s)$ $E^{\circ} = -0.14$ V
$2H_2O + 2e^- \rightarrow H_2(g) + 2OH^-(aq)$ $E^{\circ} = -0.83$ V

Tin(II) has the highest reduction potential, so we would expect it to be reduced in this environment. We expect Sn(s) at the cathode.

19.24 The number of Coulombs is: 4.00 A × 200 s = 800 C
The number of moles is:

$$\text{mol }OH^- = 800\text{ C} \times \frac{1\text{ F}}{96{,}500\text{ C}} \times \frac{1\text{ mol }OH^-}{1\text{ F}} = 8.29 \times 10^{-3}\text{ mol }OH^-$$

19.25 The number of moles of Au to be deposited is: 3.00 g Au ÷ 197 g/mol = 0.0152 mol Au. The number of Coulombs (A × s) is:

$$\text{Coulombs} = 0.0152\text{ mol Au} \times \frac{3\text{ F}}{1\text{ mol Au}} \times \frac{96{,}500\text{ C}}{1\text{ F}} = 4.40 \times 10^3\text{ C}$$

The number of minutes is:

$$\text{min} = \frac{4.40 \times 10^3\text{ A.s}}{10.0\text{ A}} \times \frac{1\text{ min}}{60\text{ s}} = 7.33\text{ min}$$

19.26 As in Practice Exercise 25 above, the number of Coulombs is 4.40×10^3 C. This corresponds to a current of:

$$A = \frac{4.40 \times 10^3 \text{ A.s}}{20.0 \text{ min}} \times \frac{1 \text{ min}}{60 \text{ s}} = 3.67 \text{ A}$$

19.27 The number of Coulombs is:
$0.100 \text{ A} \times 1.25 \text{ hr}(3600 \text{ s/hr}) = 450 \text{ C}$

The number of moles of copper ions produced is:

$$\text{mol } Cu^{2+} = 450 \text{ C} \times \frac{1 \text{ mol } e^-}{96{,}500 \text{ C}} \times \frac{1 \text{ mol } Cu^{2+}}{2 \text{ mol } e^-} = 0.00233 \text{ mol } Cu^{2+}$$

Therefore, the increase in concentration is:
$M = \text{mol/L} = (0.00233 \text{ mol } Cu^{2+})/(0.125 \text{ L}) = +\ 0.0187\ M$

Review Problems

19.50 (a)

anode: $Cd(s) \rightarrow Cd^{2+}(aq) + 2e^-$
cathode: $Au^{3+}(aq) + 3e^- \rightarrow Au(s)$
cell: $3Cd(s) + 2Au^{3+}(aq) \rightarrow 3Cd^{2+}(aq) + 2Au(s)$

(b)

anode: $Fe(s) \rightarrow Fe^{2+}(aq) + 2e^-$
cathode: $Br_2(aq) + 2e^- \rightarrow 2Br^-(aq)$
cell: $Fe(s) + Br_2(aq) \rightarrow Fe^{2+}(aq) + 2Br^-(aq)$

(c)

anode: $Cr(s) \rightarrow Cr^{3+}(aq) + 3e^-$
cathode: $Cu^{2+}(aq) + 2e^- \rightarrow Cu(s)$
cell: $2Cr(s) + 3Cu^{2+}(aq) \rightarrow 2Cr^{3+}(aq) + 3Cu(s)$

19.52 (a) $Pt(s)|Fe^{2+}(aq),Fe^{3+}(aq)||NO_3^-(aq), H^+(aq)|NO(g)|Pt(s)$
(b) $Pt(s)|Br_2(aq),Br^-(aq)||Cl^-(aq),Cl_2(g)|Pt(s)$
(c) $Ag(s)|Ag^+(aq)||Au^{3+}(aq)|Au(s)$

19.54 (a) $Sn(s)$ (b) $Br^-(aq)$ (c) $Zn(s)$ (d) $I^-(aq)$

19.56 (a) $E^\circ_{cell} = 0.96 \text{ V} - (0.77) \text{ V} = 0.19 \text{ V}$
(b) $E^\circ_{cell} = 1.07 \text{ V} - (1.36 \text{ V}) = -0.29 \text{ V}$
(c) $E^\circ_{cell} = 1.42 \text{ V} - (0.80 \text{ V}) = 0.62 \text{ V}$

19.58 The reactions are spontaneous if the overall cell potential is positive.
$E^\circ_{cell} = E^\circ_{substance\ reduced} - E^\circ_{substance\ oxidized}$
(a) $E^\circ_{cell} = 1.42 \text{ V} - (0.54 \text{ V}) = 0.88 \text{ V}$ spontaneous
(b) $E^\circ_{cell} = 1.07 \text{ V} - (0.17 \text{ V}) = 0.90 \text{ V}$ spontaneous
(c) $E^\circ_{cell} = -0.74 \text{ V} - (-2.76 \text{ V}) = 2.02 \text{ V}$ spontaneous

19.60 The half–cell with the more positive E°_{cell} will appear as a reduction, and the other half–reaction is reversed, to appear as an oxidation:
$BrO_3^-(aq) + 6H^+(aq) + 6e^- \rightarrow Br^-(aq) + 3H_2O$ reduction
$3 \times (2I^-(aq) \rightarrow I_2(s) + 2e^-)$ oxidation
$BrO_3^-(aq) + 6I^-(aq) + 6H^+(aq) \rightarrow 3I_2(s) + Br^-(aq) + 3H_2O$ net reaction

$E^\circ_{cell} = E^\circ_{substance\ reduced} - E^\circ_{substance\ oxided}$ or

$E^\circ_{cell} = E^\circ_{reduction} - E^\circ_{oxidation} = 1.44\ V - (0.54\ V) = 0.90\ V$

19.62 The half–reaction having the more positive standard reduction potential is the one that occurs as a reduction, and the other one is written as an oxidation:

$2 \times (2HOCl(aq) + 2H^+(aq) + 2e^- \rightarrow Cl_2(g) + 2H_2O)$ reduction

$3H_2O + S_2O_3^{2-}(aq) \rightarrow 2H_2SO_3(aq) + 2H^+(aq) + 4e^-$ oxidation

$4HOCl(aq) + 4H^+(aq) + 3H_2O + S_2O_3^{2-}(aq) \rightarrow$

$2Cl_2(g) + 4H_2O + 2H_2SO_3(aq) + 2H^+(aq)$

which simplifies to give the following net reaction:

$4HOCl(aq) + 2H^+(aq) + S_2O_3^{2-}(aq) \rightarrow 2Cl_2(g) + H_2O + 2H_2SO_3(aq)$

19.64 The two half–reactions are:

$SO_4^{2-}(aq) + 2e^- + 4H^+(aq) \rightarrow H_2SO_3(aq) + H_2O(\ell)$ reduction

$2I^-(aq) \rightarrow I_2(s) + 2e^-$ oxidation

$E^\circ_{cell} = E^\circ_{reduction} - E^\circ_{oxidation} = 0.17\ V - (0.54\ V) = -0.37\ V$

Since the overall cell potential is negative, we conclude that the reaction is not spontaneous in the direction written.

19.66 First, separate the overall reaction into its two half–reactions:

$2Br^-(aq) \rightarrow Br_2(aq) + 2e^-$ oxidation

$I_2(s) + 2e^- \rightarrow 2I^-(aq)$ reduction

$E^\circ_{cell} = E^\circ_{reduction} - E^\circ_{oxidation} = 0.54\ V - (1.07\ V) = -0.53\ V$

The value of n is 2: $\Delta G° = -nF E^\circ_{cell} = -(2)(96{,}500\ C)(-0.53\ J/C)$

$= 1.0 \times 10^5\ J = 1.0 \times 10^2\ kJ$

19.68 (a) $E^\circ_{cell} = E^\circ_{reduction} - E^\circ_{oxidation} = 2.01\ V - (1.47\ V) = 0.54\ V$

(b) Since n = 10, $\Delta G° = -nF E^\circ_{cell} = -(10)(96{,}500\ C)(0.54\ J/C) = -5.2 \times 10^5\ J$

$\Delta G° = -5.2 \times 10^2\ kJ$

(c)

$$E^\circ_{cell} = \frac{RT}{nF} \ln K_c$$

$$0.54\ V = \frac{(8.314\ J\ mol^{-1}K^{-1})(298\ K)}{10(96{,}500\ C\ mol^{-1})} \ln K_c$$

$\ln K_c = 210.3$

Taking the exponential of both sides of this equation:

$K_c = 2.1 \times 10^{91}$

19.70 Sn is oxidized by two electrons and Ag is reduced by two electrons:

$$E^\circ_{cell} = \frac{0.0592}{n} \log K_c$$

$-0.015\ V = (0.0592\ V/2) \times \log K_c$

$\log K_c = -0.51$

$K_c = \text{antilog}(-0.51) = 0.31$

19.72 This reaction involves the oxidation of Ag by two electrons and the reduction of Ni by two electrons. The concentration of the hydrogen ion is derived from the pH of the solution:
$[H^+]$ = antilog (–pH) = antilog (–2) = 1×10^{-2} *M*

$$E_{cell} = 2.48\text{ V} - \frac{0.0592\text{ V}}{2}\log\frac{\left[Ag^+\right]^2\left[Ni^{2+}\right]}{\left[H^+\right]^4}$$

$$= 2.48\text{ V} - \frac{0.0592\text{ V}}{2}\log\frac{\left[3.0\times10^{-2}\right]^2\left[3.0\times10^{-2}\right]}{\left[1.0\times10^{-2}\right]^4}$$

$E_{cell} = 2.48\text{ V} - 0.101\text{ V} = 2.38\text{ V}$

19.74

$$E_{cell} = E^\circ_{cell} - \frac{RT}{nF}\ln\frac{\left[Mg^{2+}\right]}{\left[Cd^{2+}\right]}$$

$$E_{cell} = 1.97 - \frac{(8.314\text{ J mol}^{-1}\text{K}^{-1})(298\text{ K})}{2(96{,}500\text{ C mol}^{-1})}\ln\frac{[1.00]}{\left[Cd^{2+}\right]}$$

$$1.54\text{ V} = 1.97\text{ V} - 0.01284\ln\frac{1}{\left[Cd^{2+}\right]}$$

$\ln(1/[Cd^{2+}]) = 33.489$

Taking (e^x) of both sides:
$1/[Cd^{2+}] = 3.50 \times 10^{14}$
$[Cd^{2+}] = 2.86 \times 10^{-15}$ *M*

19.76 In the iron half–cell, we are initially given:
$0.0500\text{ L} \times 0.100\text{ mol/L} = 5.00 \times 10^{-3}\text{ mol } Fe^{2+}(aq)$

The precipitation of $Fe(OH)_2(s)$ consumes some of the added hydroxide ion, as well as some of the iron ion: $Fe^{2+}(aq) + 2OH^-(aq) \rightarrow Fe(OH)_2(s)$. The number of moles of OH^- that have been added to the iron half–cell is:

$0.500\text{ mol/L} \times 0.0500\text{ L} = 2.50 \times 10^{-2}\text{ mol } OH^-$

The stoichiometry of the precipitation reaction requires that the following number of moles of OH^- be consumed on precipitation of 5.00×10^{-3} mol of $Fe(OH)_2(s)$:
$5.00 \times 10^{-3}\text{ mol } Fe(OH)_2 \times (2\text{ mol } OH^-/\text{mol } Fe(OH)_2) = 1.00 \times 10^{-2}\text{ mol } OH^-$

The number of moles of OH^- that are unprecipitated in the iron half–cell is:
$2.50 \times 10^{-2}\text{ mol} - 1.00 \times 10^{-2}\text{ mol} = 1.50 \times 10^{-2}\text{ mol } OH^-$

Since the resulting volume is 50.0 mL + 50.0 mL, the concentration of hydroxide ion in the iron half–cell becomes, upon precipitation of the $Fe(OH)_2$:

$[OH^-] = 1.50 \times 10^{-2}\text{ mol}/0.100\text{ L} = 0.150\ M\ OH^-$

We have assumed that the iron hydroxide that forms in the above precipitation reaction is completely insoluble. This is not accurate, though, because some small amount does dissolve in water according to the following equilibrium:

$Fe(OH)_2(s) \rightleftharpoons Fe^{2+}(aq) + 2OH^-(aq)$

This means that the true [OH^-] is slightly higher than 0.150 *M* as calculated above. Thus we must set up the usual equilibrium table, in order to analyze the extent to which $Fe(OH)_2(s)$ dissolves in 0.150 *M* OH^- solution:

	[Fe^{2+}]	[OH^-]
I	–	0.150
C	+x	+2x
E	+x	0.150+2x

The quantity x in the above table is the molar solubility of $Fe(OH)_2$ in the solution that is formed in the iron half–cell.

$K_{sp} = [Fe^{2+}][OH^-]^2 = (x)(0.150 + 2x)^2$

The standard cell potential is:

$$E^\circ_{cell} = E^\circ_{reduction} - E^\circ_{oxidation} = 0.3419\text{ V} - (-0.447\text{ V}) = 0.7889\text{ V}$$

The Nernst equation is:

$$E_{cell} = E^\circ_{cell} - \frac{RT}{nF}\ln\frac{\left[Fe^{2+}\right]}{\left[Cu^{2+}\right]}$$

$$1.175 = 0.7889 - \frac{(8.314\text{ J mol}^{-1}\text{K}^{-1})(298\text{ K})}{2(96{,}500\text{ C mol}^{-1})}\ln\frac{\left[Fe^{2+}\right]}{[1.00]}$$

$$1.175 = 0.7889 - 0.01284\ln\left[Fe^{2+}\right]$$

$\ln[Fe^{2+}] = -30.07$
$[Fe^{2+}] = 8.72 \times 10^{-14}\ M$

This is the concentration of Fe^{2+} in the saturated solution, and it is the value to be used for x in the above expression for K_{sp}.

$K_{sp} = (x)(0.150 + 2x)^2 = (8.72 \times 10^{-14})[0.150 + (2)(8.72 \times 10^{-14})]^2$
$K_{sp} = 1.96 \times 10^{-15}$

19.78 $$E^\circ = E^\circ_{cell} - \frac{RT}{nF}\ln\frac{\left[Ag^+\right]_{dilute}}{\left[Ag^+\right]_{conc}}$$

$$E^\circ = 0 - \frac{(8.314\text{ J mol}^{-1}\text{ K}^{-1})(298\text{ K})}{(1\text{ mol})(9.65\times10^4\text{ C mol}^{-1})}\ln\frac{[0.015]}{[0.50]}$$

$E^\circ = 0.090$ V

19.80 (a) $Fe^{2+}(aq) + 2e^- \rightarrow Fe(s)$
0.20 mol Fe^{2+} × 2 mol e^-/mol Fe^{2+} = 0.40 mol e^-

(b) $Cl^-(aq) \rightarrow 1/2Cl_2(g) + e^-$
0.70 mol Cl^- × 1 mol e^-/mol Cl^- = 0.70 mol e^-

(c) $Cr^{3+}(aq) + 3e^- \rightarrow Cr(s)$
1.50 mol Cr^{3+} × 3 mol e^-/mol Cr^{3+} = 4.50 mol e^-

(d) $Mn^{2+}(aq) + 4H_2O(l) \rightarrow MnO_4^-(aq) + 8H^+(aq) + 5e^-$
1.0×10^{-2} mol Mn^{2+} × 5 mol e^-/mol Mn^{2+} = 5.0×10^{-2} mol e^-

19.82 $Fe(s) + 2OH^-(aq) \rightarrow Fe(OH)_2(s) + 2e^-$
The number of Coulombs is: 12.0 min × 60 s/min × 8.00 C/s = 5.76 × 10^3 C. The number of grams of $Fe(OH)_2$ is:

$$\text{g Fe(OH)}_2 = \left(5.76 \times 10^3 \text{ C}\right)\left(\frac{1 \text{ mol e}^-}{96500 \text{ C}}\right)\left(\frac{1 \text{ mol Fe(OH)}_2}{2 \text{ mol e}^-}\right)\left(\frac{89.86 \text{ g Fe(OH)}_2}{1 \text{ mol Fe(OH)}_2}\right)$$

$$= 2.68 \text{ g Fe(OH)}_2$$

19.84 $Cr^{3+}(aq) + 3e^- \rightarrow Cr(s)$
The number of Coulombs that will be required is:

$$\text{Coulombs} = \left(75.0 \text{ g Cr}\right)\left(\frac{1 \text{ mol Cr}}{52.00 \text{ g Cr}}\right)\left(\frac{3 \text{ mol e}^-}{1 \text{ mol Cr}}\right)\left(\frac{96{,}500 \text{ C}}{1 \text{ mol e}^-}\right) = 4.18 \times 10^5 \text{ C}$$

The time that will be required is:

$$\text{hr} = \left(4.18 \times 10^5 \text{ C}\right)\left(\frac{1 \text{ s}}{2.25 \text{ C}}\right)\left(\frac{1 \text{ hr}}{3600 \text{ s}}\right) = 51.5 \text{ hr}$$

19.86 $Mg^{2+} + 2e^- \rightarrow Mg(l)$
The number of Coulombs that will be required is:

$$\text{Coulombs} = \left(60.0 \text{ g Mg}\right)\left(\frac{1 \text{ mol Mg}}{24.31 \text{ g Mg}}\right)\left(\frac{2 \text{ mol e}^-}{1 \text{ mol Mg}}\right)\left(\frac{96{,}500 \text{ C}}{1 \text{ mol e}^-}\right) = 4.76 \times 10^5 \text{ C}$$

The number of amperes is: 4.76 × 10^5 C ÷ 7200 s = 66.2 amp

19.88 The electrolysis of NaCl solution results in the reduction of water, together with the formation of hydroxide ion: $2H_2O + 2e^- \rightarrow H_2(g) + 2OH^-(aq)$. The number of Coulombs is: 2.00 A × 20.0 min × 60 s/min = 2.40 × 10^3 C. The number of moles of OH^- is:

$$\text{mol OH}^- = \left(2.40 \times 10^3 \text{ C}\right)\left(\frac{1 \text{ mol e}^-}{96{,}500 \text{ C}}\right)\left(\frac{2 \text{ mol OH}^-}{2 \text{ mol e}^-}\right) = 0.0249 \text{ mol OH}^-$$

The volume of acid solution that will neutralize this much OH^- is:

$$\text{mL HCl} = \left(0.0249 \text{ mol OH}^-\right)\left(\frac{1 \text{ mol HCl}}{1 \text{ mol OH}^-}\right)\left(\frac{1000 \text{ mL HCl}}{0.620 \text{ mol HCl}}\right) = 40.1 \text{ mL HCl}$$

19.90 The electrolysis of NaCl solution results in the reduction of water, together with the formation of hydroxide ion: $2Cl^- + 2e^- \rightarrow Cl_2(g)$. The number of Coulombs is: 2.50 A × 15.0 min × 60 s/min = 2.25 × 10^3 C. The number of moles of Cl_2 gas collected is:

$$\text{mol Cl}_2 = \left(2.25 \times 10^3 \text{ C}\right)\left(\frac{1 \text{ mol e}^-}{96{,}500 \text{ C}}\right)\left(\frac{1 \text{ mol Cl}_2}{2 \text{ mol e}^-}\right) = 0.0117 \text{ mol Cl}_2$$

The volume of Cl_2 gas that is collected is:

$$V = \frac{nRT}{P} = \frac{(0.0117 \text{ mol})\left(0.0821 \frac{\text{L atm}}{\text{mol K}}\right)(298 \text{ K})}{(750 \text{ torr} - 23.76 \text{ torr})\left(\frac{1 \text{ atm}}{760 \text{ torr}}\right)} = 0.299 \text{ L} = 299 \text{ mL}$$

19.92 Possible cathode reactions:

$Al^{3+} + 3e^- \rightleftharpoons Al(s)$ E° = –1.66 V

$2H_2O + 2e^- \rightleftharpoons H_2(g) + 2OH^-(aq)$ E° = –0.83 V

Possible anode reactions:

$S_2O_8^{2-} + 2e^- \rightleftharpoons 2SO_4^{2-}$ E° = +2.05 V

$O_2 + 4H^+ + 4e^- \rightleftharpoons 2H_2O$ E° = +1.23 V

Cathode reaction: $2H_2O + 2e^- \rightleftharpoons H_2(g) + 2OH^-(aq)$ E° = –0.83 V

Anode reactions: $2H_2O \rightleftharpoons O_2 + 4H^+ + 4e^-$ E° = –1.23 V

Net cell reaction: $2H_2O \rightleftharpoons 2H_2(g) + O_2(g)$ E° = –2.06 V

19.94 The answers to the previous Review Questions guide us here:

Possible cathode reactions:

$K^+ + e^- \rightleftharpoons K(s)$ E° = –2.92 V

$Cu^{2+} + 2e^- \rightleftharpoons Cu(s)$ E° = +0.34 V

$2H_2O + 2e^- \rightleftharpoons H_2(g) + 2OH^-(aq)$ E° = –0.83 V

Cathode reaction: $Cu^{2+} + 2e^- \rightleftharpoons Cu(s)$

Possible anode reactions:

$2SO_4^{2-} \rightleftharpoons S_2O_8^{2-} + 2e^-$ E° = –2.01 V

$2Br^- \rightleftharpoons Br_2 + 2e^-$ E° = –1.07 V

$2H_2O \rightleftharpoons O_2(g) + 4H^+(aq) + 4e^-$ E° = –1.23 V

Anode reaction: $2Br^- \rightleftharpoons Br_2 + 2e^-$

Overall reaction: $Cu^{2+} + 2Br^- \rightleftharpoons Br_2 + Cu(s)$

Practice Exercises

20.1 $^{226}_{88}\text{Ra} \rightarrow {}^{222}_{86}\text{Rn} + {}^{4}_{2}\text{He} + {}^{0}_{0}\gamma$

20.2 $^{90}_{38}\text{Sr} \rightarrow {}^{90}_{39}\text{Y} + {}^{0}_{-1}\text{e}$

20.3 Using the value of k for Pu from Example 20.2:
Activity = kN
Activity = 6.22×10^{11} Bq = 6.22×10^{11} disintegrations/second
k = 2.50×10^{-10} seconds^{-1}
6.22×10^{11} disintegrations/second = (2.50×10^{-10} seconds^{-1})N
N = 2.49×10^{21} atoms Pu
Now find the mass of Pu

$$\text{mass Pu} = (2.49 \times 10^{21}\text{ atoms Pu})\left(\frac{1\text{ mole Pu}}{6.022 \times 10^{23}\text{ atoms Pu}}\right)\left(\frac{244\text{ g Pu}}{1\text{ mole Pu}}\right) = 1.01\text{ g Pu}$$

The percentage of Pu in the sample:

$$\frac{1.01\text{ g Pu}}{2.00\text{ g sample}} \times 100\% = 50.4\%\text{ Pu}$$

20.4 Half life for Rn-222 = 4 days(24 h/day)(3600 s/hr) = 345,500 s
$k = \ln 2/t_{1/2} = (0.6931)/(345{,}500\text{ s}) = 2.01 \times 10^{-6}\text{ s}^{-1}$

Activity = 6 pCi = 6×10^{-12} Ci (3.7×10^{10} dps/1Ci)
= 0.222 Bq
= 0.222 disintegrations per second
Activity = disintegrations/sec = kN
0.222 = (2.01×10^{-6} s^{-1})N
N = 1.10×10^{5} atoms Rn-222

20.5 We make use of the Inverse Square Law:

$$\frac{I_1}{I_2} = \frac{d_2^2}{d_1^2}$$

$$\frac{4.8\text{ units}}{0.30\text{ units}} = \frac{(x\text{ m})^2}{(5.0\text{ m})^2}$$

x m = 20 m

20.6 We make use of the Inverse Square Law:

$$\frac{I_1}{I_2} = \frac{d_2^2}{d_1^2}$$

$$\frac{1.4\text{ units}}{I_2} = \frac{(1.2\text{ m})^2}{(10\text{ m})^2} = 100\text{ units (assuming 1 significant figure)}$$

Review Problems

20.45 Solve the Einstein equation for Δm:
$m = \Delta E/c^2$
$1\ kJ = 1.00 \times 10^3\ J = 1.00 \times 10^3\ kg\ m^2\ s^{-2}$
$\Delta m = 1.00 \times 10^3\ kg\ m^2\ s^{-2} \div (3.00 \times 10^8\ m/s)^2 = 1.11 \times 10^{-14}\ kg = 1.11 \times 10^{-11}\ g$

20.47 The joule is equal to one kg m^2/s^2, and this is employed directly in the Einstein equation: $\Delta m = \Delta E/c^2$, where ΔE is the enthalpy of formation of liquid water, which is available in Table 6.2.

$H_2(g) + O_2(g) \rightarrow H_2O(l)$, $\Delta H = -285.9$ kJ/mol
$\Delta m = (-285.9 \times 10^3\ kg\ m^2/s^2) \div (3.00 \times 10^8\ m/s^2)^2 = -3.18 \times 10^{-12}\ kg$
$(-3.18 \times 10^{-12}\ kg) \times 1000\ g/kg \times 10^9\ ng/g = -3.18\ ng$
The negative value for the mass implies that mass is lost in the reaction.

The percent of mass that is lost is
$$\left(\frac{3.18\times10^{-12}\ kg}{18.02\times10^{-3}\ kg}\right)\times 100\% = 1.77 \times 10^{-8}\%$$

20.49 The mass of the deuterium nucleus is the mass of the proton (1.0072764669 u) plus that of a neutron (1.0086649156 u), or 2.015941381 u. The difference between this calculated value and the observed value is equal to Δm:
$\Delta m = (2.015941 - 2.0135) = 2.44 \times 10^{-3}\ u$
$\Delta E = \Delta mc^2 = (2.44 \times 10^{-3}\ u)(1.6606 \times 10^{-27}\ kg/u)(3.00 \times 10^8\ m/s)^2$
$\Delta E = 3.65 \times 10^{-13}\ kg\ m^2/s^2 = 3.65 \times 10^{-13}\ J$

Since there are two neucleons per deuterium nucleus, we have:
$\Delta E = 3.65 \times 10^{-13}$ J/2 nucleons $= 1.8 \times 10^{-13}$ J per nucleon

20.51 (a) ${}^{211}_{83}Bi$ (b) ${}^{177}_{72}Hf$ (c) ${}^{216}_{84}Po$ (d) ${}^{19}_{9}F$

20.53 (a) ${}^{242}_{94}Pu \rightarrow {}^{4}_{2}He + {}^{238}_{92}U$
(b) ${}^{28}_{12}Mg \rightarrow {}^{0}_{-1}e + {}^{28}_{13}Al$
(c) ${}^{26}_{14}Si \rightarrow {}^{0}_{1}e + {}^{26}_{13}Al$
(d) ${}^{37}_{18}Ar + {}^{0}_{-1}e \rightarrow {}^{37}_{17}Cl$

20.55 (a) ${}^{261}_{102}No$ (b) ${}^{211}_{82}Pb$ (c) ${}^{141}_{61}Pm$ (d) ${}^{179}_{74}W$

20.57 ${}^{87}_{36}Kr \rightarrow {}^{86}_{36}Kr + {}^{1}_{0}n$

20.59 The more likely process is positron emission, because this produces a product having a higher neutron–to–proton ratio: ${}^{38}_{19}K \rightarrow {}^{0}_{1}e + {}^{38}_{18}Ar$

20.61 Six half–life periods correspond to the fraction 1/64 of the initial material. That is, one sixty–fourth of the initial material is left after 6 half lives: 3.00 mg × 1/64 = 0.0469 mg remaining.

20.63 ${}^{53}_{24}Cr^*$; ${}^{51}_{23}V + {}^{2}_{1}H \rightarrow {}^{53}_{24}Cr^* \rightarrow {}^{1}_{1}p + {}^{52}_{23}V$

20.65 ${}^{80}_{35}Br$

20.67 $^{55}_{26}Fe$; $^{55}_{25}Mn + ^{1}_{1}p \rightarrow ^{1}_{0}n + ^{55}_{26}Fe$

20.69 $^{70}_{30}Zn + ^{208}_{82}Pb \rightarrow ^{278}_{112}Uub \rightarrow ^{1}_{0}n + ^{277}_{112}Uub$

20.71 $\text{Radiation} \propto \frac{1}{d^2}$

$$\frac{I_1}{I_2} = \frac{d_2^{\,2}}{d_1^{\,2}}$$

$$d_2 = d_1\sqrt{\frac{I_1}{I_2}} = 2.0\text{m}\sqrt{\frac{2.8}{0.28}} = 6.3\text{ m}$$

20.73 This calculation makes use of the Inverse Square Law:

$$\frac{I_1}{I_2} = \frac{d_2^{\,2}}{d_1^{\,2}}$$

$$\frac{8.4\text{ rem}}{0.50\text{ rem}} = \frac{d_2^{\,2}}{(1.60\text{ m})^2}$$

$d_2 = 6.6$ m

20.75 Activity = kN and $t_{1/2} = \frac{\ln 2}{k}$ or $k = \frac{\ln 2}{t_{1/2}}$

$$\text{Activity} = \frac{\ln 2}{t_{1/2}}N$$

$$N = 0.20\text{ mg}\left(\frac{1\text{ g }^{241}\text{Am}}{1000\text{ mg}}\right)\left(\frac{1\text{ mol }^{241}\text{Am}}{241\text{ g }^{241}\text{Am}}\right)\left(\frac{6.022\times10^{23}\text{ atoms }^{241}\text{Am}}{1\text{ mol }^{241}\text{Am}}\right) = 5.0\times10^{17}\text{ atoms }^{241}\text{Am}$$

$$\text{Activity} = \left(\frac{\ln 2}{1.70\times10^5\text{ d}}\right)\times(5.0\times10^{17}\text{ atoms }^{241}\text{Am})$$

$$= \left(\frac{1.0\times10^{12}\text{ Am decays}}{\text{d}}\right)\left(\frac{1\text{ d}}{24\text{ h}}\right)\left(\frac{1\text{ h}}{3600\text{ s}}\right) = 2.4\times10^7\text{ s}^{-1}$$

$$2.4\times10^7\text{ s}^{-1} = 2.4\times10^7\text{ Bq} = 2.4\times10^7\text{ Bq}\left(\frac{1\text{ Ci}}{3.7\times10^{10}\text{ Bq}}\right)\left(\frac{1000\text{ mCi}}{1\text{ Ci}}\right) = 6.5\times10^{-1}\text{ mCi}$$

$$= 6.5\times10^2\ \mu\text{Ci}$$

20.77 Activity = kN $\qquad k = \dfrac{\text{activity}}{N}$

$$N = \text{number of } ^{131}\text{I atoms} = 1.00 \text{ mg } ^{131}\text{I}\left(\frac{1 \text{ g } ^{131}\text{I}}{1000 \text{ mg } ^{131}\text{I}}\right)\left(\frac{1 \text{ mol } ^{131}\text{I}}{131 \text{ g } ^{131}\text{I}}\right)\left(\frac{6.022 \times 10^{23} \text{ g } ^{131}\text{I}}{1 \text{ mol } ^{131}\text{I}}\right)$$

$$= 4.60 \times 10^{18} \text{ atoms } ^{131}\text{I}$$

$$k = \frac{4.6 \times 10^{12} \text{ Bq}}{4.60 \times 10^{18} \text{ atoms } ^{131}\text{I}}\left(\frac{1 \text{ s}^{-1}}{1 \text{ Bq}}\right) = 1.0 \times 10^{-6} \text{ s}^{-1}$$

$$t_{1/2} = \frac{\ln 2}{k} = \left(\frac{\ln 2}{1.0 \times 10^{-6} \text{ s}^{-1}}\right) = 6.9 \times 10^{5} \text{ s (This is about 8 days.)}$$

20.79 The chemical product is $BaCl_2$.Recall that for a first order process $k = \dfrac{0.693}{t_{1/2}}$

So $k = 0.693/30 \text{ yr} = 2.30 \times 10^{-2}/\text{yr}$. Also,

$$\ln\frac{[A]_0}{[A]_t} = kt$$

$$[A]_t = [A]_0 \exp(-kt)$$

$$\frac{[A]_0}{[A]_t} = \exp[-(2.30 \times 10^{-2} /\text{yr})(150 \text{ yr})]$$

$$\frac{[A]_0}{[A]_t} = 3.13 \times 10^{-2}$$

so 3.1% of the original sample remains.

20.81 This calculation makes use of the first order rate equation, where knowing $[A]_t$, we need to calculate $[A]_0$:

$$\ln\frac{[A]_0}{[A]_t} = kt.$$

$k = 0.693/t_{1/2} = 0.693/8.07 \text{ d} = 8.59 \times 10^{-2} \text{ d}^{-1}$

$$\ln\frac{[A]_0}{\left(25.6 \times 10^{-5} \text{ Ci/g}\right)} = \left(8.59 \times 10^{-2} \text{ d}^{-1}\right)(28.0 \text{ d})$$

Taking the exponential of both sides of the above equation gives:

$$\frac{[A]_0}{\left(25.6 \times 10^{-5} \text{ Ci/g}\right)} = e^{2.41} = 11.1$$

Solving for the value of $[A]_0$ gives: $[A]_0 = 2.84 \times 10^{-3}$ Ci/g

20.83 In order to solve this problem, it must be assumed that all of the argon–40 that is found in the rock must have come from the potassium–40, i.e., that the rock contains no other source of argon–40. If the above assumption is valid, then any argon–40 that is found in the rock represents an equivalent amount of potassium–40, since the stoichiometry is 1:1. Since equal amounts of potassium–40 and argon–40 have been found, this indicates that the amount of potassium–40 that remains is exactly half the amount that was present originally. In other words, the potassium–40 has undergone one half–life of decay by the time of the analysis. The rock is thus seen to be 1.3×10^{9} years old.

20.85 $\ln\left(\frac{^{14}C}{^{12}C}\right) = \left(1.2 \times 10^{-4}\right)t$

Taking the natural log we determine:

$$\ln\left(\frac{1.2 \times 10^{-12}}{4.8 \times 10^{-14}}\right) = \left(1.2 \times 10^{-4}\right)t$$

$$t = \left(\frac{1}{1.2 \times 10^{-4}}\right)\ln\left(\frac{1.2 \times 10^{-12}}{4.8 \times 10^{-14}}\right) = 2.7 \times 10^{4}\ \text{yr}$$

The tree died 2.7×10^4 years ago. This is when the volcanic eruption occurred.

20.87 $^{235}_{92}U + {}^{1}_{0}n \rightarrow {}^{94}_{38}Sr + {}^{140}_{54}Xe + 2\,{}^{1}_{0}n$.

Practice Exercises

21.1 $Au_2O_3(s) \rightarrow 2Au(s) + \frac{3}{2}O_2(g)$

$\Delta H° = -80.8 \text{ kJ mol}^{-1}$

$\Delta S° = (2\,S°_{Au}\ S°_{Au} + \frac{3}{2}\,S°_{O_2}) - S°_{Au_2O_3}$

$\Delta S° = [(2 \text{ mol} \times 47.7 \text{ J mol}^{-1} \text{ K}^{-1}) + (\frac{3}{2} \text{ mol} \times 205 \text{ J mol}^{-1} \text{ K}^{-1})] - (1 \text{ mol} \times 125 \text{ J mol}^{-1} \text{ K}^{-1})$

$\Delta S° = +278 \text{ J K}^{-1} = +0.278 \text{ kJ K}^{-1}$

$\Delta G°_T \approx \Delta H°_{298} - T\Delta S°_{298}$

$\Delta G°_T \approx -80.8 \text{ kJ} - T(0.278 \text{ kJ K}^{-1})$

Note that regardless of temperature, $\Delta G°_T$ will be negative. This is because the absolute temperature is always a positive quantity. This means that Au_2O_3 is unstable with respect to decomposition at any temperature.

21.2 $\Delta H = -\Delta H_f° = 601.7 \text{ kJ}$

$\Delta S° = S°(Mg(s)) + 1/2S°(O_2(g)) - S°(MgO(s))$

$= 32.5\text{J/K} + 1/2(205 \text{ J/K}) - 26.9 \text{ J/K}$

$= 108.1 \text{ J/K}$

$= 0.108 \text{ kJ/K}$

$\Delta G_T° = \Delta H° - T\Delta S°$

$= 601.7 \text{ kJ} - T(0.108 \text{ kJ/K})$

Decomposition occurs when $\Delta G<0$. Solve for $\Delta G° = 0$

$$T = \frac{601.7 \text{ kJ}}{0.108 \text{ kJ/K}}$$

$= 5570 \text{ K}$

21.3 The net charge on the complex ion must first be determined. Two $S_2O_3^{2-}$ ions contribute a charge of 4–; the metal contributes a charge of 1+. The sum of these is 3–. The formula of the complex ion is therefore $[Ag(S_2O_3)_2]^{3-}$. The ammonium salt of this ion would have the formula $(NH_4)_3[Ag(S_2O_3)_2]$.

21.4 The salt must include the six hydrated water molecules. We know that Al exists as a 3+ ion and that chloride has a charge of 1–. The hydrate would have the formula $AlCl_3{\cdot}6H_2O$. The complex ion most likely has the formula $[Al(H_2O)_6]^{3+}$.

21.5 (a) $[SnCl_6]^{2-}$
(b) $(NH_4)_2[Fe(CN)_4(H_2O)_2]$

21.6 (a) potassium hexacyanoferrate(III)
(b) dichlorobis(ethylenediamine)chromium(III) sulfate

21.7 Since there are three ligands and $C_2O_4^{2-}$ is a bidentate ligand, the coordination number is six.

21.8 (a) The coordination number is six. There are two bidentate ligands and two unidentate ligands.
(b) The coordination number is six. Both $C_2O_4^{2-}$ and ethylenediamine are bidentate ligands. Since there are three bidentate ligands, the coordination number must be six.
(c) EDTA is a hexadentate ligand so the coordination number is six.

Review Problems

21.102 First we need $\Delta H°$ for this reaction.

$\Delta H° = 2\Delta H_f°(Hg_{(g)}) + \Delta H_f°(O_{2(g)}) - 2\Delta H_f°(HgO_{(s)})$

$= 2(61.3 \text{ kJ}) + 0 - 2(-90.8 \text{ kJ}) = 304 \text{ kJ}$

Similarly;

$\Delta S° = 2(175 \text{ J/K}) + 205 \text{ J/K} - 2(70.3 \text{ J/K}) = 414 \text{ J/K}$

Determine the temperature when $\Delta G = 0$

$\Delta G° = \Delta H° - T\Delta S°$

$$T = \frac{\Delta H°}{\Delta S°} = \frac{304 \text{ kJ}}{0.414 \text{ kJ/K}} = 734 \text{ K}$$

21.104 The net charge is –3, and the formula is $[Fe(CN)_6]^{3-}$. The IUPAC name for the complex is hexacyanoferrate(III) ion

21.106 $[CoCl_2(en)_2]^+$

21.108 (a) $C_2O_4^{2-}$ oxalato (b) S^{2-} sulfido or thio
(c) Cl^- chloro (d) $(CH_3)_2NH$ dimethylamine

21.110 (a) hexaamminenickel(II) ion
(b) triamminetrichlorochromate(II) ion
(c) hexanitrocobaltate(III) ion
(d) diamminetetracyanomanganate(II) ion
(e) trioxalatoferrate(III) ion or trisoxalatoferrate(III) ion

21.112 (a) $[Fe(CN)_2(H_2O)_4]^+$
(b) $[Ni(C_2O_4)(NH_3)_4]$
(c) $[Fe(CN)_4(H_2O)_2]^-$
(d) $K_3[Mn(SCN)_6]$
(e) $[CuCl_4]^{2-}$

21.114 The coordination number is six, and the oxidation number of the iron atom is +2.

21.116

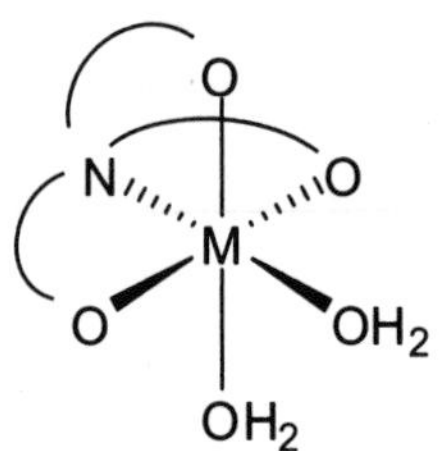

The curved lines represent $-CH_2-C(=O)-$ groups

21.118 Since both are the *cis* isomer, they are identical. One can be superimposed on the other after simple rotation.

21.120

cis:

Br, NH3, Pt, Cl, NH3

trans:

Br, NH3, Pt, H3N, Cl

21.122

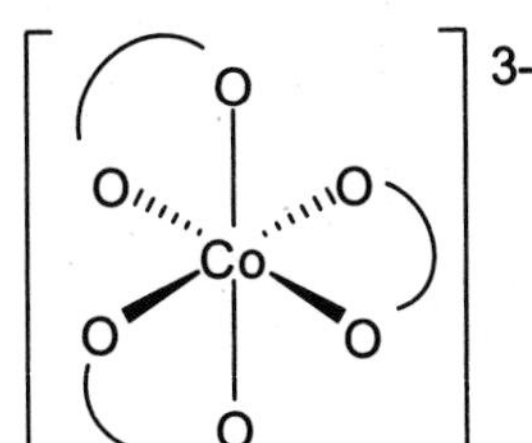

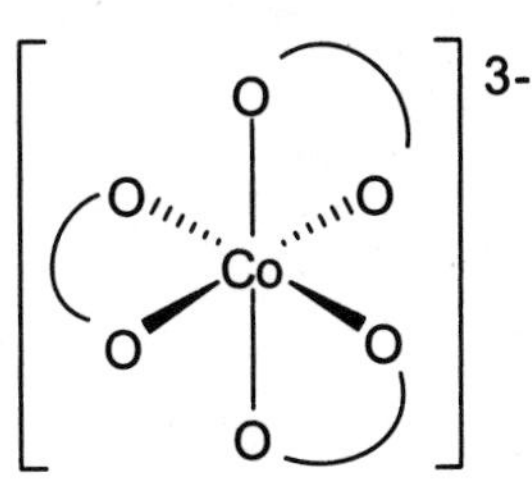

21.124 (a) $[Cr(H_2O)_6]^{3+}$ (b) $[Cr(en)_3]^{3+}$

21.126 $[Cr(CN)_6]^{3-}$

21.128 (a) The value of Δ increases down a group. Therefore, we choose: $[RuCl(NH_3)_5]^{3+}$
(b) The value of Δ increases with oxidation state of the metal. Therefore, we choose: $[Ru(NH_3)_6]^{3+}$

21.130 This is the one with the strongest field ligand, since Co^{2+} is a d^7 ion: CoA_6^{3+}

21.132 This is a weak field complex of Co^{2+}, and it should be a high–spin d^7 case. It cannot be diamagnetic; even if it were low spin, we would still have one unpaired electron.

Practice Exercises

22.1 2,2–dimethylpropane

22.2 (a) 3–methylhexane
(b) 4–ethyl–2,3–dimethylheptane
(c) 5–ethyl–2,4,6–trimethyloctane

22.3

$$CH_3CH_2\text{–}\underset{\displaystyle |}{\overset{\displaystyle OH}{C}}H\text{–}CH_2CH_3$$

22.4 (a)

$$CH_3\text{–}\overset{\displaystyle O}{\overset{\|}{C}}H \quad \text{or} \quad CH_3\text{–}\overset{\displaystyle O}{\overset{\|}{C}}OH$$

(depending on the strength of the oxidizing agent)

(b)

$$CH_3CH_2\overset{\displaystyle O}{\overset{\|}{C}}CH_2CH_3$$

22.5

$$CH_3CH_2\text{–}\overset{\displaystyle O}{\overset{\|}{C}}\text{–}NHCH_2CH_2CH_3$$

The functional groups are the acid and the amine which condense to form an amide

22.6 (a)

$$CH_3CH_2\text{–}\overset{\displaystyle O}{\overset{\|}{C}}\text{–}O^- + HOCH(CH_3)_2$$

(b) $CH_3CH{=}CH_2$

22.7

$$\text{—}\underset{\underset{\displaystyle CH_3}{|}}{\overset{\overset{\displaystyle H}{|}}{C}}\text{—}\underset{\underset{\displaystyle H}{|}}{\overset{\overset{\displaystyle CH_3}{|}}{C}}\text{—}\underset{\underset{\displaystyle CH_3}{|}}{\overset{\overset{\displaystyle H}{|}}{C}}\text{—}\underset{\underset{\displaystyle H}{|}}{\overset{\overset{\displaystyle CH_3}{|}}{C}}\text{—}\underset{\underset{\displaystyle CH_3}{|}}{\overset{\overset{\displaystyle H}{|}}{C}}\text{—}\underset{\underset{\displaystyle H}{|}}{\overset{\overset{\displaystyle CH_3}{|}}{C}}\text{—}$$

Repeat Unit (first two carbons)

22.8

$$\text{—}\underset{\underset{\displaystyle F}{|}}{\overset{\overset{\displaystyle F}{|}}{C}}\text{—}\underset{\underset{\displaystyle F}{|}}{\overset{\overset{\displaystyle F}{|}}{C}}\text{—}\underset{\underset{\displaystyle F}{|}}{\overset{\overset{\displaystyle F}{|}}{C}}\text{—}\underset{\underset{\displaystyle F}{|}}{\overset{\overset{\displaystyle F}{|}}{C}}\text{—}\underset{\underset{\displaystyle F}{|}}{\overset{\overset{\displaystyle F}{|}}{C}}\text{—}\underset{\underset{\displaystyle F}{|}}{\overset{\overset{\displaystyle F}{|}}{C}}\text{—}$$

Repeat Unit (first two carbons)

Review Problems

22.97

(a)

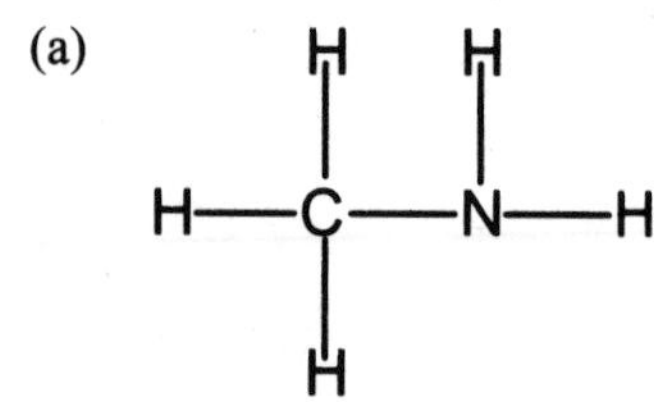

(b)

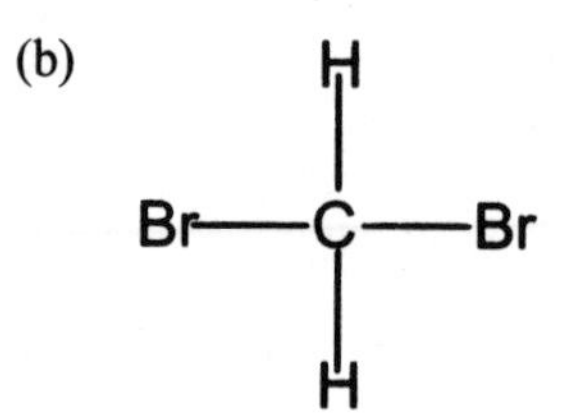

(c)

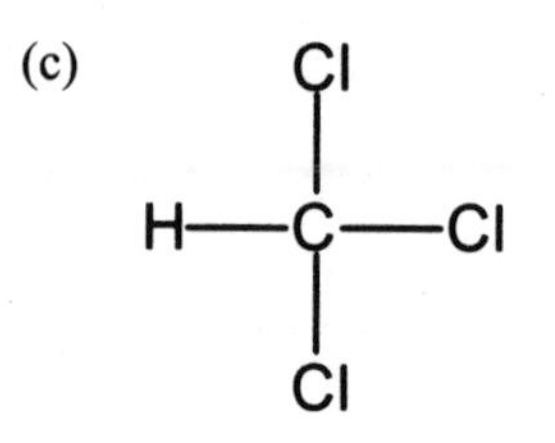

(d)

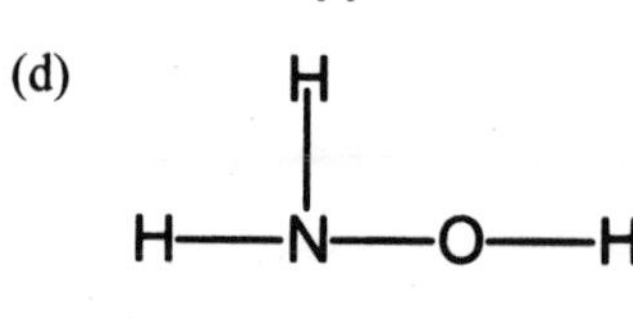

(e)

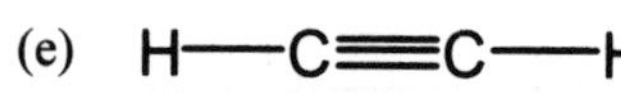

(f)

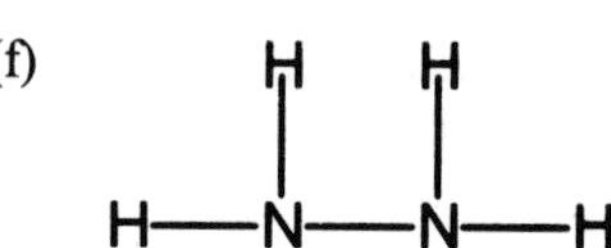

22.99 (a) alkene (d) carboxylic acid
(b) alcohol (e) amine
(c) ester (f) alcohol

22.101 The saturated compounds are b, e, and f.

22.103 (a) amine (b) amine (c) amide (d) amine, ketone

22.105 (a) These are identical, being oriented differently only.
(b) These are identical, being drawn differently only.
(c) These are unrelated, being alcohols with different numbers of carbon atoms.
(d) These are isomers, since they have the same molecular formula, but different structures.
(e) These are identical, being oriented differently only.
(f) These are identical, being drawn differently only.
(g) These are isomers, since they have the same molecular formula, but different structures.

22.107 (a) pentane
(b) 2–methylpentane
(c) 2,4–dimethylhexane

22.109 (a) No isomers

(b)

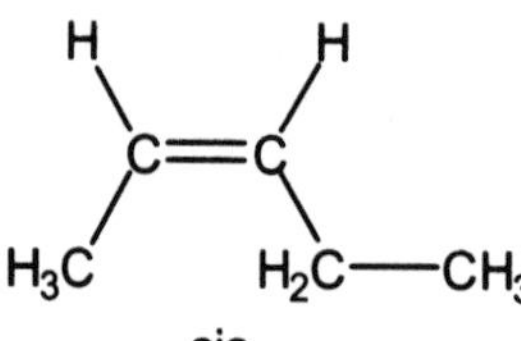

(c)

cis: Br and H_3C on one carbon, Cl and H on the other (Br–C=C–Cl, H_3C–C=C–H)

trans: Br and H_3C on one carbon, H and Cl on the other (Br–C=C–H, H_3C–C=C–Cl)

22.111 (a) CH_3CH_3
(b) $ClCH_2CH_2Cl$
(c) $BrCH_2CH_2Br$
(d) CH_3CH_2Cl
(e) CH_3CH_2Br
(f) CH_3CH_2OH

22.113 (a) $CH_3CH_2CH_2CH_3$
(b) H_3C—CH(Cl)—CH(Cl)—CH_3

(c) H_3C—CH(Br)—CH(Br)—CH_3

(d) H_3C—CH_2—CH(Cl)—CH_3

(e) H_3C—CH_2—CH(Br)—CH_3

(f) H_3C—CH_2—CH(OH)—CH_3

22.115 This sort of reaction would disrupt the π delocalization of the benzene ring. The subsequent loss of resonance energy would not be favorable.

22.117

CH_3OH	IUPAC name = methanol;	common name = methyl alcohol
CH_3CH_2OH	IUPAC name = ethanol;	common name = ethyl alcohol
$CH_3CH_2CH_2OH$	IUPAC name = 1–propanol;	common name = propyl alcohol
$H_3C—CH(CH_3)—OH$	IUPAC name = 2–propanol;	common name = isopropyl alcohol

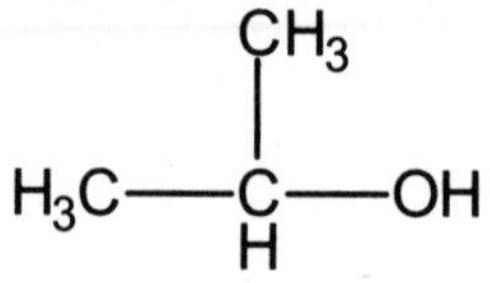

22.119

$CH_3CH_2CH_2–O–CH_3$	methyl propyl ether
$CH_3CH_2–O–CH_2CH_3$	diethyl ether
$(CH_3)_2CH–O–CH_3$	methyl 2–propyl ether

22.121 (a)

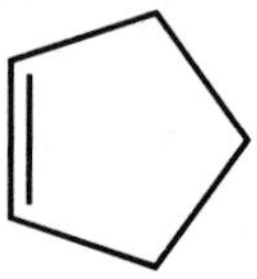

(b)

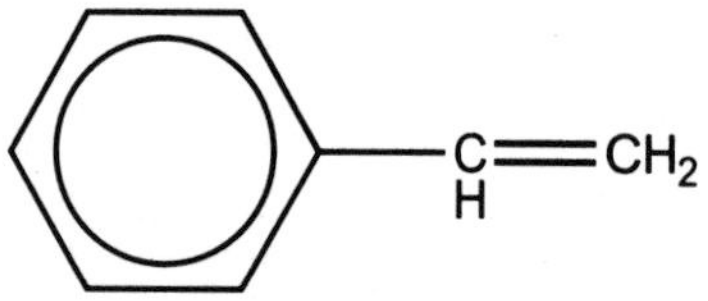

$C_6H_5—CH{=}CH_2$

(c)

$C_6H_5—CH{=}CH_2$

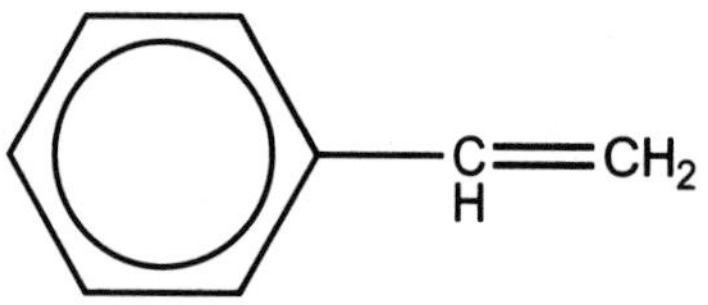

22.123 (a)

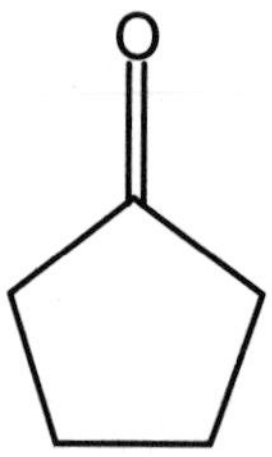

(b)

$C_6H_5—C(=O)—CH_3$

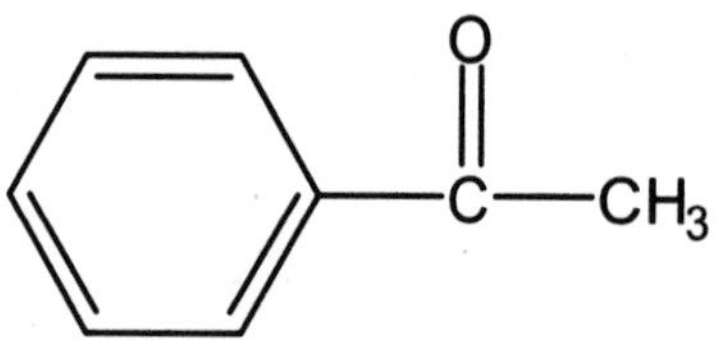

(c)

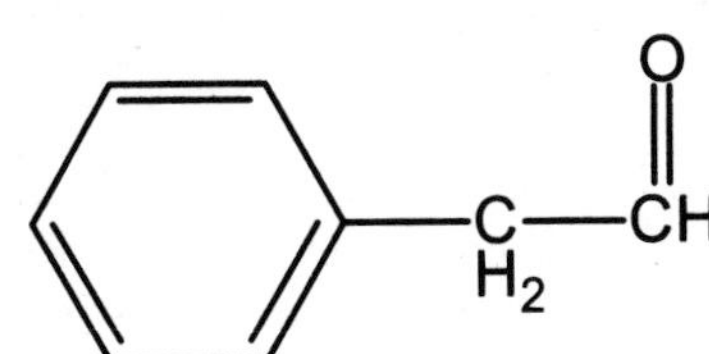

22.125 The elimination of water can result in a C=C double bond in two locations:

$CH_2=CHCH_2CH_3$	$CH_3CH=CHCH_3$
1–butene	2–butene

22.127 The aldehyde is more easily oxidized. The product is:

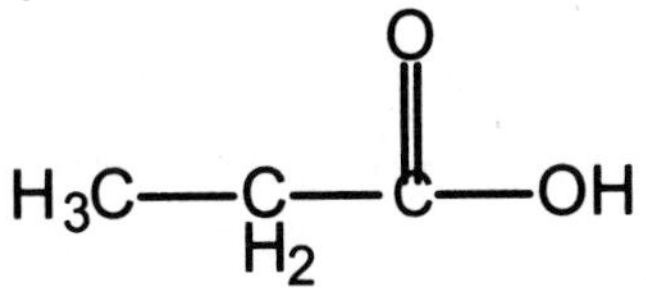

$H_3C—CH_2—C(=O)—OH$

22.129 (a) $CH_3CH_2CO_2H$
(b) $CH_3CH_2CO_2H\ +\ CH_3OH$
(c) $Na^+ + CH_3CH_2CH_2CO_2^-\ +\ H_2O$

22.131 $CH_3CO_2H\ +\ CH_3CH_2NHCH_2CH_3$

22.133

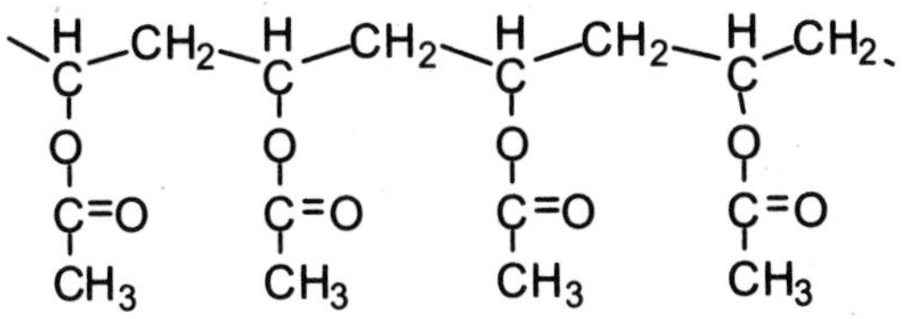

H CH2 H CH2 H CH2 H CH2
C C C C
O O O O
C=O C=O C=O C=O
CH3 CH3 CH3 CH3

22.135

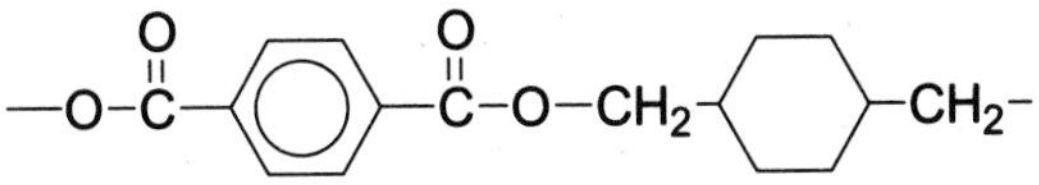

—O–C(=O)–C6H4–C(=O)–O–CH2–C6H10–CH2–

22.137

$H_2C—O—C(=O)—(CH_2)_7—CH=CH(CH_2)_7CH_3$

$HC—O—C(=O)—(CH_2)_7·CH=CH—CH_2—CH=CH(CH_2)_4CH_3$

$H_2C—O—C(=O)—(CH_2)_{14}CH_3$

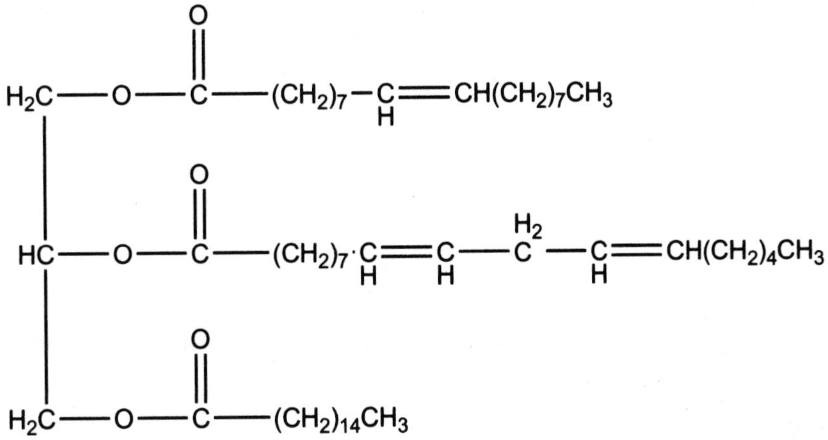

22.139

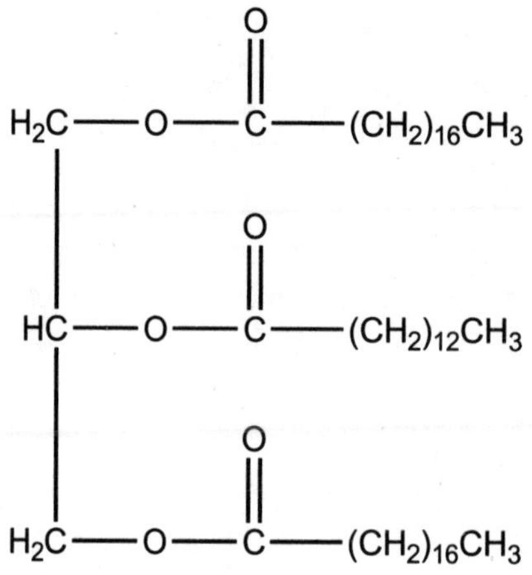

$$H_2C-O-\overset{O}{\overset{\|}{C}}-(CH_2)_{16}CH_3$$
$$HC-O-\overset{O}{\overset{\|}{C}}-(CH_2)_{12}CH_3$$
$$H_2C-O-\overset{O}{\overset{\|}{C}}-(CH_2)_{16}CH_3$$

22.141 Hydrophobic sites are composed of fatty acid units. Hydrophilic sites are composed of charged units.

22.143

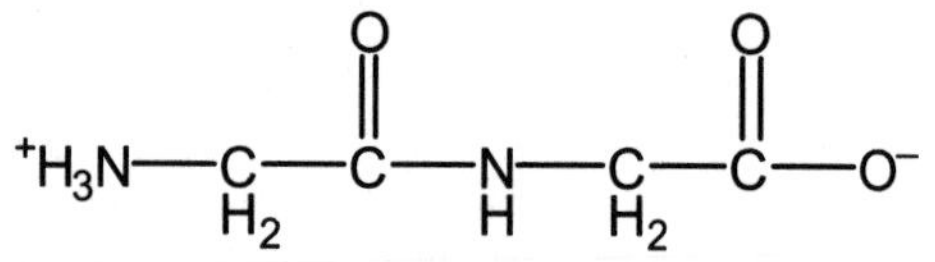

$$^{+}H_3N-\underset{H_2}{C}-\overset{O}{\overset{\|}{C}}-\underset{H}{N}-\underset{H_2}{C}-\overset{O}{\overset{\|}{C}}-O^{-}$$

22.145

$$^{+}H_3N-\underset{H_2}{C}-\overset{O}{\overset{\|}{C}}-\underset{H}{N}-\underset{CH_2C_6H_5}{CH}-\overset{O}{\overset{\|}{C}}-O^{-}$$

$$^{+}H_3N-\underset{CH_2C_6H_5}{CH}-\overset{O}{\overset{\|}{C}}-\underset{H}{N}-\underset{H_2}{C}-\overset{O}{\overset{\|}{C}}-O^{-}$$

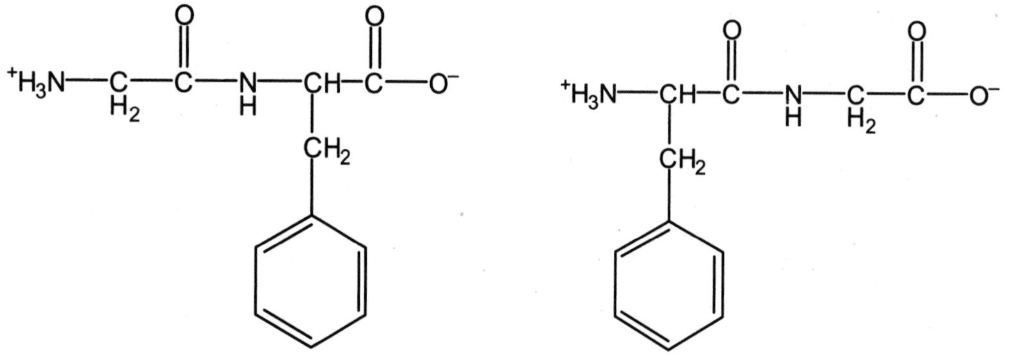